COLLECTED WORKS OF RICHARD J. CHORLEY

Volume 1

FRONTIERS IN GEOGRAPHICAL TEACHING

FRONTIERS IN
GEOGRAPHICAL TEACHING

Edited by
RICHARD J. CHORLEY
AND PETER HAGGETT

Routledge
Taylor & Francis Group

LONDON AND NEW YORK

First published in 1965 by Methuen and Co. Ltd

Second edition published 1970

This edition first published in 2019
by Routledge
2 Park Square, Milton Park, Abingdon, Oxon OX14 4RN

and by Routledge
52 Vanderbilt Avenue, New York, NY 10017

Routledge is an imprint of the Taylor & Francis Group, an informa business

British Library Cataloguing in Publication Data
A catalogue record for this book is available from the British Library

ISBN: 978-0-367-22096-9 (Set)
ISBN: 978-0-429-27321-6 (Set) (ebk)
ISBN: 978-0-367-22099-0 (Volume 1) (hbk)
ISBN: 978-0-367-22175-1 (Volume 1) (pbk)
ISBN: 978-0-429-27330-8 (Volume 1) (ebk)

Publisher's Note
The publisher has gone to great lengths to ensure the quality of this reprint but points out that some imperfections in the original copies may be apparent.

Disclaimer
The publisher has made every effort to trace copyright holders and would welcome correspondence from those they have been unable to trace.

Frontiers in Geographical Teaching

EDITED BY

Richard J. Chorley and Peter Haggett

METHUEN & CO LTD

LONDON

First published 1965
by Methuen & Co Ltd
11 New Fetter Lane, London EC4
© *1965 and 1970 R. J. Chorley and P. Haggett*
Reprinted twice in 1967
Second Edition 1970
Printed in Great Britain by
The Camelot Press Ltd
London and Southampton
SBN 416 16840 X

First published as a University Paperback 1970
SBN 416 18340 9

Distributed in the U.S.A. by Barnes & Noble, Inc.

Contents

Contents vii

PART II TECHNIQUES

VIII THE APPLICATION OF QUANTITATIVE METHODS
TO GEOMORPHOLOGY   *page* 147

R. J. Chorley (Lecturer in Geography, University of Cambridge)

The Weakening of Denudation Chronology   148
The Distrust of Quantification   150
The Promotion of Quantification   153
Aspects of Quantification   154
References   161

IX SCALE COMPONENTS IN GEOGRAPHICAL PROBLEMS 164

P. Haggett (Professor of Geography, University of Bristol)

The Scale Coverage Problem   164
The Scale Linkage Problem   170
The Scale Standardization Problem   177
Conclusion   182
References   183

X FIELD WORK IN GEOGRAPHY, WITH PARTICULAR
EMPHASIS ON THE ROLE OF LAND-USE SURVEY 186

C. Board (Lecturer in Geography, London School of Economics)

Wooldridge and 'Real Field Work'   186
The Method of Field Teaching   189
The Nature of Field Research   191
The Role of the Visible Landscape   192
Regional Surveying and Field Work   194
Field Work and Land-Use Surveys   195
The Effect of Regional Survey on Land-Use Studies 198
The Importance of Visible Features in Land-Use
Studies   201
The Search for Speed and Efficiency in American
Land-Use Studies   202
British Studies take up the Search   204
The Interpretation of Land-Use Patterns   204
Farm Surveys as an Alternative to Land-Use Surveys 206
Conclusion   208
References   208

Contents ix

PART III   TEACHING

Foreword

This volume of essays emerged from the first two residential courses for teachers of pre-university geography held at Madingley Hall near Cambridge under the aegis of Cambridge University's Extra-Mural Board. Although modest in its scope and level of treatment, its publication in the autumn of 1965 highlighted a polarization of attitudes by teachers of geography at a time when rapid methodological changes in the discipline were beginning to place them under considerable stress. Understandably reactions to the book varied, with early reviews in British journals reflecting a cautious and often hostile reaction to the book's diagnosis and prognosis. However, the trends foreshadowed in it accelerated so quickly that later reviews found it only 'mildly iconoclastic' and in current perspective it seems clear to the editors that we erred not in being radical, but in not being radical enough.

In North America the reception to the original volume was encouraging and the substantive critical reviews in a number of volumes – notably in the *Professional Geographer* (1966, pp. 341–5), *Geographical Review* (1967, pp. 443–6), *Canadian Geographer* (1967, pp. 59–60) and *Economic Geography* (1967, pp. 182–4) – suggested the essays were less isolationist than we had feared. In preparing this new edition for press we have therefore incorporated chapters on geographical teaching in the United States, as well as revising the materials on Britain. We are grateful to Professor Clyde F. Kohn of Iowa University, Professor Placido LaValle of the University of Windsor, Ontario, and Dr A. R. H. Baker of Cambridge University, for their additional contributions to this revision.

Successive Madingley seminars lead to a series of later volumes – *Socio-economic Models in Geography* (1968), *Integrated Models in Geography* (1969) and *Physical and Information Models in Geography* (1969) – and a number of the early contributors have established in *Progress in Geography*, I- (1969–) a regular outlet for research reviews. However, the original '*Frontiers*' volume remains the only source work from the series specifically aimed at discussing for teachers in schools the questions 'What is happening in geography? What impact does this have on school geography?' We hope that this revised volume will continue to provide a real link for students on both sides of the Atlantic who are making the increasingly difficult transition from school to university geography.

R. J. C.
P. H.

Acknowledgements

The authors gratefully acknowledge the permissions by the following individuals and organizations to reproduce figures: The Warden, Madingley Hall, Cambridge, for the print used for the dust-jacket; The Director, U.S. Geological Survey, for figures 2.1, 2.2, and 8.2; G. B. Masefield, Esq., M.A., Editor, *Tropical Agriculture*, for figures 3.3, 3.5, 3.6, 3.7, and 3.10; Professor A. N. Strahler, Columbia University, New York, for figure 8.1; The Editor, Royal Geographical Society, for figures 9.7, 13.2, 13.4, 13.6, 13.7, 13.10, and 17.4; Her Majesty's Stationery Office (Crown Copyright Reserved), for figures 13.1, 13.5, 13.8, 13.9, 13.12, 13.13, and 13.14; Longmans, Green & Co. Ltd., for figures 17.1, 17.3, 17.5, and 17.6; The Petroleum Information Bureau, for figure 17.2; The Editor, Institute of British Geographers, for figure 13.3; The Editor, *Town Planning Review*, for figure 13.11.

PART ONE
CONCEPTS

CHAPTER ONE

Changes in the Philosophy of Geography

E. A. WRIGLEY

Lecturer in Geography, University of Cambridge

It is perhaps a platitude that a perennial problem of geography, true of many subjects but felt very acutely in geography, has been to find a satisfactory way of organizing the welter of observational material with which geographers commonly deal. It may prove helpful to look back to some solutions to this problem offered in the past as a background to contemporary developments, and to pay especial attention to the writings of Vidal de la Blache, since both his successes and his final failure are very instructive.

'CLASSICAL' GEOGRAPHY

Modern geography is often said to begin with two important early nineteenth-century German scholars, von Humboldt and Ritter.[1] These two men saw eye to eye in many things and agreed in pouring scorn on their predecessors because they dealt with geographical information in such a haphazard and unsystematic fashion. They gave clear expression to a view of geographical methodology which remained dominant for most of the nineteenth century. It was still the guiding principle of Friedrich Ratzel, at least in his earlier writings in the 1880's, and was shared by many who were not geographers. It is indeed a pointless exercise at this comparatively early date to distinguish between those who were geographers and those who were not. Humboldt would not fit any label conveniently. Buckle in the early chapters of his *History of Civilisation in England* (1857–61) gives a succinct account of the new attitude.

This conception of the organization of geography, which may

[1] This claim is conventional and perhaps convenient, but it is quite possible to argue the case for a later date; for example for the work of Vidal de la Blache.

conveniently be labelled the 'classical'[1] view since it held sway during the formative period of modern geography in the nineteenth century, was straightforward and simple. Men like Humboldt and Ritter considered the writings of earlier geographers to be defective because they were largely descriptive and were in their view very ill-organized.[2] They considered the scientific organization of knowledge to be a two-stage affair: a first stage which consisted of the careful assembly of detailed and accurate factual material; and a second in which the material was given coherence and made intelligible by being subsumed under a number of laws which should express the relationships of cause and effect to be found in the phenomena as simply and concisely as possible.[3] The vital feature of any science was this second stage. Without it any branch of learning was simply pigeon-holing and antiquarianism. This was a first main characteristic of 'classical' geography. Status in the world of learning depended on the successful formulation of laws enabling the material to be organized and made intelligible. If geography were to be worthy of ranking with the sciences it must succeed in establishing such laws. It must go on, as Ratzel put it, airing his Latin tags, *rerum cognoscere causas*,[4] to know the causes of things.

A second chief prop of the 'classical' view of geography was the

[1] The term 'classical' is used here in a different sense from that employed by Hartshorne for whom the deaths of Humboldt and Ritter in 1859 mark the end of the 'classical' period.

[2] E.g. 'A systematic organization of material is seldom to be found in them (the older type of geographies). . . . They contain at bottom only an arbitrary, unorganized and unsystematic compilation of all sorts of noteworthy phenomena, which in the different parts of the globe appear to be especially striking. . . . The description of Europe is begun with either Portugal or Spain because Strabo began his narration in this order. The facts are arranged like the pieces of a patchwork quilt, now one way, now another, as if each disconnected piece could stand by itself' (Ritter, 1862, pp. 21–2).

[3] E.g. 'In proportion as laws admit of more general application, and as sciences mutually enrich each other, and by their extension become connected together in more numerous and more intimate relations, the developments of general truths may be given with conciseness devoid of superficiality. On being first examined, all phenomena appear to be isolated, and it is only by the result of a multiplicity of observations, combined with reason, that we are able to trace the mutual relations existing between them' (Humboldt, 1849, Vol. 1, p. 29).

[4] 'But this deeper conception (of geography) cannot possibly remain content with description but must, following the irresistible example of all natural sciences, within whose sphere it developed in the most intimate relationship, go on from description to the higher task of "Rerum cognoscere causas"' (Ratzel, 1882, Vol. 1, p. 5).

conviction[1] that in the final analysis there was no difference methodologically between what would now be called the social and the physical sciences. In both cases the ultimate aim was the formulation of laws expressing the universal operation of cause and effect.[2] It was widely agreed that the subject matter was vastly more diverse and complicated in the study of societies than, say, in physics and that it might be much longer before satisfactory formulations of laws could be made, but the Newtonian model was assumed, though often implicitly rather than explicitly, to be appropriate. The denial of the methodological unity of all knowledge, and particularly the assertion of a special position for the study of social change and functioning in the writings of men like Dilthey and Max Weber, still lay in the future. This is important because it largely obviated a difficulty which has been much more keenly felt by geographers in the last two generations, namely the problem of running in harness, as it were, physical geography and social geography. Since the methodology of the social sciences is now very commonly held to be different from that of the physical sciences (for example, the possibility of formulating universally valid laws descriptive of social functioning is often denied, and Weberian 'ideal' types and middle-order generalizations advocated in their stead), there are clearly problems in asserting the unity of geographical knowledge and particularly of geographical methodology. During the period of 'classical' geography no difficulties could arise on this score because the general methodology of all branches of the subject could be held to be the same.

A third general point on which the writings of the 'classical' geographers show agreement in the main is that a prime object of geographical study is to investigate the ways in which the physical environment affects the functioning and development of societies.[3]

[1] Shared of course by many contemporaries, for example A. Comte and J. S. Mill.

[2] 'In regard to nature, events apparently the most irregular and capricious have been explained, and have been shown to be in accordance with certain fixed and immutable laws. This has been done because men of ability and, above all, men of patient, untiring thought have studied natural events with the view of discovering their regularity: and if human events were subjected to a similar treatment, we have every right to expect similar results' (Buckle, 1857, Vol. 1, p. 6).

[3] 'Only having a firm methodological principle can protect it (geography) from going astray: the clear commitment to the central theme of the relationship between the forms of terrestrial phenomena and mankind' (Ritter, 1862, p. 28).

The *Erdkunde* gained Humboldt's warm approbation because of its success

This is not to suggest that this was the main theme of all their works. There were many in which it appeared rarely; some of a specialist nature in which it did not appear at all; and in some works of geographical methodology it was firmly rejected. But, from a general view, this appears to be the third necessary support of the 'classical' attitude to geography.[1] It gave point to all the subsidiary lines of investigation and it made it perfectly clear why geography must have both a physical and social side. Some knowledge of both is obviously vital to any attempt to understand this matter. Given the first two main characteristics of 'classical' geography, the belief in the necessity of formulating laws expressing relationships of cause and effect and the conviction that the basic methodology of both social and physical sciences is the same, it is very reasonable to feel that one of the most promising lines of investigation to pursue is the explanation of social change and function by reference to features of the physical environment. In the early decades after the publication of the *Origin of Species* this type of work, while it might be modified, was also encouraged and strengthened by a flow of new concepts, as may be seen in much of Ratzel's large output.

The story of the decline and fall of the 'classical' conception of geography is a most interesting chapter in intellectual history. By the end of the century it was widely attacked for its rigidity, because it was wedded to what came to be called geographical determinism (although this was not an accurate criticism to level at the major figures of the group), and because the new ideas about methodology in the social sciences made its whole approach to the understanding of social action and social change seem unrewarding.[2] 'Classical' geography has left a considerable legacy. It can be seen, for example, in the continuing arguments about what are usually termed 'possibilism' and 'probabilism', which are rooted in the 'classical' attitude

in showing the influence of the environment 'on the migrations, laws, and manners, of nations, and on all the principal events enacted upon the face of the earth' (Humboldt, 1849, Vol, 1, p. 28).

Ratzel subtitled his great work *Anthropo-geographie: oder Grundzüge der Anwendung der Erdkunde auf die Geschichte.*

[1] It is interesting that Hettner once wrote to Joseph Partsch that what first attracted him to geography was the idea of the dependence of man on nature and described his surprise at discovering how little this entered into his teaching when he went to study under Kirchoff at his first university, Halle (*Heidelberger Geographische Arbeiten*, 1960, p. 77).

[2] It is interesting to note that Hettner knew Max Weber in the years when both men were at Heidelberg.

to the subject, and are open to much the same objections as the 'classical' system. Or again, it can be seen in the layout of many textbooks which begin with such things as solid geology and climate and progress through vegetation and soils to settlement, agriculture, industry and transport—a perfectly logical sequence of exposition in 'classical' terms,[1] but less so if the 'classical' view is abandoned.

'REGIONAL' GEOGRAPHY

The reaction against 'classical' geography took many forms: some essentially a development from it, like the writings of Hettner; some in opposition to it, like Brunhes' ideas;[2] some developed along new lines without close reference to it. The work of Vidal de la Blache falls best perhaps in the last category.[3] It is of the greatest importance and may serve as an introduction to the 'post-classical' world, and to what may be termed the 'regional' view of the nature of geography.

Vidal saw that it made little sense to set the physical and social environments of man over against one another, as it were, and examine the way the former influenced the latter, still less to do this in a systematic fashion in the hope that general laws describing the relationships could be discovered. Instead he propounded a different idea. Whereas in the 'classical' view the study of the physical environment and the study of society were riveted to each other because a

[1] One great strength and beauty of the 'classical' system was that the sequence was at one and the same time a coherent method of conveying information and of explaining it stage by stage. Description and explanation were so neatly bound up together that the second could hardly be distinguished from the first.

[2] Brunhes admired Ratzel's work very greatly, considering that he had conceived a fresh and fruitful approach to human geography, but his views on method stand in strong contrast with the 'classical' school. 'Between the facts of physical nature there are sometimes causal relations; between those of human geography there are really only relations of connexion. To force, as it were, the bond which connects phenomena to each other is to produce a work of false science; and a critical attitude is very necessary to allow one to specify with good judgement those complex cases where interconnexion (*connexité*) does not in the least imply causation' (Brunhes, 1925, Vol. 2, p. 877).

[3] This is not, of course, to say that Vidal de la Blache was not acquainted with the works of his predecessors and contemporaries. On the contrary, few men have drawn with such facility from the whole corpus of geographical literature from Greek times onwards.

main purpose of geographical study was to investigate the condition-
ing of society by environment, in Vidal's scheme they were linked
because they were inseparable. Any physical environment in which a
society settles is greatly affected by the presence of man, the more
so if the society has an advanced material culture. The plant and
animal life of France, to take an obvious example, was vastly different
in the nineteenth century from what it would have been if there had
been no settlement there by man. Equally, the adjustment of each
society to the peculiarities of the local physical environment, taking
place over many centuries, produces local characteristics in that
society which are not to be found elsewhere. Man and nature become
moulded to one another over the years rather like a snail and its
shell. Yet the connexion is more intimate even than that, so that it is
not possible to disentangle influences in one direction, of man on
nature, from those in the other, of nature on man. The two form a
complicated amalgam. Vidal often reiterated that he was not studying
a people but a landscape, yet he chose to do this in a way which came
close to denying the distinction between the two. The area within
which an intimate connexion between man and land had grown up in
this way over the centuries formed a unit, a region, which was a
proper object for geographical study. Since each region was so much
a product of local circumstance, both social and physical, what was
significant in one area might prove to be irrelevant in another. His
conception of the subject weighted it strongly in favour of the
regional and against the systematic treatment of material.

It was essential to the best flowering of Vidal's method that the
society living in an area should be 'local' and that it should be basically rural. It must be local in the sense that the bulk of the materials
used as food, for building, for fuel, in the manufacture of tools and
machines and so on, should be of local origin. Each small region might
conduct a trade with other areas in special commodities but the basic
stuff of life and work was to be local. This naturally gave rise to
typical regional foods and dishes, styles of domestic and farm
architecture, clothes, and so on. Equally, it must be rural, rooted in
the land, with the bulk of the population either working on the land or
servicing those who did. Even the bourgeoisie and the local landed
gentry might find it difficult to break free from local, rural patterns
of life. The peasant was deeply imbedded in them (see e.g. Vidal de
la Blache, 1911, pp. 384–5). Hence Vidal's interest in the minutiae
of the material culture of each small area. He dwelt at some length
on the importance to geographers of ethnological museums in this

connexion. Within them might be preserved the whole range of tools with which a society went about its daily business, not merely the instruments used for grinding corn, of working wood, or ploughing the soil, but also the houses, the types of clothing, the methods of heating and lighting which were or had been in use (Vidal de la Blache, 1922, pp. 119-21). These would all be made from local materials and designed to overcome local difficulties or take advantage of local opportunities. With such a display before him a geographer should be able to read off many both of the main and the more detailed characteristics of the environment in which they were developed.

To emphasize the continuity of traits in the life of societies of this sort Vidal made use of an arresting image. He reminded his readers that the surface of a pond may be ruffled by a passing breeze so that the eye of the watcher is unable any longer to see the bottom of the pond through the clear water, but that as soon as the wind dies down and the waters are again calm the bottom is once more visible and all the old contours may again be seen undisturbed by the movements of the surface water (Vidal de la Blache, 1911, p. 386). In just the same way the advent of war, pestilence, famine, rebellion, may appear to disrupt the life of a region and throw its steady routines of action into chaos, but once the crisis is over the same long-established pattern of life, of working and holding the land, of building, of clothing, of feeding, even of trade, will reappear. Change may come in such communities. Vidal showed much change in Alsace and Lorraine in *La France de l'Est* in the centuries before the French Revolution, but it was change within a continuing dialectic of man and land, the elaboration of a pattern set by the exigencies of life and soil.

All this was rooted in a very apt appreciation of a truth of great importance. The communities of Europe throughout the medieval and early modern period were rural, were local, and were the result of a long interplay between man and land. There were important local variations in material culture which gave to each region characteristic styles of domestic architecture, clothing and food. In some cases and for some classes in the community there were national rather than local characteristics, but Vidal's was a powerful and legitimate vision of the functioning of societies during most of European history. It was, however, ironically, a vision of things past or about to pass, not a vision of things present or to come. The method developed by Vidal de la Blache is admirably suited to the historical geography of Europe before the Industrial Revolution, or indeed to the limited and shrinking areas of the world today whose economies are still based on peasant agriculture and local self-sufficiency in most of the material things of life, but it is not applicable to a country which has undergone industrial revolution.

What gives to the work of Vidal de la Blache its special interest is not just the fact that he conceived a new attitude to the organization of geographical material and founded a very influential 'school', but also that the rigour of his argument led him to recognize that his method could not cope with the aftermath of the Industrial Revolution. This can be seen clearly in the *France de l'Est* (1917), perhaps his most original and important work. The *France de l'Est* is devoted to the study of the formation of the landscapes and rural societies of Alsace and Lorraine. It covers a period of two millenia during which the landscapes and societies emerged. As a result a large fraction of the book is taken up with the period before 1789. The method of treatment is chronological in the main and there comes a point in the later stages of the work where the finely developed dialectic between man and land[1] which he had been at such pains to pursue over many centuries suddenly begins to fail to comprehend and make sense of the course of events. In the middle of the nineteenth century – Vidal actually found a date, 1846, for it (1917, p. 126) – the waters of the pond were ruffled by something much more disturbing than the storms of the past, something which did not leave the contours on the bottom of the pond as they had always been but caused them to adopt quite a new configuration. The advent of the steam-engine, the railways, the coal-carrying canals and of the new Alsatian cotton industry did not mean the superimposition of a few new strands upon an old-established pattern, but was the first stage in the dissolution of the traditional, rural, local, regional pattern of life. Vidal was too shrewd and conscientious not to recognize this. Whereas the industries of the past were easily assimilated into the model he proposed, the new industries represented a new type of society. The new society was able to produce industrial goods on a vastly bigger scale, and was formed round cheap and speedy communications. Food, clothing, building materials, tools, all soon ceased to be locally made and different in one region from its neighbour. Instead, from the

[1] 'A people, great or small, possesses a personality, whose appearance, like anything else, must submit to the erosion of time, but yet it keeps through the ages the fundamental traits which it acquired as it developed in the region where it settled. . . . It is in conjunction with time, and face to face with the soil, that their traits became fixed once and for all, and thus a personality became established which cannot but be noticed and deserves respect' (Vidal de la Blache, 1917, p. 43).

mid-nineteenth century onwards in Alsace and Lorraine, and from a much earlier date in England, the leading characteristics of the traditional society and economy crumbled slowly away. As a result geographical methodology has been obliged to abandon the 'regional' concept of the subject, just as half a century earlier it had to abandon the 'classical' scheme. The changes in society and economy which produced this situation did not of course come overnight, nor did they affect all areas equally or equally quickly, but the Industrial Revolution began a process which today leaves only one man in twenty on the land in this country[1] and a small and falling fraction on the land in all other materially advanced countries; which has left almost no typical local foodstuffs, clothes or house types; which leaves us all dependent upon a network of communications covering the whole globe. The basic stuff out of which Vidal fashioned his analyses is no longer to be found in Europe and North America, though it is still to be seen in parts of Africa, Asia and South America. Advanced communities are no longer local, no longer fundamentally rural, no longer characterized in their material culture by a host of features which are not to be seen elsewhere.

Vidal de la Blache regretted what he could not help but observe. He considered that much that was best in the life of France arose out of the range and balance of regional communities to be found there. He considered, like many of his French contemporaries, that the moral qualities of rural life were important to the nation and feared their decay. It was only natural also that he should regret the onset of a train of events which was in time to make nonsense of his life's work, but it is a measure of his stature as a scholar that he not only saw that the change was coming, but also suggested how sense would come to be made of the new order of things. He noted in the *France de l'Est*, for example, that the organizing principle of economic life in the future would be the relationship of an area to the metropolitan centre to which it was subservient, that is a relationship born of ease of access to an urban centre, rather than a rural relationship between man and land such as had been so long the case.[2]

The work of Vidal de la Blache was, of course, only one of a range

[1] In December 1963 there were 862,000 persons employed in agriculture, forestry and fishing out of a national total of 24,234,000, or about 3·6 per cent.

[2] 'The idea of the region in its modern form is a conception to do with industry; it is associated with that of the industrial metropolis' (Vidal de la Blache, 1917, p. 163).

of variants within the general 'regional' view of geography. This view gained widest currency between the wars under the umbrella title of landscape or *Landschaft* geography.[1] At the end of the inter-war period Dickinson (1939), argued that landscape geography was the main recent development in the subject and that Britain was out of step in being much less fully committed than the continent to this view of the subject whose acceptance he strongly urged. Landscape geography developed many forms, and it is necessarily hazardous to generalize, but in outline the starkest form of landscape geography consisted in the examination of all that was visible on the surface of the earth (Brunhes' use of the idea of surveying the surface of the earth from a balloon illustrates both what was intended and the vintage of the concept), and the investigation of the characteristic associations of phenomena which existed there. A part of that which is visible consists of the works of man, part may be chiefly natural, a great deal is, as it were, an admixture of the two – fields, crops, animals, afforested areas, etc. It was frequently remarked that landscape geography provided the discipline with its own peculiar subject matter. This 'classical' geography had not possessed. 'Moreover, a science cannot be defined on the basis of particular causal relationships; it must have a definite body of material for investigation. The recognition of this fact, and the direction of research to the study of landscape and society on the lines here presented, is the most significant and widespread trend of post-war geography in other countries' (Dickinson, 1939, p. 8). Landscape geography developed a scholasticism of its own so that there were arguments about whether a house as a more permanent and fixed feature of the landscape was more 'geographical' than a man; and in a broader context there was much argument about how far it was proper to take into account things which were not visible. Landscape geographers of a purist inclination either excluded economic geography altogether or wished to see it relegated to the status of an associated subject rather than an integral part of the discipline, because it was by nature systematic and involved the consideration of abstract and general principles rather than the concrete reality of the landscape.[2] Landscape

[1] The terms are used here loosely and in a very general sense. Hartshorne devotes a chapter to the problems of the correct use of the terms; see Hartshorne, *The Nature of Geography* (1939), pp. 149–74.

[2] See, e.g. M. A. Lefèvre, *Principes et problèmes de géographie humaine* (1945), esp. pp. 29–30 and 195–6. The book in general is an interesting and typical example of landscape geography of a rather strict type.

geography had its strengths – the inter-war generation of continental scholars produced much of interest and value in conformity with its precepts – but it suffered from many weaknesses. Above all it suffered from the same fundamental shortcoming which afflicted Vidal's work. All variants of the 'regional' view of geography are at their best when dealing with areas of rural, local economies. All are ill at ease when dealing with areas thoroughly caught up in the Industrial Revolution. Yet the 'regional' period of geographical methodology, like the 'classical', has left many traces, some of which will perhaps prove permanent, on the methods used in organizing and presenting geographical material. Any discipline is both the product and the victim of its own past successes and these were two of the most important successes thrown up by geographical scholarship.[1]

WHAT REPLACES 'REGIONAL' GEOGRAPHY?

The view that the study of the region and regional life was the peculiar crown and peak of geographical work, that which held the subject together, that which solved most of the methodological difficulties which had become apparent in 'classical' geography by the turn of the present century, is no longer tenable. There has been, however, little, if any, retreat from regional geography, if by that one means the study of things in association in area, which still affords endless opportunities for *ad hoc* studies. The regional method thus remains the means for much geographical work but is no longer its end. One may say that much geography is still regional, but no longer that geography is about the region. What we have seen is a concept overtaken by the course of historical change. 'Regional' geography in the great mould has been as much a victim of the Industrial Revolution as the peasant, landed society, the horse and the village community, and for the same reason.

Granted that the 'regional' concept of geographical methodology has lost its general appeal because the western world has changed so rapidly, one might argue that it is possible to adopt two attitudes to

---

[1] These are not, of course, the *only* two general views of the subject which have gained wide currency. There is, for example, the inversion of the 'classical' system, the view that geography is the study of the effect of man on land, of the effect of the material culture of societies on local ecological, hydrological and other systems. This has supported and still supports a very interesting literature.

the question of the place of regional analysis in geographical work when dealing with communities which have been drawn into the Industrial Revolution. One may hold either that with the final disappearance of the old local, rural, largely self-sufficient way of life the centrality of regional work to geography has been permanently affected – that in one respect we are back with the 'classical' early nineteenth-century position where regional study was important but on the whole less important than systematic study. Or, secondly, one may argue that what departed with an older type of economic life was only one type of regional study and that the general significance of the region to geographical work remains unchanged. On this view all that has happened is that there are now different fish to catch in the sea, fish which escape the older sort of net which Vidal made so well, but which are worth catching and can be had with different types of net or methods of fishing.[1]

On the whole the first of the two alternatives seems the truer. The line of intellectual descent which began with von Thünen and J. G. Kohl and leads down through Alfred Weber, Christaller, Lösch and Isard has perhaps supplied the most fruitful of the ideas which have enabled geographers to tackle the question of the regional ordering and functioning of economy and society in post-industrial communities, and their thinking is, of course, systematic in nature, though very flexible for use in special studies. It is true that in a sense the Industrial Revolution has made possible a degree of regional differentiation of economic activity which was not possible earlier and in this way brought out regional distinctiveness with a sensitivity not previously seen. In the days of substantial local self-sufficiency, for example, corn was grown for local consumption in many parts of England where today the land is largely down to grass. Each agricultural area of Britain can realize advantages of site and soil which remained potential only as long as transport was expensive and uncertain and markets little developed. Or again, to contrast the agricultural economy of California, with its specialization on a range of crops peculiarly suited to its climate and irrigation possibilities, with the agricultural economy of an area physically similar, such as central Chile, where a wider range of basic foodstuffs is grown, is to see the implication of modern high-speed, cheap communications and easy access to great consuming centres. But these changes and the

---

[1] Either view is consistent with the assertion that regional and systematic studies are both necessary. This was recognized in the main by both 'classical' and 'regional' geographers. The difference is a matter of emphasis.

contemporary pattern of regional specialization are only intelligible in terms not of one region but of a whole congerie of interlocked economies. Furthermore, the great bulk of employment in modern industrial countries is to be found in secondary and tertiary occupations, not on the land, and in these days when in almost all industries the most important locating factor is accessibility to the major markets, this means that a systematic treatment alone holds out hope of understanding.

It is notoriously difficult to be clear about the trend of contemporary events, but the question which naturally suggests itself at this point in the argument is, of course, whether, granted that the two great earlier conceptions of the subject are inadequate, some new over-arching conception of the nature and methodology of the subject is emerging. Since the situation is alive and changing, any observation about this is likely to contain an element of advocacy as well as analysis. This section of the essay will have served its purpose, therefore, if it helps to stimulate further thought.

One may begin by remarking that geography and geographers do not live in intellectual isolation. Both 'classical' and 'regional' methodologies of geography were closely related to the general intellectual history of their day. The 'classical' view of geography was a natural avenue to explore given the contemporary assumptions about the nature of scientific knowledge and the way in which it should be organized. Equally the 'regional' idea in the hands of a man like Brunhes has many affinities with 'functional' social anthropology[1] and with the whole argument at the turn of the last century about the methodology appropriate to the social as opposed to the physical sciences. In the same way many contemporary developments in geographical technique and in ideas about geographical methodology are linked to thought in the social and physical sciences in a wider context.

One aspect of this which appears to be of singular importance is the application of statistical concepts and devices to many new areas of study. This is true of social sciences like economics where in the last generation econometrics had greatly flourished; of the biological sciences; and of more utilitarian branches of study like operational

---

[1] The meaning which Brunhes attached to the word *connexité* and its importance in his organization of material make this clear. There are many points of resemblance between what he was attempting to do for geography and what men like Malinowski or Raymond Firth wished to do for social anthropology.

research where the use of statistical techniques and computers has made possible rational planning of such things as the holding of stocks or the optimal use of machinery.

The use of statistical techniques has spread rapidly in geography in the last generation. Statistical methods are now commonly used in dealing with questions like the testing of regional boundaries; the spacing, size and areas of influence of settlements; locational theory; migratory movements; characteristic crop combinations and plant associations; and a host of geomorphological and hydrological questions. Sometimes such studies make use of available statistical techniques; sometimes modifications are used designed to help especially with the measurement of association in area.[1] Although geographers have been rather slow to make use of the opportunities offered, statistical techniques are peculiarly well suited to geographical problems for two reasons. In the first place some statistical techniques are capable of bringing into a meaningful relationship to one another a large number of variables which may be only rather weakly correlated with one another, or which may be significantly related only when combined and considered in groups. This characteristic is a godsend to geography because geographers have often wished to hold within the focus of their attention a large number of rather disparate factors, all of which are thought to be of some importance, but they have commonly been unable to establish the nature or degree of intensity of relationship between the many elements in the situation. Attempts to overcome the difficulty by intuitive assessments have often been strikingly unsuccessful, leaving the impression that more had been bitten off than could conveniently be chewed. If the welter of possible interconnections can be examined statistically and an accurate measurement of correlation made a much firmer foundation for analysis is available. It needs to be said repeatedly that statistics is an aid to good judgement rather than a substitute for it, but it is a very powerful aid and enables many problems to be examined again or for the first time which could not be tackled previously, or tackled only perfunctorily without statistical aid.

Secondly, statistical techniques are likely to be attractive to geography because they help with one of the subject's most intractable methodological problems of the last three-quarters of a century since the decay of the 'classical' school, the recurring worry about the best way to accommodate in a single discipline a physical and a social

[1] See Chapter 9.

side.[1] This issue was always at its most sensitive when the question of the influence of the physical environment on social change and functioning was raised, but arose in other connexions also. Experience made men wary of treating these questions in an '*A* caused *B*' way yet geographers continued to be interested in questions which demanded a knowledge both of society and environment. It was in this connexion that Brunhes advocated the idea of *connexité*, interconnection, functional relationship. The use of statistical techniques permits greater precision in these matters and, if properly used, helps in avoiding some pitfalls frequently visited by the unwary in the past. In short both the practical and methodological problems of geography are such that the use of statistical techniques is likely to exercise a strong attraction.

It may be remarked that it is one thing to say that statistical techniques are very useful and another to say that this is a development equivalent to the rise of the 'classical' or 'regional' conceptions of the subject; that statistical techniques are only a tool; that they cannot be a methodology in themselves; that they cannot supply a general vision of the subject or give it the sort of unity that the older conceptions, whatever their defects, provided. This is true. Geographical writing and research work has in recent years lacked any generally accepted, overall view of the subject even though techniques have proliferated. This, where it has been recognized, has been widely regarded as a bad thing. A unifying vision is a very comforting thing, but one may perhaps question whether it is as vital a thing as is sometimes supposed. Without it there is always a danger of a slow drifting apart of the congerie of interests which together make up a subject. With it, on the other hand, there is also danger from rigidity and from the creation of an orthodoxy. At all events it is arguable that the best sign of health is the production of good research work rather than the manufacture of general methodologies, though perhaps the two together are to be preferred to either singly if they form a fruitful harmony, as in the work of Vidal de la Blache.

Perhaps the most sensible attitude now as at other times to adopt towards the question of method in geography is to be eclectic – to use whichever method of analysis, Blachian or systematic, landscape or Löschian, appears to offer the best hope of dealing with the problem

[1] This is one of the most fundamental questions in the recent history of geographical methodology and deserves study on a much larger scale. It is perhaps the best point of departure for a history of geographical writing in the last hundred years since it is central to so much else.

in hand. There is no reason, for example, why a study similar to that of the *France de l'Est* or others done by the followers of Vidal should not be written for many parts of the underdeveloped world today where life is still essentially local and rural. However, for reasons already touched upon, geographers tend to find it hard to leave it at that. Both the older 'classical' school, because of the question to which they addressed themselves, and the followers of Vidal de la Blache, because of the way in which they conceived of the region, held the physical and social halves of geography very firmly together, indeed in the second case they they were held to form a seamless robe which could not be perceived except as a unity. If both the view of geography as the investigation of the effects of the physical environment upon social functioning and development, and the view of it as the examination of the region in the manner of Vidal de la Blache are rejected there is an evident danger that the links between the two halves of the subject will be weakened. This makes little or no difference to individual pieces of research. There are now and have always been since the days of Humboldt and Ritter, and indeed much earlier, particular pieces of work in which a knowledge of both social and physical geography have been essential, and others which were purely physical or social. But in the general sense, viewed overall, the connexion is weakened.

It would be mistaken to suppose that this is a new situation, that we are moving away from a period when there were no uncertainties, seeing the breaking up of something which in times past was a firm whole. Even at the beginning of modern geography the same problem was present. When Ritter wrote the introduction to the *Erdkunde* he intended to survey each of the earth's continents in turn. In the event he was able to deal only with Africa and parts of Asia. This is a pity because he also said that he considered Europe to be different from other continents (following Hegel) because hers was not a static but a developing civilization, and he suggested that whereas the study of the local physical environment might provide many clues to the functioning of the static civilizations of other continents it would be of much less utility in Europe.[1] In other words he foresaw that when dealing with Europe there would be difficulty in holding the two sides of the subject in the same close conjunction that was possible elsewhere. From Ritter's day to the present this and similar questions have

[1] See e.g. Ritter, *Allgemeine Erdkunde* (1862, p. 229). Buckle has much of interest to say on this theme: see especially *Civilization in England*, Vol. 1, Chap. 2.

given rise to discussion, at some times desultory, at others urgent. The degree of overlap between the two halves of the subject has always varied from topic to topic. This made the existence of an overall vision the more important if a clear connexion between the two were judged of supreme importance. In general the more backward in material culture, and the more rural in nature, the closer the evident connexion: while the connexion is least obvious in the industrial and urban countries of today. An eclectic attitude towards geographical method in work on the 'advanced' countries today will tend to underline the comparatively slight degree of overlap which exists.[1]

Handsome is as handsome does. The final test of the value of any intellectual labour is its ability to help men to understand questions in which they are interested. Men such as Ritter and Vidal de la Blache in their day succeeded notably in this. Questions of method in geography in the future as in the past will be decided by the quality of the work produced by men of different methodological persuasions. Progress lies in rejecting conceptions which are no longer fruitful in favour of those which can help the understanding. All such schema are provisional: in time they will be replaced by others which meet contemporary needs better. Intellectual development is a continuing process of modification, rejection, addition and replacement of conceptual tools. The more fully past experience in geography is digested the more likely it is that contemporary discussions will be productive. *Reculer pour mieux sauter* is good advice here as in other connexions. Only when the merits of long-standing methods are seen in their original setting can their present utility be adequately judged. The most complete prisoners of the past are those who are unconscious of it.

[1] Forde contributed an interesting article to the little burst of methodological writing which occurred in the *Scottish Geographical Magazine* in 1939, and in dealing with a similar issue came to the conclusion that 'Actually the human geographer stands in need of a knowledge of physical environment to precisely the same degree as do the archaeologist, the ethnographer, and the economic historian' (Forde, 1939, pp. 229–30). And the published work of many human geographers bears him out in his contention.

References

BRUNHES, J., 1925, *La Géographie humaine*; 2 vols, 3rd ed. (Paris).

BUCKLE, H. T., 1857–61, *History of Civilization in England*; 2 vols (London).

DICKINSON, R. E., 1939, 'Landscape and Society', *Scot. Geog. Mag.*, 55, 1–15.

FORDE, C. D., 1939, 'Human Geography, History and Sociology', *Scot. Geog. Mag.*, 55, 217–35.

HARTSHORNE, R., 1939, *The Nature of Geography* (Lancaster, Pa.).

HEIDELBERGER GEOGRAPHISCHE ARBEITEN, 1960, *Alfred Hettner: Gedenkschrift zum 100 Geburtstag* (Heidelberg).

HUMBOLDT, A. von, 1849, *Cosmos*; Transl. by E. C. Otté (London).

LEFEVRE, M. A., 1945, *Principes et problèmes de géographie humaine* (Brussels).

RATZEL, F., 1882–91, *Anthropogeographie*; 2 vols (Stuttgart).

RITTER, K., 1862, *Allgemeine Erdkunde* (Berlin).

VIDAL DE LA BLACHE, P., 1911, *Tableau de la géographie de la France* (Paris).

— 1917, *La France de l'Est* (Paris).

— 1922, *Principes de la géographie humaine* (Paris).

CHAPTER TWO

A Re-evaluation of the Geomorphic System of W. M. Davis

R. J. CHORLEY

Lecturer in Geography, University of Cambridge

As the work of William Morris Davis recedes further into the past and becomes more and more identified with the basic structure of what might be termed 'classical geomorphology' its outlines blur and the impression which it commonly produces is one of intangible strength. His voluminous, repetitive, but often subtly-modulated writings appear to us as through a haze of secondary interpretation producing unreal optical effects, not the least striking of which is that the cycle of erosion concept seems large enough to embrace the whole of geomorphological reality. An added distortion results from Davis' avowedly prime position as a teacher, for the achievement of a teacher must be judged largely by the effects which his teachings produce. Thus, an evaluation of the cycle of erosion theory requires that one distinguishes between the stated intention of Davis as conveyed by his writings, the implicit and often unstated assumptions underlying his work, the sometimes distorted interpretations of Davis' work stemming from his students, and, finally, the effect of his teaching on succeeding generations of geomorphologists.

Increasingly, however, since the death of Davis in 1934, criticisms of the cyclic approach to landform study have been intensifying, but the vague and confused form taken by these criticisms and, in particular, the form in which they have reached teachers has not permitted the true character of these objections to appear clearly. It is true to recognize also that the obvious teaching qualities possessed by the cycle concept have hardened the resistance of teachers to these criticisms, and it must be stressed at the outset that modern objections to the Davisian approach have not developed because the cycle has been found to be a totally inappropriate vehicle for geomorphic thought or teaching, but because its restrictive and highly-specialized, built-in characteristics have been high-lighted by recent investigations.

c

The cycle of erosion is thus now being recognized as merely one framework within which geomorphology may be viewed, wherein those aspects of landforms which are susceptible to progressive, sequential and irreversible change through time are especially stressed (just as the system of Euclid is now considered as merely one of many 'geometries'). The cycle is no more a complete and exclusive definition of geomorphic reality than the pronouncement by the proverbial Indian blind man on feeling an elephant's leg that the animal is like a tree. What has happened in the last thirty years or so is that, to continue the metaphor, other blind geomorphologistshave been feeling the geomorphological elephant's trunk and sides and are variously describing it as being like a snake or a wall. It is under-standable that the equally-blind 'onlookers' should have become confused and vaguely resentful, particularly because they have been brought up to believe that trees are much more rational, beautiful and believable things than either snakes or walls!

THE MODEL

The strongest and most compelling feature of the cycle of erosion concept is that it presents many of the features of a theoretical model (Chorley, 1964). As distinct from classification, which merely involves the dissection and categorizing of information in some convenient manner, model-building requires the identification and association of some supposedly significant aspects of reality into a working system which seems to possess some special properties of intellectual stimula-tion. This stimulative quality, often resulting from the special juxtaposition of information which is the very foundation of the structure (one is almost tempted to say the 'artistic form') of the model, finds expression in an enlargement of what is thought of as 'reality' (i.e. involving the kind of scientific prediction which resulted from the construction of Newton's model). It was thus very character-istic of Davis' intellectual achievement that after he developed his cyclic model he was able to say that he could think of many more landforms than he could find examples of in the field! A moment's reflection on the magnitude of possible combinations of structure, process and stage in landforms shows exactly what he meant. It is therefore in the bringing together of certain aspects of the 'web of reality', stripped of other considerations, into a clear-cut theoretical model that much of the intellectual attractiveness and teaching strength of the cycle lies. Davis (1909, pp. 253-4) knit certain aspects of landforms together into a meaningful association both in space within a given landscape and in time throughout an assumed evolutionary history. In order to understand many of the special characteristics of the cyclic theory, and in particular to recognize both its strengths and limitations, it is profitable to consider it in the light of three of the properties common to all such models – their essentially theoretical character, the inherent need for the discarding or 'pruning' of much information, and the fact that no part of reality can be uniquely and completely built into any one model. Davis (1909, p. 281) himself recognized and defended the theoretical nature of his model, writing '. . . the scheme of the cycle is not meant to include any actual examples at all, because it is by intention a scheme of the imagination and not a matter for observation; yet it should be accompanied, tested, and corrected by a collection of actual examples that match just as many of its elements as possible'. Rather than being surprised by such a statement, one should recognize this as a very characteristic state of mind for the model builder in a natural science, where the subject matter of even a small part of reality has usually to be accepted in uncontrollable mutual associations. The result is that there is an attendently large 'elbow room' within which the researcher may select, organize and interpret his material, introducing a subjective bias into all work – good or bad. This brings one to the second model property, that much possible information relating to even a small part of reality has to be rejected in order that the rest (i.e. that information and those relationships which appear especially significant or interesting) may be presented in sharp outline. All models caricature reality by this pruning, but the most successful ones (e.g. that of Newton) are those wherein that which remains still retains some observable or testable significance as far as the 'real world' is concerned. Davis' cyclic model is heavily pruned, such that changes in the geometry of erosional landforms through time emerge as the central theme. When he excluded the possible effects of climatic change or of progressive movements of base level from his scheme, Davis was not (as he asserted, e.g. 1909, p. 283) doing so to facilitate and simplify his *explanation* but to make the cyclic scheme *possible at all*. One can only imagine what would have remained of the cycle if Davis had permitted the possible effects of continuous movements of base level to have been superimposed upon those associated with the progressive subaerial degradation of the landmass. This is, in fact, just what Walther Penck attempted to do, and is the reason why his model is much more confused and unsatisfactory than the cycle. The third property of models which I think is appropriate here follows directly from the second. If discrimination and selection operate in terms of the building of information into a model structure then no single model can form a universally appropriate approximation to a segment of reality. It is interesting that it was one of Davis' most faithful supporters, Nevin Fenneman (1936), who, almost inadvertently one feels, stated this most cogently: '. . . the cycle itself is not a physical process but a philosophical conception. It contemplates erosion in one of its aspects, that of changing form. But erosion does not always and everywhere present this aspect. . . .' 'Cycles have parts and the parts make wholes, and the wholes may be counted like apples. Non-cyclic erosion can only be measured like cider. There is neither part nor whole, only much or little.' It is around the essentially non-cyclic model of Grove Karl Gilbert (Gilbert, 1877, Chapter 5; Chorley, 1962) that much modern thought is centring.

THE DOGMA OF PROGRESSIVE, IRREVERSIBLE AND SEQUENTIAL CHANGE

Another feature of the cycle of erosion concept, one which is basic to the whole reasoning underlying it, is the tacit assumption that the amount of energy available for the transformation of landforms is a simple and direct function of relief or of angle of slope. This unformulated, but nevertheless real, assumption on the part of Davis (and one which seems so logical in the abstract as to be unquestionably accepted as an axiom) is that rates of mass transfer by all agencies are greater on steeper slopes than on less steep ones. From this assumption many others are deduced – some apparently true and others, often not so apparently, untrue. The ideas, for example, that steep slopes are eroded faster than less steep ones and that stream velocity is solely dependent on bed slope, derive from this axiom, and lead inevitably to the conclusion (Davis, 1909, pp. 255–6) that rates of change of landforms, as well as their geometrical magnitude, are direct functions of local relief. It follows, therefore, that the progressive changes of relief during the consumption of a landmass by erosion are held to be universally associated with a progressive landscape evolution wherein the geometry of individual landforms and the rates of their erosional change are both subject to sequential transformations through time. Considering individual valley-side slope elements and stream reaches, for example, this reasoning leads to the assumption that they are progressively transformed into lower and lower energy (i.e. gradient) forms as the general relief is reduced following late youth (Davis, 1909, pp. 268–9). The study of change is therefore the guiding purpose of the 'geographical cycle', wherein a sequence of sketchily-treated changes leading to ill-defined conditions of 'grade' (Chorley, 1962) in stream channels and slopes, is followed by a progressive, sequential and irreversible transformation of virtually all aspects of landforms as the potential energy (i.e. relief) of the system is dissipated. Although it is obvious that the general reduction of a landmass by erosion must be associated with broad changes involving in the long run the replacement of steeper slopes by less steep ones, modern research is indicating that the relationship between gradient and rate of mass transfer is more partial and complex than Davis assumed (this word is deliberately employed in that Davis never attempted to test this axiom), such that some aspects of landscape geometry may be relatively unchanging throughout large segments of 'cyclic time', whereas the detailed pattern of change of others may be neither progressive nor continual. Recent works relating gradient and process by Leopold (1953) and Young (1960) have respectively demonstrated, for example, that both mean velocity and bed velocity of rivers *increase* downstream (the increasing depth of flow more than offsetting the decrease of bed slope, as embodied in the eighteenth-century Chézy flow formula), and that the rate of soil creep on some slopes seems to be more strongly controlled by the frequency and amount of moisture changes than by the slope angle.

In order to illustrate both how the built-in sequential assumptions of Davis inevitably lead to highly stylized and restricted concepts of change of form through time and how modern research is questioning these concepts, we can profitably examine three aspects of landscape geometry – drainage density, erosional slopes and river meanders.

A cyclic interpretation of the development of drainage density (the total length of stream channels in a given area divided by the area) was given by Glock (1931; see also Wooldridge and Morgan, 1959, p. 173) wherein drastic changes through time were inferred, but Melton (1957) and Strahler (1958) have shown that the factors which seem to exercise the most important control over drainage density are those of rainfall characteristics, infiltration, surface resistance to erosion and runoff intensity, rather than those (e.g. relief) which might dictate significant or progressive changes of drainage density through time. It seems that, virtually independent of relief or 'stage' in any cyclic scheme, drainage density (which is probably the most important single parameter of landscape geometry) is most characteristic of the rainfall/infiltration characteristics of a region and thus may be relatively unconnected with stage throughout long periods of erosional history.

Valley-side slope (which combines with drainage density, relief and upper slope curvature to virtually define the geometry of erosional landscapes) was treated by Davis (1909, pp. 266–9) as beginning steep and irregularly covered with coarse debris in youth and as getting progressively less steep with age as it becomes composed of a thickening mantle of finer and finer debris. As has been mentioned above, there can be no dispute as to the decrease of *average* erosional slope angle as relief is lowered, but modern work is shedding some interesting light on the detailed pattern of the development of slope elements within the broad framework of surface degradation (which is, after all, a concept which predates Davis by several hundreds of years). It is now patently apparent that it is completely unrealistic to hold rigid views as to any unique pattern of slope development, even within a given region of uniform structure, lithology or climate. The geometry of valley-side slopes is controlled by a number of interlocked variables which may operate in very different combinations and magnitudes. Thus, within a single climatic environment some slopes may recline whereas others retreat parallel (Schumm, 1956 B); in the same lithological environment and 'stage' of dissection adjacent recline and parallel retreat can be deduced (Strahler, 1950 A, p. 804); and in some limiting cases a steepening of slopes can be deduced through part of their history (Carter and Chorley, 1961).

Davis (1909, p. 265) associated the development of meanders with the practical cessation of downcutting at 'grade' when the continuance of outward cutting changes the curves of youth into systematic meanders of radius proportional to the river's discharge, which increase in size progressively as the gradient of the flood plain lowers during the subsequent progress of the cycle. Thus the existence of meanders, together with their magnitude, is viewed as having some time-significance in terms of 'stage' within the cycle. The vexed question of grade has been touched on elsewhere (Chorley, 1962), but it is profitable here to examine briefly how modern research bears on Davis' interpretation of meanders. At the outset it is important to recognize two elements of the problem which are interconnected but quite distinct – the question of the initiation of meanders and that of the relationship between meander form and fluvial processes once the meanders have developed. Obviously the first question is the more difficult one, but it is now reasonably certain that the regularly spaced pools and shallows (riffles) of meandering rivers are comparable with similar features in straight streams (Leopold and Wolman, 1957; Leopold, Wolman and Miller, 1964), that the deposition of these riffles probably occurs during the time of falling discharge (when the decreasing thread of water becomes more and more deflected), and that meandering can develop in streams which are in the broadest sense, aggrading, degrading, or 'poised' (Matthes, 1941).

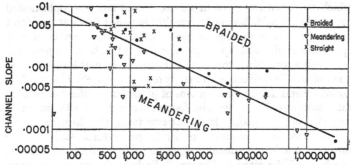

FIG. 2.1. *The control over the meandering or braided condition of rivers exercised by bankfull discharge and channel slope (after Leopold and Wolman, 1957).*

The relationships between meander geometry and discharge have been exhaustively examined in actual rivers, irrigation canals and in model flumes, but some recent work by Leopold and Wolman (1957), involving observations directed towards the second of the questions identified above has had an important bearing on the first question. Leopold and Wolman found that whether streams meander or braid seems to depend on a fairly simple multiple relationship between bankfull discharge and channel slope (the latter being partly a function of the calibre of the bed material) (Figure 2.1). It is thus apparent that whether a stream develops a sinuous meandering course on unconsolidated flood plain material is really determined by *chance* relationships between discharge and slope (calibre) and does not *per se* have any time or 'stage' significance. Having said this, it is only fair to add that with the passage of time the decrease of bed calibre and channel slope makes the *chance* of obtaining a meandering condition (as distinct from braiding or straightness) *more likely*, but this is a rather different statement from that of Davis. In short, many geomorphic form changes are now being viewed from a more *ad hoc* standpoint than they were by Davis.

When one compares the assumptions of Davis with the findings of more recent workers regarding the three examples given above, an outstanding feature of Davis' reasoning becomes abundantly clear. This is that given aspects of landscape result simply from a small number of given causes – often from one single cause – which operate through time in a progressive and sequential manner. This reasoning, which was often employed in a more blatant and less sophisticated manner by Davis' followers, has resulted in two of the most significant features of the reasoning processes which have been commonly applied to cyclic geomorphology. The first is that both the processes responsible for a given topographic form and its past history can be unambiguously deduced from a study of the form itself. Thus modern notions that different combinations of processes or different histories can result in similar topographic forms are at variance with much that underlies cyclic thinking. In its extreme forms this disregard of the highly ambiguous character of landscape features leads to disturbingly uncritical treatments of geomorphic problems, an example of which will be given in Chapter 8. Perhaps the most outstanding instance of this is the naïve view that river terraces and nickpoints are almost invariably associated with movements of base level, and one has only to turn to the work of Lewis (1944) and Yatsu (1955) – the former showing that terraces can be produced by the varying of stream load/discharge relationships, and the latter that breaks of stream slope can be associated with changes in calibre of the bed load – to see the limitations of this assumption. The second result of the lack of a multivariate view of reality nourished by the cyclic approach is that geomorphic processes tend often to be viewed in an over-simplified manner. Everyone is familiar with such arguments as 'more rain means more erosion' (i.e. desert wadis must have been excavated during more pluvial past conditions) or that 'the intense rainstorms of desert areas are associated with high rates of erosion'. Recent work by Langbein and Schumm (1958) has indicated that rates of sediment yield and erosion tend to be at a maximum in climates having a rainfall of about 12 inches per year (i.e. where rainfall is high enough to cause substantial erosion and vegetation not dense enough to prevent it) and possibly only rising to another maximum at rainfalls exceeding 50 or 60 inches where the impeding effect of vegetation cannot be increased by further increase of precipitation (Figure 2.2).

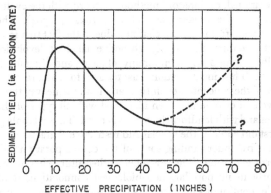

FIG. 2.2. *A schematic suggestion of variations in sediment yield (i.e. erosion rate) associated with variations in precipitation (after Langbein and Schumm, 1958, and Fournier).*

THE INTELLECTUAL SETTING OF THE 'CYCLE'

A proper understanding of the specialized character of the cycle of erosion concept can be achieved only by considering it in the context of nineteenth-century thought in the natural sciences, for, despite the deceptively youthful appearance which its teaching facility has maintained, the cyclic notion was first mentioned by Davis as long ago as 1884 and first stated in fairly complete terms only five years later (Davis, 1889). At this time the writings of Herbert Spencer and others were extending the concept of evolution from the biological into the physical, social and mental spheres such that it seemed to form a basic organizational framework for the whole world of experience. Although prominent Harvard philosophers of the time were resisting this extension (Leighly, 1955, p. 312), there is no doubt that the idea of organic evolution was one of the most important mainsprings of the cycle of erosion theory. In his first statement of the cycle notion Davis (1884) termed it a *cycle of life* in which, as he later wrote, 'land forms, like organic forms, shall be studied in view of their evolution', (Davis, 1909, p. 279), such that the cyclic concept has the 'capacity to set forth the reasonableness of land forms and to replace the arbitrary, empirical methods of description formerly in universal use, by a rational, explanatory method in accord with the evolutionary philosophy of the modern era' (Davis, 1922, p. 594). The main problem arising from this association is due to the fact that in the later nineteenth century the highly attractive label of 'evolution' had practically become a synonym for any 'change', and often for 'history' in general. This identification has tended to obscure the *special character* of the concept of evolution, and in the same way the concept of the cycle of erosion has been identified with all types of change in landforms and with landform history in general. Thus it is only possible to understand some of the special and restrictive characteristics of the cyclic idea by understanding some of the contemporary implications of the term 'evolution'. The late-nineteenth-century view of evolution, particularly in its popular, non-biological connotation, implied an inevitable, continuous and irreversible process of change producing an orderly sequence of transformations, wherein earlier forms could be considered as stages in a sequence leading to later forms. 'Time' thus became, at least for many of those concerned with adapting the evolutionary notion to wider fields, almost synonymous with 'development' and 'change', such that it was viewed not merely as a temporal framework within which events occur but *a process itself*. It was in this sense that Davis employed the concept of evolution as a basis for the cycle of erosion, and it is easy to see why what Fenneman termed 'non-cyclic erosion' seems just as inconsequential to the cyclic concept as the lack of sequential development of certain biological organisms through long time periods seemed to the theory of evolution.

In other ways, too, Davis' synthesis was typical of nineteenth-century scholarship in general and of geographical scholarship in particular. The emphasis upon historical sequences rather than functional associations, the reconnaissance and artistic basis of his field work, and the stress laid upon causal description are features of Davis' work which make it appear most antique to the modern student of landforms. Davis followed Ritter in his concept of the scope and nature of geography, wherein human activities were subordinated to, and largely based upon, the main features of the physical environment. There are many overtones of the 'landschaft' concept of geography implicit in Davis' reasoning, in that it is assumed that the landscape features contain within themselves the unambiguous evidence of their origin. While stressing the Victorian character of much of Davis' work it is only fair to note that he departed from the characteristic standards of much nineteenth-century work in the natural sciences in three important particulars; his lack of detailed field observations, his unconcern with details of processes prompting change, and the entirely qualitative nature of his methods.

This last characteristic leads us to a further feature of the geographical cycle which militates against its popularity with modern workers – its highly dialectical and semantic quality. Anyone at all familiar with the voluminous writings of Davis cannot but be struck with the essentially verbal logic which he employed, characterized by his obsessive concern over terminology. Much of this resulted, of course, from the theoretical basis of his work, but in its extreme form this preoccupation resulted in 'research by debate' (e.g. Symposium, 1940). Speaking of the reaction of the Davisian geomorphologist confronted by a radically different approach to the subject, Bryan (1940, p. 254) wrote . . . 'Slightly bemused by long, though mild intoxication on the limpid prose of Davis' remarkable essays, he wakes with a gasp to realize that in considering the important question of slope he has always substituted words for knowledge, phrases for critical observation.' Again, to be fair to Davis, it must be recognized that he never intended his cyclic theory to be 'scientific' – at least in the sense that the term is currently employed, but there can be no doubt that the long-term effect of his work was to take a whole branch of natural science often intimately concerned with mass, force, resistance, rates of change, and many of the other basic parameters of physics and to effectively divorce it from the main stream of scientific thought. Until the Second World War geomorphology developed very much like a private game, played by comparatively few initiates most of whom were unable or unwilling to draw upon the general body of scientific experience. Thus, although the avowed aim of the cyclic approach to landforms was to provide a general view of the degradational succession of erosional forms, its effect (particularly when yoked to the concept of denudation chronology) was to throw the emphasis upon historical studies of special regions. This idiographic attitude has always found favour with geographers since the collapse of nineteenth-century determinism, but its application in geomorphology successfully isolated the subject from every other science except a small segment of historical geology. Largely deprived of the stimulus of cross-fertilization, geomorphology in the half century after 1890 developed by in-breeding into a highly-stylized discipline wherein the keen edge of research was blunted, and lacking the active professional echelon concerned with practical problems which during the same period, for example, transformed the sister science of meteorology.

THE DEVELOPMENT OF THE CYCLE

In the academic life cycle of William Morris Davis old age was succeeded by rejuvenation and, even as atrophy was setting in with regard to the subject in general, he was during the period 1920–34 revising many of his earlier views. The irony is that what is most easily available to students today as the 'essential' teaching of Davis are certain of his essays written prior to 1909 and the writings of his most influential students. Pre-eminent in both these respects was the editor of *Geographical Essays*, Douglas Wilson Johnson, the Professor of Geomorphology at Columbia University. In his teaching Johnson, who until his death in 1944 held a foremost – almost dictatorial – position in American geomorphology, followed in detail the approach to the subject adopted by Davis in his middle years (Strahler, 1950 B), differing from him only on the spelling of 'peneplain' (peneplane). Davis, however, showed remarkable versatility after the age of seventy, modifying his views on peneplanation and the youthful stage (1922), recognizing the lack of real differences between many humid and arid landforms (1930 A), and acknowledging after setting up permanent residence in California in 1928 the difficulty of applying simple cyclical notions to an area of active orogeny.

However, the scheme of the cycle represented such a compelling geomorphic framework that in a few decades it was extended into all branches of geomorphology, usually much less satisfactorily than had been its application to 'normal' fluvial features. Davis was variously responsible for these extensions: the arid cycle was wholly worked out by Davis (1905, 1909, pp. 296–322), as was the cycle of upland glacial erosion (Davis, 1900; 1909, pp. 658–66); and the suggestion by Davis (1896; 1909, p. 709) regarding the stages of shoreline development was inflated by Johnson (1919) into cycles of submergence and emergence. In contrast, however, he viewed karstic features as developing only as an early mature stage in the normal cycle (1930 B) although a whole cycle of karst development had been deduced in the meantime (Cvijić, 1918, following Sawički and others). The reasons for the lack of current popularity for these cyclic extensions are precisely those which account for the decline of the 'normal' cycle, although in the former instances they are usually more obvious than in the latter. The most important reason has been the specialized and restrictive nature of the initial assumptions, for example, that the arid cycle is referred to block-faulted basin and range structures and that the destruction of glacial mountains is almost entirely attributed to cirque-cutting. The other problem is the one to which I have already referred with regard to the normal cycle, that of the misinterpretations regarding the successive development of topographic forms attendant upon a very imperfect knowledge of process. Thus Shepard (1960) has shown, for example, that barrier beaches (offshore bars), far from being criteria for coastal emergence, are now being observed to develop in association with stationary and rising sea-levels.

DENUDATION CHRONOLOGY AND THE CYCLE

Of all the apparent developments of the cyclic approach to the study of landforms none has been more important than that relating to 'denudation chronology'. During the first half of the twentieth century studies in the sequential development of erosional forms referred to changes in baselevel formed the mainstream of geomorphological work in Britain, the United States and France, under the influence of S. W. Wooldridge, D. W. Johnson and H. Baulig, although there was a different emphasis on either side of the Atlantic regarding the eustatic or diastrophic nature of the baselevel changes (Chorley, 1963). The relationship of the concept of the cyclic development of landforms to denudation chronology is at once complex and ambiguous, such that the two are commonly confused. It is important to realize that studies of denudation chronology, using the term in precisely the same way as it is currently employed, preceded or accompanied the formulation of the cyclical approach by Davis. In the 1860's and 70's Jukes and Ramsay proposed sequences of landscape development involving changes of baselevel; in the 1880's in the United States Joseph Le Conte interpreted breaks in stream profiles as indicative of the discontinuous lowering of baselevel; and one year before the first really important statement of the cycle McGee (1888) developed an erosional chronology for that part of the

Appalachians later to be made classic by Davis (1889), Johnson (1931) and many others. The notion of cyclic change fitted so well into the interpretation of landforms directed pre-eminently towards an evaluation of baselevel changes (which is the real aim of students of denudation chronology) that the two approaches merged, and it is with surprise that we now realize that, in terms of research (as distinct from teaching) all that remains of the cyclic concept can be measured largely in terms of its reinforcement of denudation chronology. The closeness of this association tells us a great deal both about the character of denudation chronology and the cycle. The former, relying often upon highly ambiguous evidence, assumed like the cycle the character of a highly stylized game indulged in by a free-masonry who after commiting themselves to certain basic initial steps of faith (e.g. topographic flat means stillstand; higher is older and lower is younger; uplift is generally discontinuous, etc.) reached conclusions which seem often to be more a product of the means of analysis rather than a physical reality. To adapt an expression of Sauer's (1925, p. 52), many studies of denudation chronology look like the products of men set out to 'bag their own decoys'. However, the best and most convincing studies of this type (e.g. Wooldridge and Linton, 1955) rely for their conviction and satisfaction upon evidence provided by deposits of known origin and date. This permits the true character of denudation chronology to emerge, and it becomes patently apparent that it represents, as McGee stated quite clearly, a branch of *historical geology* in which the central theme is the interpretation of past forms rather than the full understanding of the present landscape. Of course, these two aims may occasionally coincide, particularly when landforms are changing at a slow rate, but an historical pre-occupation for its own sake cannot provide a universal basis for the discipline of geomorphology. Despite Davis' protestations as to the essentially geographical nature of his cyclic approach, the same preoccupation with the deduction of past forms is apparent, and the expulsion of geomorphology from American geography some forty years ago was largely due to this feature of Davis' emphasis. It is significant that, according to Johnson (1929, p. 209), in his last years Davis came to the view that most of his writings were not strictly geographical in character.

THE 'GEOGRAPHICAL CYCLE' AND GEOGRAPHY

This brings us to the last, and undoubtedly the most presently appropriate question which must be asked: What is the relevance of the cyclic basis of geomorphology to current geographical teaching? It has, I think, become apparent that the supposed geographical significance of the cycle assumed by Davis was based upon his essentially antique view of the nature of the subject. In terms of the modern, man-oriented geographical synthesis the explanatory description of landforms as a function of deduced origins has never been satisfactorily assimilated, despite the efforts of S. W. Wooldridge. Even the most sophisticated treatments of geographical 'regions' or 'landscapes' have not satisfactorily circumvented this difficulty, while the standard geographical text commonly presents the reader with a ritualistic introduction of dead, undigested and largely irrelevant physical information. Even when this is presented within a cyclic framework the relevance of this material does not increase (usually the reverse!), and this is largely the result of the difference in the time scales which are involved. Commonly the 'yesterday' of historical geography is still the 'today' of geomorphology. Russell (1949) put some of the difficulties of integrating a past-oriented geomorphology into the body of geography when he wrote: 'Geographers ordinarily find difficulty in discovering useful information in the conclusions of the pure morphologist. That a particular river is a consequent stream with an obsequent extension, or that some part of a river is superimposed rather than antecedent, or that a windgap suggests a cause of stream piracy, really means little to the person working on the problems of some specific cultural landscape. . . . The geomorphologist may concern himself deeply with questions of structures, process, and time, but the geographer wants specific information along the lines of what, where, and how much' (1949, pp. 3–4). Neither can it be contended, however, that the modern 'quantitative' approach to geomorphology which I shall treat in a later chapter is any more 'geographical' in character than its predecessor but, almost paradoxically, these studies which have been attacked as 'removing geomorphology from geography' are in the process of providing *as a by-product* just that basically relevant geographical material called for by Russell. A geographer does not want to know that a stream is 'mature' but what discharges have been recorded for it, not that a river terrace may possibly represent an interglacial event but information regarding its dimensions and composition, not that a slope form may indicate the poly-cyclic origin of the valley but details of its geometry and soil characteristics. This is not to degrade either geography or geomorphology, for both are quite distinct disciplines and are each proceeding to higher and different syntheses. It is no more the exclusive aim of the geomorphologist to provide physical data for the geographer, than it is for the geographer to content himself with a mass of physical information unintegrated into his human theme.

References

BRYAN, K., 1940, 'The Retreat of Slopes', *Ann. Assn. Amer. Geog.*, **30**, 254–68.

— 1941, 'Physiography 1888–1938', *Geol. Soc. Amer.*, *50th Anniversary Vol.*, 1–15.

CARTER, C. S. and CHORLEY, R. J., 1961, 'Early Slope Development in an Expanding Stream System', *Geol. Mag.*, **98**, 117–30.

CHORLEY, R. J., 1962, 'Geomorphology and General Systems Theory', *U.S. Geol. Survey, Prof. Paper*, *500–B*, 10 pp.

— 1963, 'Diastrophic Background to Twentieth-century Geomorphological Thought', *Bull. Geol. Soc. Amer.*, **74**, 953–70.

— 1964, 'Geography and Analogue Theory', *Ann. Assn. Amer. Geog.*, **51**, 127–37.

CVIJIĆ, J., 1918, 'Hydrographie souterraine et évolution morphologique du karst', *Rec. des Trav. de l'Inst. de Géog. alpine* (Grenoble), 6(4), 56 pp.

DAVIS, W. M., 1884, 'Geographic Classification, Illustrated by a Study of Plains, Plateaus and their Derivatives', *Proc. Amer. Assn. Adv. Sci.*, **33**, 428–32.

— 1889, 'The Rivers and Valleys of Pennsylvania', *Nat. Geog. Mag.*, **1**, 183–253 (also *Geographical Essays*).

— 1896, 'The Outline of Cape Cod', *Proc. Amer. Acad. Arts and Sciences*, **31**, 303–32 (also *Geographical Essays*).

— 1899, 'The Geographical Cycle', *Geog. Jour.*, **14**, 481–504 (also *Geographical Essays*).

— 1900, 'Glacial Erosion in France, Switzerland and Norway', *Proc. Boston Soc. Nat. Hist.*, **29**, 273–322 (also *Geographical Essays*).

— 1904, 'Complications of the Geographical Cycle', *Proc. Eighth Int. Geog. Congr.* (Washington), 150–63 (also *Geographical Essays*).

— 1905, 'The Geographical Cycle in an Arid Climate', *Jour. Geol.*, **13**, 381–407 (also *Geographical Essays*).

— 1909, *Geographical Essays* (Boston), 777 pp.

DAVIS, W. M., 1922, 'Peneplains and the Geographical Cycle', *Bull. Geol. Soc. Amer.*, 33, 587–98.

— 1930 A, 'Rock Floors in Arid and Humid Climates', *Jour. Geol.*, 38, 1–27 and 136–58.

— 1930 B, 'Origin of Limestone Caverns', *Bull. Geol. Soc. Amer.*, 41, 475–628.

FENNEMAN, N. M., 1936, 'Cyclic and Non-Cyclic Aspects of Erosion', *Science*, 83, 87–94.

GILBERT, G. K., 1877, *The Geology of the Henry Mountains*, U.S. Dept. of the Interior (Washington) (Chapter 5, Land Sculpture).

GLOCK, W. S., 1931, 'The Development of Drainage Systems', *Geog. Rev.*, 21, 475–82.

JOHNSON, D. W., 1919, *Shore Processes and Shoreline Development* (New York), 584 pp.

— 1929, 'The Geographic Prospect', *Ann. Assn. Amer. Geog.*, 19, 167–231.

— 1931, *Stream Sculpture on the Atlantic Slope* (New York), 142 pp.

LANGBEIN, W. B. and SCHUMM, S. A., 1958, 'Yield of Sediment in Relation to Mean Annual Precipitation', *Trans. Amer. Geophys. Union*, 39, 1076–84.

LEIGHLY, J., 1955, 'What has happened to Physical Geography?', *Ann. Assn. Amer. Geog.*, 45, 309–18.

LEOPOLD, L. B., 1953, 'Downstream Change of Velocity in Rivers', *Amer. Jour. Sci.*, 251, 606–24.

LEOPOLD, L. B. and WOLMAN, M. G., 1957, 'River Channel Patterns: Braided, Meandering and Straight', *U.S. Geol. Survey, Prof. Paper*, 282–B, 39–85.

LEOPOLD, L. B., WOLMAN, M. G. and MILLER, J. P., 1964, *Fluvial Processes in Geomorphology* (San Francisco), 522 pp.

LEWIS, W. V., 1944, 'Stream Trough Experiments and Terrace Formation', *Geol. Mag.*, 81, 241–53.

MCGEE, W. J., 1888, 'Three Formations on the Middle Atlantic Slope', *Amer. Jour. Sci.*, 3rd Ser., 35, 120–43, 328–30, 367–88 and 448–66.

MATTHES, G. H., 1941, 'Basic Aspects of Stream Meanders', *Trans. Amer. Geophys. Union*, Pt. 3, 632–6.

MELTON, M. A., 1957, 'An Analysis of the Relations among Elements of Climate, Surface Properties, and Geomorphology', *Office of Naval Research Project NR 389–042*, Tech. Rept. 11, Dept. of Geol., Columbia Univ., New York, 102 pp.

RUSSELL, R. J., 1949, 'Geographical Geomorphology', *Ann. Assn. Amer. Geog.*, 39, 1–11.

SAUER, C. O., 1925, 'The Morphology of Landscape', *Univ. of Calif. Pubs. in Geog.*, 2, 19–53.

D

SCHUMM, S. A., 1956 A, 'Evolution of Drainage Systems and Slopes in Badlands at Perth Amboy, New Jersey', *Bull. Geol. Soc. Amer.*, **67**, 597–646.

— 1956 B, 'The Role of Creep and Rainwash on the Retreat of Badland Slopes', *Amer. Jour. Sci.*, **254**, 693–706.

SHEPARD, F. P., 1960, 'Gulf Coast Barriers', in 'Recent Sediments, North-west Gulf of Mexico', ed. by F. P. Shepard, F. B. Phleger and T. H. Van Andel, *Amer. Assn. Petroleum Geologists*, Tulsa, 197–220.

STRAHLER, A. N., 1950 A, 'Equilibrium Theory of Erosional Slopes, approached by Frequency Distribution Analysis', *Amer. Jour. Sci.*, **248**, 673–96 and 800–14.

— 1950 B, 'Davis' Concepts of Slope Development viewed in the Light of Recent Quantitative Investigations', *Ann. Assn. Amer. Geog.*, **40**, 209–13.

— 1952, 'Dynamic Basis of Geomorphology', *Bull. Geol. Soc. Amer.*, **63**, 923–38.

— 1958, 'Dimensional Analysis applied to Fluvially Eroded Landforms', *Bull Geol. Soc. Amer.*, **69**, 279–300.

SYMPOSIUM, 1940, 'Walther Penck's Contribution to Geomorphology', *Ann. Assn. Amer. Geog.*, **30**, 219–80.

WOOLDRIDGE, S. W., 1958, 'The Trend of Geomorphology', *Trans. Inst. Brit. Geog.*, No. 25, 29–35.

WOOLDRIDGE, S. W. and LINTON, D. L., 1955, *Structure, Surface and Drainage in South-east England*, 2nd Edn. (London), 176 pp.

WOOLDRIDGE, S. W. and MORGAN, R. S., 1959, *An Outline of Geomorphology*, 2nd. Edn. (London), 409 pp.

YATSU, E., 1955, 'On the Longitudinal Profile of a Graded River', *Trans. Amer. Geophys. Union*, **36**, 655–63.

YOUNG, A., 1960, 'Soil Movement by Denudational Processes on Slopes', *Nature*, **188**, 120–2.

CHAPTER THREE

Some Recent Trends in Climatology

R. P. BECKINSALE

Senior Lecturer in Geography, University of Oxford

Since 1945 all branches of meteorology have made great progress and have stimulated corresponding advances in climatology, although allowance must be made for the usual long educational time-lag. As a rule meteorological progress has been associated with advances in physics, chemistry, observational techniques and in the use of mathematical models, all of which tend to move meteorology outside the sphere of simple climatology or at least to remove many meteorological findings, unless grossly over-simplified, beyond the scope of non-scientific studies. Fortunately, since meteorology is based mainly on scientific principles, many of the narrower aspects of climatology can also be based securely on a simple knowledge of physics and chemistry. Unfortunately, much of traditional climatology demands geographical correlations and global or regional generalizations which at best are unscientific and at worst are little short of incorrect. Thus generalizations which are intended to form a fundamental prop for non-scientific students may become anathema to students with more than a nodding acquaintance with simple atmospheric physics and aerodynamics. Whereas the meteorologist writes for specialists in a specialized way, the climatologist often has the unenviable task of making meteorological data intelligible and useful to non-specialists. The climatologist today has to balance intelligibility and accuracy, and not infrequently it seems better to be intelligible and semi-accurate than unintelligible and accurate. The cardinal rule is that elementary principles of physics and chemistry should not be flouted.

The climatologist usually has the advantage of acute geographical knowledge. He can often come down to earth, though there seems no need always to parachute when a gentler descent from the attractive refuge above the bright blue sky would often lead to a happier landing. Modern meteorology has opened the way to greater climatic reality: climate has never been so real, nor so complicated. An attempt to demonstrate this will be made from recent advances in five

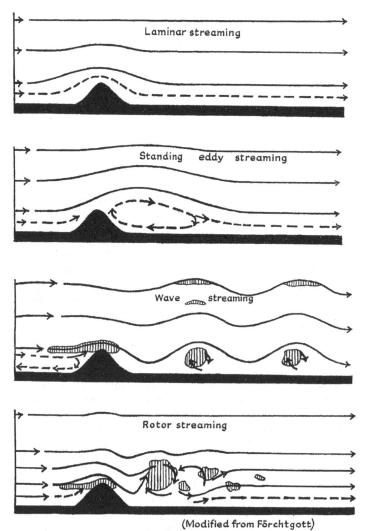

Laminar streaming

Standing eddy streaming

Wave streaming

Rotor streaming

(Modified from Förchtgott)

FIG. 3.1. *Types of airflow over a long ridge.*

interrelated aspects of climatology: airflow; precipitation; airmass definition and fronts; the general circulation; and the nature of the lower atmosphere.

AERODYNAMICS AND PLANETARY AIRFLOW

Aerodynamics

The frequent occurrence of orographically-formed cloud-masses above dip-slopes near scarps and of lenticular, banner, and arched clouds over lee-slopes has long drawn attention to airflow across relief-barriers. Today the use of gliders and of powered aircraft has brought great advances in the aerodynamics of relief-influenced airflow. These problems have been studied, for example, by Scorer (1961) and Corby (1954) (Figure 3.1).

With light winds, laminar streaming occurs and a single shallow wave is uplifted symmetrically above the obstacle. With slightly stronger airflow, the crest of the laminar streaming is displaced downwind and a lee eddy forms, often with important climatic effects. With still stronger winds, the lee eddy is replaced by a series of lee-waves which affects all the lower airflow. With very strong winds, especially if the relief barrier is high compared with the airflow, the symmetrical wave streaming breaks down into a complicated turbulence or 'rotor' streaming. An interesting and well-illustrated study of lee waves in the French Alps, published by Gerbier and Bérenger in 1961, shows clearly the appreciable departures from these typical conditions when wind-speeds do not increase regularly towards the tropopause. The characteristic features of wave-streaming airflow across a long mountain range when the wind-speed increases with altitude have been summarized by Wallington (1958; 1960) and are reproduced in a modified form in Figure 3.2.

The aerodynamic effects of smaller obstacles, such as isolated hills, buildings and tree-belts have also been summarized by Caborn (1955 and 1957) and L. P. Smith (1958, pp. 72–77). The extent of the shelter effect depends mainly on the height and permeability of the wind-break. Generally the reduction of wind velocity begins at about 9 times the shelter-belt height to windward and extends to about 30 times the shelter-belt height to leeward. A solid wind-break may in time of strong winds afford practically no shelter to windward and a marked protection (80 per cent or more wind-speed reduction) for only a short distance to leeward before vigorous turbulence (rotor or highly turbulential streaming) occurs. A slightly permeable shelter-belt will develop an eddy ('air-cushion') to windward and a marked lee eddy, which under average conditions affords a shelter of over 20 per cent wind-reduction to distances of twice the height of the windbreak to windward and of 15 to 20 times its height to leeward.

Planetary Airflow

Much progress is being made in the knowledge of the two main surface-wind systems of the globe, the so-called tropical easterlies

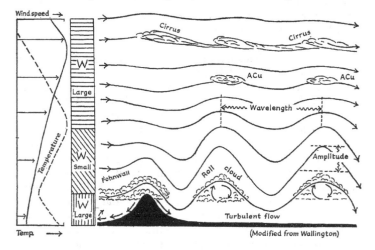

FIG. 3.2. *Features of strong wave-streaming across a long mountain range typically associated with a three-layer troposphere, when an inversion temperature layer in the middle troposphere tends to induce airflow with waves of smaller length and greater amplitude. The maximum streamwave amplitude in the more stable temperature layer is shown. The wavelength is of the order of 2 to 20 miles. ACu denotes altocumulus lenticularis or lenticular cloud. W denotes the general magnitude of the natural or characteristic wavelength determined by airflow and temperature conditions. The decrease of upper wind-speed occurs near the tropopause.*

(Riehl, 1954, pp. 210–34; Koteswaram, 1958) and the sub-tropical and polar westerlies. The former, known as 'Trades' over the oceans, usually play an important or even dominant role in climates between latitudes 32° N and S or upon nearly half the world's area. The two-tier nature of the tropical airspace was recognized a century ago but today the layer-structure and the areal extent of the Trades have been more precisely determined. Crowe (1949 and 1950) has emphasized the great seasonal expansions equatorward and westward of the oceanic Trades from 'root' areas over cold oceanic-upwellings off the west sides of continents. He and Mintz and Dean (1952) show that strong surface Trades occupy 30 million square miles in March and about 40 million square miles in July. Riehl and others (1951 and 1954) have emphasized the vertical stratification of the 'tropical'

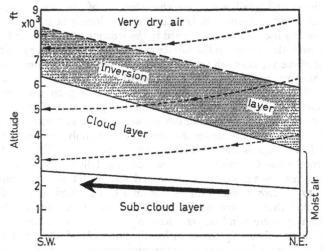

FIG. 3.3. *Schematic cross-section of NE Trades over Pacific between 32° N, 136° W and 21° N, 158° W (Honolulu). Large arrow shows main airflow; arrows on dotted lines indicate general subsidence of airmass through the inversion layer (after Riehl, Yeh, Malkus and La Séur).*

troposphere which, apart from the active portions of disturbances, generally consists of a lower moist and an upper dry layer. The refinement is carried much further. A moist surface-layer, warmed by contact with warm oceans and humidified by evaporation from rough seas, is usually overlain by a strong inversion layer marked by a sharp drop in relative humidity and an appreciable rise in temperature (Figure 3.3). Above is very dry air. The inversion-layer may be near sea-level over sub-tropical littoral deserts but normally in the north Pacific it is at about 4,000 feet in 35° N and rises steadily to

7,000 feet and more inside the tropics where it weakens sufficiently to allow the formation of high cumulo-nimbus clouds and rain. Above this inversion layer the very dry easterly airflow lessens in frequency with height over large areas and, for example, in the northern hemisphere poleward of about 10° N in January and 15° N in July above about 18,000–20,000 feet its mean zonal component changes to westerly or antitrade.

The Westerlies have been described recently by Hare (1960) and Lamb (1959) as two separate vast circumpolar vortices. Whereas steadiness and shallowness are characteristic of the Trades, the Westerlies are extremely variable in direction locally at the surface and usually increase in zonal speed to the tropopause. In winter the polar westerlies prevail at the surface at about 35°–70° N and 35°–62° S but above 16,000 or 18,000 feet (500 mb) westerly flow dominates to within 10° of the equator. The seasonal migrations are considerable; thus in the northern summer the surface or polar westerlies contract northward about 5° latitude and the upper westerlies nearly 10° latitude. The surface westerlies are well documented whereas the upper westerlies have, under the term *jet stream*, only just begun to stimulate a large literature (Riehl, 1962; Reiter, 1963). The World Meteorological Organization recommends the following definition: 'A jet stream is a more or less horizontal, flattened, tubular current, close to the tropopause, with its axis on a line of maximum windspeed and characterized not only by high wind-speeds but also by strong transverse wind shears. Generally speaking, a jet stream is some thousands of kilometres in length, hundreds of kilometres in width, and some kilometres high; the minimum wind speed is 30 m/s at every point on its axis. . . .'

It is perhaps important to emphasize that although the westerlies generally increase in speed and constancy up to the tropopause, or about 30,000 feet in 'temperate' latitudes and above 45,000 to 50,000 feet or more in the sub-tropics, the broad picture of a single upper westerly belt with a central jet stream in the upper troposphere between latitudes 25°–40° N and S is quite inadequate. Usually there are two or three jet streams or ribbons of higher velocity (Figure 3.4). In January the northern hemisphere sub-tropical jet commonly occurs at about 40,000 feet in 30° N but other ribbons of jet occur between 35° and 80° N. Moreover, the jet streams migrate seasonally with the upper (and polar) Westerlies and their horizontal position advances and retreats irrespective of the seasons. The possible role of these non-seasonal migrations has led to the idea of the *Index Cycle*

(Namias, 1950) which is discussed in many meteorological texts. The jet stream also undergoes vertical oscillations, or long waves, with an amplitude of several thousand miles. These oscillations, together with variations in jet stream velocity and in its motion relative to the

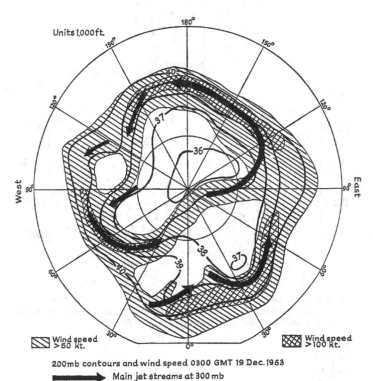

200mb contours and wind speed 0300 GMT 19 Dec.1953
➤ Main jet streams at 300 mb

FIG. 3.4. *Upper troposphere airflow at 0300 GMT on 19 December 1953. The 200 mb contours and wind-speed demonstrate the great strength of the sub-tropical jet stream (over 100 knots) at 38,000 to 40,000 feet, whereas at the same time the circumpolar jet reached its greatest development at or near the 300 mb level (28,000 to 30,000 feet) (after Sawyer).*

earth's rotation, are thought to be associated with cyclogenesis and with the position of 'blocking anticyclones' (Rex, 1951; Sanders, 1953; Sumner, 1959). In fact, the jet stream has become a sort of synoptic panacea, which the causal climatologist will find most attractive.

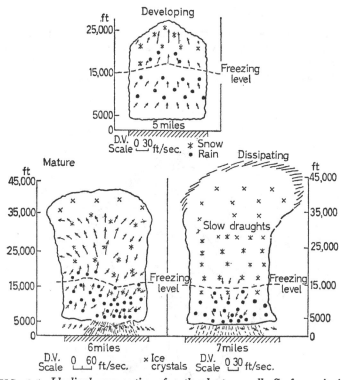

FIG. 3.5. *Idealized cross-section of a thunderstorm cell. Surface rain is shown by dashes; D. V. denotes draught-vector scale, length of arrows being roughly proportional to air-speed (after Byers).*

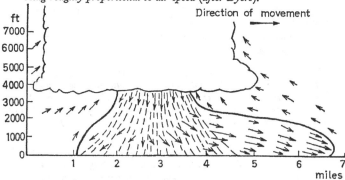

FIG. 3.6. *Schematic section showing the spreading of the downdraught from a thunderstorm cell (after Byers).*

PRECIPITATION

Extraordinary strides have been made recently in the knowledge of precipitation physics, including the processes of formation of cloud-droplets, raindrops and of hail.

Cloud Microphysics: Where the tropopause is high, as in the tropics generally, and in, for example, some warm-air sectors in Britain, the formation of rain at above zero temperatures and without ice-nuclei is firmly established. Here the essential requirement is the presence of some cloud droplets so much above average in size that they fall fast enough to collide and coalesce with smaller droplets. Presumably the larger cloud-drops form on larger or more hygroscopic nuclei.

Outside the tropics, and commonly within, the ice-crystal nuclei mechanism operates. At temperatures below $-40°$ C all cloud droplets freeze automatically or spontaneously but at temperatures between zero and $-40°$ C ice crystals form among supercooled water droplets. These ice crystals grow by accretion at a much faster rate than the water droplets and soon reach small raindrop size. The crystals coalesce as they fall and form snowflakes which nearer earth may melt into raindrops. Details of these cloud microphysical processes, with further references, are given by Mason (1957 and 1959) and Durbin (1961).

Cloud Physics: Modern studies of the thunderstorm and of hail are of great climatological value and prove a real educational asset when such phenomena occur locally. The convectional thunderstorm consists of a collection of cells, each of which may experience a typical life-cycle (Byers and Braham, 1949). The normal cell commonly has a horizontal dimension of about ½ mile up to 6 miles and when triggered off in an unstable airmass may develop at a rapid rate and complete its cycle within 20 to 60 minutes. In the developing stage a slow ascent of air proceeds, usually until ice crystals form (Figure 3.5). Then in the mature stage, precipitation spreads rapidly throughout the cell and the accumulation of water aloft eventually overcomes the ascensional impulses in some lower parts of the cell; a mass of rain-laden air rushes earthward (Figure 3.6) and thrusts before it a violent gust of cold air or a 'line-squall' (Wallington, 1961). Thus the cell is overturned and it now slowly dissipates. However, the sudden down-draught of rain-cooled air from the mature cell usually triggers off the overturning of other cells especially if they are 3 to 5 miles distant. So the thunderstorm rolls across the sky. Although in a

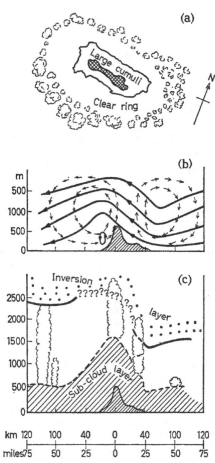

FIG. 3.7. *Schematic reconstruction of day-time orographic-convection cell of tradewind island (Puerto Rico), showing ascent over island and subsidence in surrounding ring on 25 June 1952 (after Malkus). Crosshatching denotes land over 2000 feet; in (b) streamlines of the main airflow are shown in thick lines and those of the local convection cell in thin lines; in (c) the inversion layer was not measured in parts.*

practically uniform air-mass the cells may have a random or irregular pattern, it seems that they usually group themselves into bands or lines. In some areas these 'progressions' may be due to potential cells passing successively across a more or less stationary trigger-action; in others, there may be 'squall-lines' in the lower atmosphere (Soane and Miles, 1954).

The formation of large hailstones has also attracted much attention and also involves the nature of ascensional impulses in convective or ascensional systems (Ludlam, 1961). Large hail stones, in Europe at least, are often associated with some kind of minor front with a steady but sloping updraught, the slope being opposite to the general direction of the storm's movement. Small hail forms in the usual manner in the upward currents and drops out of the top of the cumulus anvil protruding in front of the storm; in falling, it is caught again in the main ascent and again carried up until it becomes so big that it falls as large hail out of the rear side of the uplift and squall-line. To produce really large hailstones the hail particles that re-enter the storm centre must be the right size to grow as to match closely the increase of speed of the updraught and to stay for a relatively long time in the upward currents.

Recently another form of convective air-movement, the orographic cell, has received revived attention. Although the nature of mountain- and valley-breezes has long been imputed to regional as well as local causes, sufficient credit has not hitherto been given to the potency of orographic cells in the tropics and sub-tropics. The diurnal thermal pulsation of mountainous areas, especially of isolated uplands, greatly encourages local convectional over-turnings. High islands even in Trade-wind areas build up orographic cells which weaken and pierce the inversion layer so that rain-clouds form over higher slopes and, to a lesser extent, a cloud-ring develops some distance offshore (Figure 3.7). This convective influence is often generated in spite of the aero-dynamic influences of the relief on the steady Trade airflow (Malkus. 1955).

AIRMASS DEFINITION AND FRONTS

The airmass concept is today accepted generally in climatology and is proving of great geographical value, especially when depicted on equal-area maps instead of on the grossly-misleading Mercator projection. But the airmass is no more rigidly definable than many other 'major regions' used in regional studies. Owing to surface friction and influence the airmass does not move as a strictly rigid unit although it does transpose many of its 'core' characteristics. The idea of airmass depth and of airmass stability or instability aloft seems a necessary addition. The extended shorthand for an airmass then becomes as follows, where Polar maritime airmasses are used as an example:

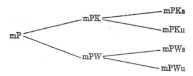

K denotes the airmass is heated from below, W cooled from below, s stable aloft and u unstable aloft. However, the same purpose would also be served if the cross-section of the whole troposphere was always studied in any area.

The airmass concept has already been tied to world climatic classifications, loosely by Flohn (1950 and 1952), and more securely by Miller (1953), Borchert (1953) and Strahler (1960, pp. 189–91). In addition the proportion of airmass and airflow in each month or season is becoming a vital part of regional climatic studies.

It must be admitted that in some areas the airmass which sustains the climate dominates over the fronts which supply the weather. Yet the seasonal change in insolation at the earth's surface and the associated migration of the thermal equator and the fluctuations in areal extent of the main airmasses are so great that the horizontal movement of frontal surfaces can hardly be overemphasized. It seems inevitable that the nature of fronts should provide a constant stimulus to meteorological research, particularly in extra-tropical latitudes where horizontal-mixing of airmasses is prevalent. With an increase of instrumental observations, the nature of temperate-latitude depressions acquires an increasing complexity. As an ideal the simple Abercrombie–Bjerknes–Solberg scheme of a typical 'low' was not likely to be found more frequently than any other ideal, but it was rather oversimplified. To give two examples, cold fronts usually depart from the simple pattern; and correlations with the middle and upper troposphere could not be included scientifically in the earlier patterns as the details were not then available.

Cold fronts were early recognized as being of two main types dependent on whether the warm air was for the most upsliding (anafront) or descending (katafront) at the cold wedge. Subsequently it was increasingly realized that 'there is no *average* cold front', the weather being determined 'by the nature of the warm airmass which is being lifted and by the degree of lift to which the warm air is subjected by the advancing cold air wedge' (*Weather Ways*, 3rd. edn. 1961, p. 80. Met. Branch, Dept. of Transport, Ottawa).

Cold fronts have been studied by Sansom (1961) and Miles (1962). The former showed that the anafront produced greater precipitation and underwent more abrupt changes in temperature and wind, which backed rapidly with height. The katafront revealed only a slight backing of wind with height and brought a rapid clearance followed by fine weather. Miles (1962) has shown that the implicit suggestion that a cold front consists of two distinct airmasses separated by a narrow transition zone is frequently not upheld by observations. Many cold fronts do have a comparatively narrow temperature transition zone aloft but they fail to develop any cloud system sloping up the frontal surface. The type of cold front with moist, cloud-filled air lying above a well-marked cold wedge is, in fact, rare. The commonest type of cold front has a convex air shear zone or temperature transition zone overlain by very dry warm air. Frequently, this warm dry air also protrudes, at 10,000 feet or so, above the warm moist air in the frontal zone which normally extends about fifty miles ahead of the surface cold front. Thus the air above the cold-wedge is very dry and warm while the air ahead of the frontal zone is warm and moist and the change in humidity from very dry to moist commonly occurs in a zone about fifty miles wide almost vertically above the convex snout of the cold-air wedge (Figure 3.8). This humidity transition zone is often the rear edge of the main rain- and cloud-belt associated with the front. Thus the change aloft is in humidity and wind-speed and not in temperature. 'With many well marked surface cold fronts the temperature difference at 700 mb or 500 mb is spread over a distance of at least 200 miles, often without significant horizontal displacement from one level to another' (Miles, 1962, p. 286). It seems that the warm dry air is not necessarily descending in the frontal zone and that it is not very realistic to postulate warm air subsiding down the cold wedge at a katafront.

The search for some causal relationship between surface weather phenomena and upper troposphere conditions in a temperate-latitude depression continues. Frontogenesis or the growth of the depressional

wave is being increasingly related to the jet stream which may encourage or discourage, as the case may be, convergence (subsidence)
or divergence (ascent) at levels in the middle and upper troposphere.
The early concept of frontal depressional models was, in fact, three-
dimensional in outlook but its upper air conditions were based largely

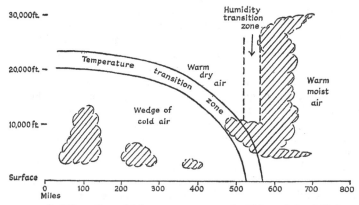

FIG. 3.8. *Schematic model for a common type of cold front (after Miles).*

on theory. Today upper-air soundings allow reliable charts to be
drawn for selected levels over many continental areas at least; the
frontal zones can be studied at various heights and frontal contour
charts compiled. The possibility of linking depressions with the circumpolar jet stream is discussed by Murray and Johnson (1952) and
by Sawyer (1958). Modern ideas of incorporating the possible influence of upper-air conditions on frontogenesis, are expressed in the
three-front model, 'a logical extension of the Bjerknes' model to
include upper air data'. Galloway (1958–60) supplies details of this
scheme which shows the position of the fronts of Arctic, Polar, and
Tropical airmasses at various levels in the troposphere outside the
tropics.

The failure of the simple application of the Bjerknes' depression
model to weather disturbances in the tropics is not surprising. The
dynamic or synoptic meteorology of the tropics remains an academic
battlefield. The old term Tropical Front is upheld by some who
consider it has 'all the properties of a front' (Bergeron, 1954) while
others consider it best to replace the term by the expression 'equatorial
air-stream boundaries' when applied to within 10° N and 10° S of the

equator (Watts, 1955). However, it is generally agreed that various forms of lows, often non-frontal, and of frontal surfaces operate widely and frequently within the tropics. Perturbations or waves in the easterlies are stressed by Palmer (1951) and Riehl (1954), while most studies of the tropical airspace stress the great distances travelled seasonally by 'troughs' and 'fronts'. The rotating tropical cyclone is today well documented (Malkus, 1958; Dunn and Miller, 1960; Neiburger and Wexler, 1961), and a brief geographical commentary on the complexity of causes of tropical rainfall will be found in Beckinsale (1957).

It happens that the climates of tropical territories demonstrate well the great value and the great difficulties inherent in the airmass concept. As these have been summarized by Trewartha (1961) we need only use the Indian monsoon as an example of modern meteorological trends. The meteorological explanations of the burst of the Indian summer monsoon are highly contradictory but they are strongly unified by insistence on the influence of middle and upper tropospheric conditions, including the jet stream (Yin, 1949; Lockwood, 1965, pp. 2–8; Symposium, 1957–8). Normally the sub-tropical jet stream would migrate northward in summer in Asia, but, according to some theories, it lingers unduly along the southern side of the Himalayan–Tibetan mountain arc. Not until the jet stream migrates over or north of the mountain belt does the monsoon burst in with its typical 'frontal' weather. On the other hand, other suggestions impute the trigger-action to thermal changes in tropopause conditions and consider that the onset of the monsoon slightly precedes the main jet stream migration. Whichever theory proves acceptable, the significance of the new explanation is that a synoptic or upper air concept has replaced an oversimplified climatological generalization. Yet recent studies also reveal that land- and sea-breezes are a scientific reality and differential heating of land and sea may be relegated to a minor status but it cannot be entirely eliminated!

CHANGING MODELS OF THE GENERAL CIRCULATION

A large literature has appeared on aspects of the general circulation since Rossby in his famous essay of 1949 suggested that the observed mean zonal wind profile could be satisfactorily accounted for by complete lateral mixing north of a certain latitude and that the zone of maximum westerly flow (now known as the sub-tropical jet stream) would be near or at the equatorward edge of the lateral-mixing region. The eddies involved in the advection were thought to add energy to the mean zonal airflow.

Subsequently, numerous studies were made of the energizing or 'balance requirements' of the general circulation. These studies emphasized the balance of angular momentum, the balance of energy or the transfer of energy from low to high latitudes, and the water balance. Sheppard (1958) has summarized the balance of angular momentum and Tucker (1960, 1961 and 1962) has discussed in detail various aspects of dynamical climatology which he considers 'the major development of climatology during the last decade'. The general trend has been to raise the importance of dynamic processes at the expense of surface or geographical influences.

However, meteorologists do not agree upon the relative importance of solar heating, with its meridional implications, and of lateral mixing or eddy-action in the energizing of the global circulation as a whole. Starr (1956) suggests that the primary meridional circulation caused by solar radiation is disrupted by the Earth's rotation into large-scale eddies (cyclones and anti-cyclones) and that the rotation acts further in channelling these turbulent motions into prevailing east and west winds. These major circulations are thought to derive their energy of motion from the large eddies, which convert potential energy into kinetic energy. The actual circulation is considered to be almost opposite to that often postulated, as air is presumed to be carried downwards in lower latitudes and upward in higher latitudes. However, many will recall a rather similar three-cell general circulation proposed by Bergeron in 1928, although its latitudinal extent was very different (Figure 3.9).

It must be noticed, however, that no meteorologist has satisfactorily explained the balance of the general circulation without assuming some form of mean meridional circulation, at least in the tropics. Thus, wind observations demonstrate a zonal circulation that consists broadly of tropical easterlies, temperate westerlies, and polar easterlies at the surface; and strong temperate-subtropical westerlies and tropical easterlies in the middle and upper troposphere. At the same time, dynamic analyses, theoretical, observational and experimental, indicate certain broad circulation patterns, of which the chief are:

1. A tropical cell which remains a prime, and to some the prime, feature of the general circulation. It is a convective Hadley cell of

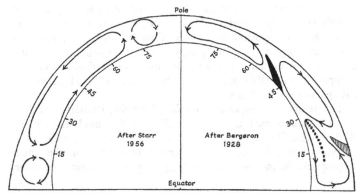

FIG. 3.9. *Old and new ideas on the general meridional circulation of the troposphere.*

remarkable depth for the troposphere, as has been clearly expressed by Palmén (1951 and 1963) (Figure 3.10);

2. In middle latitudes a zone of predominantly lateral or horizontal mixing or in other words a zone where the air-space is usually dominated by large-scale eddies (turbulence) and advection prevails over convection. Here the troposphere is relatively shallow;

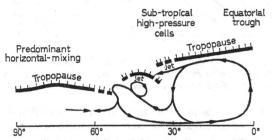

FIG. 3.10. *Scheme for mean meridional circulation in winter (after Palmén). The Polar front, which migrates vast distances, is omitted.*

3. A belt aloft near the junction of (1) and (2) where the upper air westerlies frequently accelerate to form the sub-tropical jet; and

4. Near the poles, some form of air subsidence.

These and other associated features migrate and expand or contract meridionally or latitudinally, or both, with the seasons, allowance being made for the usual time-lags. Many of them vary at least slightly according to surface thermal conditions as well as to upper tropospheric or dynamic conditions. Vertically their components or energy may perhaps be crudely expressed as closed-circuit or cellular zones but horizontally at the surface they can be depicted only as semipermanent cells which frequently allow meridional transfer in the spaces between them.

An attempt to incorporate the ideas of Rossby, Palmén and others into a diagrammatic form useful for climatologists has been made by Birot (1956). Most modern climatological texts show a similar interpretation but a few omit any reference to motions in the upper troposphere in middle latitudes (Figure 3.11). It seems, however,

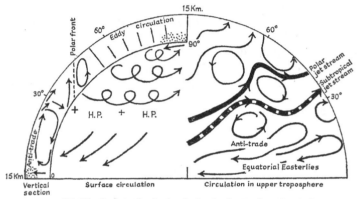

FIG. 3.11. *Model of tropospheric circulation in the northern hemisphere, compiled by Birot after Palmén. The dotted areas in the vertical cross-section denote zones of relative air accumulation.*

unfortunate that some of these generalized diagrams fail to show that the thickness of the troposphere averages about 10 to 12 miles in the tropics and only about 7 miles in mid-latitudes. Needless to say, if the cross-sections of the troposphere shown on global quadrants were drawn true to scale the existence of any general meridional circulation would appear little short of miraculous! Probably the finest exhibition of visual diagrammatic representations of the general atmospheric circulation ever shown was that of the Royal Society (described by Lamb, 1960).

THE NATURE OF THE LOWER ATMOSPHERE

To the progress of the knowledge of the troposphere discussed above must be added the tremendous advances in the knowledge of the lower atmosphere generally. As the ionosphere seems rather outside climatology, reference will be made here only to the ozone region and the stratosphere, or heights between about 60 km and 12 km. The tropopause is now known to be ill-defined and often almost absent above equatorial areas. The lower stratosphere has been shown to have small weather-qualities and to possess many signs of a general meridional circulation. From helium and water-vapour distribution, Brewer (1949) postulated such a circulation sufficient to make 'a significant contribution to the energy of the general circulation'. Subsequently his contentions have been supported by the study of the movement of other tracer substances, such as ozone and radioactive debris (Goldsmith and Brown, 1961; Murgatroyd and Singleton, 1961-2). Newell discusses the problem in detail (1963 and 1964) and, although favouring considerable lateral-mixing, thinks that the effects of 'mean meridional motions cannot be ignored entirely'.

The 'ozonosphere' or region of ozone concentration at about 20 km to 50 km seems to undergo definite seasonal and even diurnal oscillations as well as a general circulation. This layer of atomic oxygen, O_3, absorbs about 5 per cent of the sun's total radiant energy, including nearly all the ultra-violet rays, and forms a warm region overlying the lower stratosphere. Its screening effect is of paramount importance to mankind. Details of its qualities and movements are described fully by Dobson (1963).

There is perhaps no need to warn the climatologist that the effect of changes in the stratosphere and ozone region on the weather of the troposphere is far from being understood. Some of the many secular variations in the circulation pattern of the troposphere are tentatively imputed to stratospheric influences but the extreme complexity of possible correlations are at once evident, as is shown clearly in the relevant chapters of *Changes of Climates* (UNESCO, 1963).

References

GENERAL

BERRY, F. A., BOLLAY, E. AND BEERS, N. R. (eds), 1945, *Handbook of Meteorology* (New York).

BYERS, H. R., 1959, *General Meteorology* (New York).

DOBSON, C. M. B., 1963, *Exploring the Atmosphere* (Oxford).
H.M.S.O., 1961, *A Course in Elementary Meteorology* (London).
MALONE, T. F. (ed.), 1951, *Compendium of Meteorology* (Boston).
PETTERSSEN, S., 1956, *Weather Analysis and Forecasting* (2 vols) (New York).
— 1958, *Introduction to Meteorology* (New York).
SAWYER, J. S., 1957, *The Ways of Weather* (London).
SUTTON, O. G., 1962, *The Challenge of the Atmosphere* (London).
WILLETT, H. C. and SANDERS, F., 1959, *Descriptive Meteorology* (New York).

AERODYNAMICS

CABORN, J. M., 1955, 'The Influence of Shelter-belts on Microclimate', *Quart. Jour. Roy. Met. Soc.*, **81**, 112–15.
— 1957, 'Shelterbelts and Microclimate', *Forestry Commission Bulletin*, **29** (H.M.S.O.).
CORBY, G. C., 1954, 'The Airflow over Mountains', *Quart. Jour. Roy. Met. Soc.*, **80**, 481–521.
GERBIER, N. and BÉRENGER, M., 1961, 'Experimental Studies of Lee Waves in the French Alps', *Quart. Jour. Roy. Met. Soc.*, **87**, 13–23.
SCORER, R. S., 1961, 'Lee Waves in the Atmosphere', *Scientific American*, **204**, 124–34.
SMITH, L. P., 1958, *Farming Weather* (London) (*see especially pp. 72–77*).
WALLINGTON, C. E., 1958, 'Orographic waves . . .', *Met. Mag.*, **87**, 80–87.
— 1960, 'An Introduction to Lee Waves in the Atmosphere', *Weather*, **15**, 269–76.

PLANETARY AIRFLOW

Trades:

CROWE, P. R., 1949, 'The Trade-wind Circulation of the World', *Trans. Inst. Brit. Geog.*, **15**, 37–56.
— 1950, 'The Seasonal Variation in the Strength of the Trades', *Trans. Inst. Brit. Geog.*, **16**, 23–47.
MINTZ, Y. and DEAN, G., 1952, 'The Observed Mean Field of Motion of the Atmosphere', *Geophysical Research Papers*. No. 17, 37–42 (Cambridge, Mass.).
RIEHL, H., 1954, *Tropical Meteorology* (New York).
RIEHL, H. *et al*, 1951, 'The North-east Trade of the Pacific Ocean', *Quart. Jour. Roy. Met. Soc.*, **77**, 598–626.

Westerlies:

HARE, F. K., 1960, 'The Westerlies', *Geog. Rev.*, **50**, 345–67.
LAMB, H. H., 1959, 'The Southern Westerlies', *Quart. Jour. Roy. Met. Soc.*, **85**, 1–23.

Jet Stream:

BERGGREN, R. *et al.*, 1958, 'Observation Characteristics of the Jet Stream', *World Meteor. Organization*, Tech. Note 19, No. 71, T.P. 27 (Geneva).

KOTESWARAM, P., 1958, 'The Easterly Jet Stream in the Tropics', *Tellus*, **10**, 43–57.

NAMIAS, J., 1950, 'The Index Cycle and Its Role in the General Circulation', *Jour. Met.*, **7**, 130–9.

REITER, E. R., 1963, *Jet-Stream Meteorology* (Chicago).

REX, D. F., 1951, 'The Effect of Atlantic Blocking Action Upon European Climate', *Tellus*, **3**, 100–11.

RIEHL, H., 1962, *Jet Streams of the Atmosphere*, Tech. Paper No. 32, Colorado State Univ., 117 pp.

SANDERS, R. A., 1953, 'Blocking Highs Over the Eastern North Atlantic', *Monthly Weather Rev.*, **81**, 67–73.

SAWYER, J. S., 1957, 'Jet Stream Features of the Earth's Atmosphere', *Weather*, **12**, 333–44.

— 1959, 'The Jet Stream', *New Scientist*, **6**, 947–9.

SUMNER, E. J., 1959, 'Blocking Anticyclones in the Atlantic–European Sector of the Northern Hemisphere', *Met. Mag.*, **88**, 300–11.

PRECIPITATION

Cloud Physics:

DURBIN, W. G., 1961, 'An Introduction to Cloud Physics', *Weather*, **16**, 71–82 and 113–25.

MASON, B. J., 1957, *The Physics of Clouds* (Oxford).

— 1959, 'Recent Developments in the Physics of Rain and Rain-making', *Weather*, **14**, 81–97.

Thunderstorm:

BYERS, H. R. and BRAHAM, R. R., 1949, *The Thunderstorm*, U.S. Weather Bureau.

SOANE, C. M. and MILES, V. G., 1955, 'On the Space and Time Distribution of Showers in a Tropical Region', *Quart. Jour. Roy. Met. Soc.*, **81**, 440–8, and **82**, 534–5.

WALLINGTON, C. E., 1961, 'Observations of the Effects of Precipitation Downdraughts', *Weather*, **16**, 35–44.

Hail:

LUDLAM, F. H., 1961, 'The Hailstorm', *Weather*, **16**, 152–62.

Orographic Cells:

MALKUS, J. S., 1955, 'The Effects of a Large Island Upon the Trade-wind Air Stream', *Quart. Jour. Roy. Met. Soc.*, **81**, 538–50, and **82**, 235–8.

AIRMASS DEFINITION AND FRONTS

Airmasses:

BORCHERT, J. R., 1953, 'Regional Differences in World Atmospheric Circulation', *Ann. Assoc. Amer. Geog.*, **43**, 14–26.

FLOHN, H., 1950, 'Neue Anschauungen über die allgemeine zirkulation der Atmosphäre . . .', *Erdkunde*, **4**, 155–9.

— 1952, 'Grundzüge der atmosphärischen zirkulation', *Deutscher Geographentag Frankfurt*, **28**, 105–18.

MILLER, A. A., 1953, 'Air Mass Climatology', *Geog.*, **38**, 55–67.

STRAHLER, A. N., 1960, *Physical Geography* (New York) (*see especially* pp. 189–91).

Fronts:

MILES, M. K., 1962, 'Wind, Temperature and Humidity Distribution at Some Cold Fronts Over S.E. England', *Quart. Jour. Roy. Met. Soc.*, **88**, 286–300.

— 1962, 'Fronts', *Weather*, Schools Suppl. No. 12, 45–48.

SANSOM, H. W., 1951, 'A Study of Cold Fronts Over the British Isles', *Quart. Jour. Roy. Met. Soc.*, **77**, 96–120.

Fronts and Upper Air Conditions:

GALLOWAY J. L., 1958, 'The Three-front Model . . .', *Weather*, **13**, 3–10.

— 1960, 'The Three-front Model, the Developing Depression and the Occluding Process', *Weather*, **15**, 293–309.

MURRAY, R. and JOHNSON, D. H., 1952, 'Structure of the Upper Westerlies', *Quart. Jour. Roy. Met. Soc.*, **78**, 186–99.

SAWYER, J. S., 1958, 'Temperature, Humidity and Cloud Near Fronts in the Middle and Upper Troposphere', *Quart. Jour. Roy. Met. Soc.*, **84**, 375–88.

Tropical Airmass Climatology and Rainfall:

BECKINSALE, R. P., 1957, 'The Nature of Tropical Rainfall', *Tropical Agriculture*, **34**, 76–98.

BERGERON, T., 1954, 'The Problem of Tropical Hurricanes', *Quart. Jour. Roy. Met. Soc.*, **80**, 131–64.

DUNN, G. E. and MILLER, B. I., 1960, *Atlantic Hurricanes* (Louisiana).

LOCKWOOD, J. G., 1965, 'The Indian Monsoon—A Review', *Weather*, **20**, 2–8.

MALKUS, J. S., 1958, 'Tropical Weather Disturbances—Why do so few become hurricanes?', *Weather*, **13**, 75–89.

NEIBURGER, N. and WEXLER, H., 1961, 'Weather Satellites', *Scientific American*, **205**, 80–94.

PALMER, C. E., 1952, 'Tropical Meteorology', *Quart. Jour. Roy. Met. Soc.*, **78**, 126–64.

SYMPOSIUM, 1957–8, 'On the General Circulation over Eastern Asia', *Tellus*, 9, 432–46 and 10, 58–75 and 299–312.

TREWARTHA, G. T., 1961, *The Earth's Problem Climates* (New York).

WATTS, I. E. M., 1955, *Equatorial Weather* (London).

YIN, M. T., 1949, 'A Synoptic-Aerologic Study of the Onset of the Summer Monsoon Over India and Burma', *Jour. Met.*, 6, 393–400.

GENERAL CIRCULATION AND DYNAMICAL CLIMATOLOGY

BIROT, P., 1956, 'Evolution des théories de la circulation atmosphérique générale', *Ann. de Géog.*, 65, 81–97.

BREWER, A. W., 1949, 'Evidence for a World Circulation provided by the Measurements of Helium and Water Vapour Distribution in the Stratosphere', *Quart. Jour. Roy. Met. Soc.*, 75, 351–63.

GOLDSMITH, P., 1962, 'Patterns of Fallout', *Discovery*, 23, 36–42.

GOLDSMITH, P. and BROWN, D., 1961, 'World-wide Circulation of Air Within the Stratosphere', *Nature*, 191, 1033–7.

LAMB, H. H., 1960, 'Representation of the General Atmospheric Circulation', *Met. Mag.*, 89, 319–30.

MURGATROYD, R. J. and SINGLETON, F., 1961–2, 'Possible Meridional Circulation in the Stratosphere and Mesosphere', *Quart. Jour. Roy. Met. Soc.*, 87, 125–35, and 88, 105–7.

NEWELL, R. E., 1963, 'Transfer Through the Tropopause and Within the Stratosphere', *Quart. Jour. Roy. Met. Soc.*, 89, 167–204.

— 1964, 'The Circulation of the Upper Atmosphere', *Scientific American*, 210, 62–74.

PALMÉN, E., 1951, 'The Role of Atmospheric Disturbances in the General Circulation', *Quart. Jour. Roy. Met. Soc.*, 77, 337–54.

PALMÉN, E. and VUORELA, L. A., 1963, 'On the Mean Meridional Circulation in the Northern Hemisphere During the Winter Season', *Quart. Jour. Roy. Met. Soc.*, 89, 131–8.

ROSSBY, C-G., 1949, 'On the Nature of the General Circulation of the Lower Atmosphere', *The Atmosphere of the Earth and Planets*, G. P. Kuiper (ed.), Chicago Univ. Press, 16–48.

SHEPPARD, P. A., 1958, 'The General Circulation of the Atmosphere', *Weather*, 13, 323–36.

STARR, V. P., 1956, 'The General Circulation of the Atmosphere', *Scientific American*, 195, 40–45.

TUCKER, G. B., 1960, 'The Atmospheric Budget of Angular Momentum', *Tellus*, 12, 134–44.

— 1961, 'Some Developments in Climatology during the Last Decade', *Weather*, 16, 391–400.

— 1962, 'The General Circulation of the Atmosphere', *Weather*, 17, 320–40.

UNESCO, 1963, *Changes of Climate*, 485 pp.

CHAPTER FOUR

Geography and Population

E. A. WRIGLEY

Lecturer in Geography, University of Cambridge

People are the stuff of all social sciences and of history, yet it is remarkable how little population as a general concept has entered into the discussion of social change and function until comparatively recently. There were population discussions of importance and some subtlety before the days of Malthus (1798), but he it was who first elevated the study of population to a central place in the social sciences. His views provoked much discussion in his own lifetime and later, but in the main population has tended to be treated as a variable of secondary importance by social scientists and economists until the post-war surge of interest in underdeveloped countries forced it much nearer to the centre of the stage and underlined the importance of some knowledge of the interplay of nuptiality, fertility and mortality and of these in turn with other sociological and economic features of a society. The same neglect of population has been evident in the writing of history. It is only in very recent years that historians have taken population seriously as an element in general history, in spite of the great importance of such things as the impact of epidemic disease, the average age at marriage, the average size of family, infant mortality, expectation of life, and the proportion of celibates in the adult population, to economic, social and general history. The work of Goubert (1960) and others in France has shown how powerfully an interest in population matters can influence the writing of history. Since the Second World War demography has developed distinctively and has produced important works of general theory as well as a mass of technical literature (e.g. Sauvy, 1956–9).

Population is as important to geography as to history, but much the same criticism can be made of geographers as of historians. Population has not perhaps been so badly neglected, but it has often been treated in a rather perfunctory and wooden way, even though the distribution and density of population, and such things as occupational structure and mobility, have long been basic elements in the study of human geography.[1] In recent years a number of prominent geographers have come to place it much higher in the heirarchy of geographical interests (see, e.g. Trewartha, 1953: also Beaujeu-Garnier, 1956 and Zelinsky, 1962). There has, in short, as so often in the past, been a movement in parallel with those taking place in other subjects. I should like to comment on two aspects of the interest in population which seem to me to make it especially fitting for geography and indeed to hold out the prospect of important new developments for the subject.

CHANGING IDEAS ABOUT THE PLACE OF POPULATION IN GEOGRAPHICAL STUDIES

The first point is connected with the same chain of events discussed above.[2] The whole range of problems involved in the attempt to understand why men come to live where they do, and in what numbers, and how they earn their living, which will perhaps do as a thumbnail definition of human geography, has been very substantially changed by the working out of the Industrial Revolution in western countries and elsewhere. The Industrial Revolution first weakened and then largely destroyed the close traditional ties between a society and the local land, and in so doing made anachronistic the vision of geography of men like Vidal de la Blache. His regional conception had been very well able to deal with the location of population, its density and the manner in which it earned its living under the conditions holding good before the Industrial Revolution, but largely failed when confronted with what came after. But though the old answer no longer satisfied, the question was as interesting and important as ever. It remained the case that human geography was very much taken up with the problem of explaining the location, the size and the economic functioning of populations. The population map remained the point of departure for much else.[3]

The striking thing about the general study of the location of populations and of the industries which afford them employment has been

[1] It is interesting to note that Hettner, having commended Süssmilch's early work on the statistical measurement of social characteristics, goes on to describe the second edition of Malthus' *Essay on Population* as almost a geographical work (Hettner, 1927, p. 70).

[2] See Chapter 1 above.

[3] It is interesting to note that Vidal de la Blache made extensive use of population material in the *France de l'Est* (1917). There are only two maps in this remarkable work. Both are population maps.

the way in which it has underlined the point that populations are nowadays more and more their own justification, to put the matter rather cryptically. In the main (the large retired populations of today are an obvious exception to the rule) men and their families live where there is work for them to do. In pre-industrial societies most people lived on the land and the most important factor in understanding the density and distribution of population was the distribution of land that was of good quality given the agricultural and pastoral techniques of the group and period in question (there were considerable changes in both the density and distribution of population in this country between, say, the late pre-Roman period and the time of the Norman Conquest, but they are intelligible within the terms just proposed in the main: new agricultural techniques turned land which had previously been difficult to use into valuable arable). Given any particular range of agricultural techniques the density and distribution of population could be treated as a dependent variable – something which could be understood in terms of the distribution of land of differing quality.

For a time after the Industrial Revolution it appeared to be possible to understand the great new masses of population growing up in areas previously thinly populated by an extension and modification of the old attitude. The large populations which appeared in places like South Wales, central Scotland, the Pittsburgh area, the Ruhr, central Belgium, the Saar and northern France grew up where they did because coal was necessary in large quantities to most industries, and because the total cost of assembling the necessary raw materials, converting them into finished or semi-finished products and delivering them to an intermediate or final consumer was normally less when the industry was on the coalfield than when it was at any other point (Wrigley, 1961, Chapter 1). Because of this it was possible to add to a treatment of the density and distribution of agricultural populations as a function of the distribution of land of different grades, a treatment of the distribution of manufacturing populations in terms of the distribution of mineral resources, and above all coal.[1] The one treatment paralleled the other and the two in combination accounted for the bulk of productive industry and most of the population. The numbers and distribution of the new manufacturing populations could be made to follow on from the consideration of the poverty or

[1] Making use of the line of reasoning which was given its classic expression in the early part of Alfred Weber's *Uber den Standort der Industrien*, available in translation as *Alfred Weber's Theory of Location of Industries* (Friedrich, 1929).

richness of mineral resources just as the distribution and density of agricultural populations followed from a discussion of soils and natural vegetation. In as much as those employed in tertiary industry (government service, retail and wholesale trading, transport and communications, banking and commercial services, and the professions) were taken into consideration at all they could quite conveniently be treated as dependent upon the populations engaged in agriculture and industry for whom they provided services. Further technological and economic change, however, has rendered this method of analysis largely obsolete. The internal combustion engine, the long-distance transmission of electricity, the use of oil and natural gas, the great gains in economy of fuel use, the triumph of the lorry, the tremendous growth in the durable consumer goods industries, and the associated changes in economic life, have meant that in the twentieth century the best location for more and more industries has become a point close to the largest market. A great city grows still greater like a snowball rolling downhill under its own momentum. Only a few industries continue to be raw material orientated – wood and food processing and sometimes iron and steel. The presence of large tertiary populations may stimulate the growth of secondary, manufacturing industry, rather than vice versa.

The model of the great French social scientist Frédéric Le Play may be invoked at this point to restate this argument in a helpful way. Le Play wrote during the middle decades of the nineteenth century a number of works, of which one, *Les Ouvriers européens* (1855) is of especial interest in this connexion. He formulated three heads under which he felt that information should be brought together and analysis undertaken, popularly known in English as Place, Work and People, or in modern terminology, physical environment, material technology and economic organization, and social characteristics (of which one is the distribution and number of population). He felt, just as the 'classical' geographers felt (see Chapter 1), that it was natural to progress from Place through Work to People. Most geographical textbooks have followed this pattern, beginning with a description of the physical environment, then tackling the characteristics of the economy of the area (agriculture and industry), and finally passing on to deal with population distribution and density and with such things as transport, trade and cities (those industries or aspects of the economy peculiarly bound up with tertiary industry). Grouping material under heads like Place, Work and People and dealing with it in this sequence is the time-honoured method.

The great change which the increasing market orientation of industry has involved is this – that it now makes better sense to invert this three-part sequence and to begin with population distribution in order to explain the distribution of industry rather than vice-versa. It is in this sense that one can justify the remark that populations nowadays are their own justification. It is a variation on the old theme that to them that have shall be given. The presence of ten million people in London constitutes a huge market which is very attractive to most types of manufacturing industry given the economic structure of contemporary society. This causes a large proportion of new industrial plant to be built in the London area and this in turn by affording still more employment reinforces the pre-eminence of London. The tendency to market orientation has long been true of some industries, such as the baking of bread or the garment industry. It is discernible in a very wide range of durable consumer goods industries from motor cars to radios, and is to be seen even in industries which in the past have shown a tendency to raw material orientation because they used a great tonnage of raw materials and wished to minimize total transport costs.[1]

The characteristic modern locational pattern can be seen most clearly in countries which escaped the nineteenth-century Industrial Revolution but have subsequently developed a full range of manufacturing industry. In such countries there are no great coalfield industrial agglomerations of the type which were once so typical to complicate matters. An unusually clear-cut illustration of the modern pattern is afforded by the state capitals of Australia which dominate the Australian manufacturing scene. Melbourne, and Sydney, for example, each account for more than three-quarters of the industrial production by value of their respective states. Each houses more than half the state population but has an even higher share of manufacturing output. A few industries, notably those engaged in processing raw materials, like the N.S.W. dairy factories or the Broken Hill lead and zinc refineries, which reduce the bulk of agricultural or mineral raw materials and so lessen the cost of transporting them, are still located close to raw materials or at

---

[1] This is true even of the production of primary iron and steel as the history of the Fontana plant near Los Angeles shows (though it is to be remembered, of course, that large markets for steel are also important sources of steel scrap and are therefore major sources of raw material, as it were).
See e.g. Isard and Cumberland (1950).

places determined by the availability of raw materials, but such industries are unusual nowadays. The iron and steel industry is the only major exception to the rule in Australia today and it is perhaps unfortunate that it should attract so much attention in many geographical analyses of Australian industry. The prime fact which any geography of Australia should seek to drive home and explain is the remarkable concentration of industry and population in the state capitals.

The reason for the change in industrial locations over the last two centuries will be familiar. Put rather simply it may be suggested in this way. If transport costs were astronomically high per ton-mile, it would clearly be essential for each manufacturer to seek that location which reduced his transport costs to a minimum since what he would save in this way would far outweigh any additional expenses in other directions to which he might be put. If, on the other hand, transport costs per ton-mile were nil the consideration of transport costs would not enter into any calculation of best location, though other things like differential wage costs in different areas would continue to do so. What has happened in the last two centuries has produced a movement along the spectrum of possibilities between the two extremes, not of course from one end to the other but a good way along the band.[1] Therefore the old constrictions upon industrial location have weakened, those, for example, which a century ago obliged so many industries to seek a coalfield location, and other factors have a relatively greater play. In these circumstances industry has gone towards the great markets, as, other things being equal, it will always tend to do. There the manufacturer is not only aided by close contact with his main consumers and competitors, but is also usually in the largest and most flexible labour market, is close to specialized banking, commercial and professional services, is at the hub of a great transport network, and so on. Some continuing regional differences, like those between London and provincial wage rates, seem rather to reflect the greater attraction of places like London than to counteract it.

An important additional feature of the modern situation is that tertiary employment is of much greater importance than in the past. The tertiary industries between them often nowadays account for more of the working population than secondary, manufacturing

[1] A more economical and effective use of raw materials (e.g. a sharp fall in the amount of coal needed to produce a ton of finished iron and steel products) has, of course, the same effect as a fall in transport costs.

industries; more sometimes than industry and agriculture combined (see Clark, 1940 and Table 1, p. 72). Rising real incomes per head accompanied this change to tertiary employment in the past and are tending to accentuate the importance of tertiary industries today. If the presence of people and the purchasing power of which they dispose is a prime determinant of industrial location it follows that the existence of large tertiary populations instead of being treated simply as a result of the presence of industry and agriculture should now be seen also as a cause of the growth of industry and indeed of the intensification of agriculture. Because Paris is the seat of the government of France, of the chief university, of many important financial and business houses, of some of the major hospitals, of great entertainment industries, and so on, all affording tertiary employment, it is an attractive location for many manufacturing industries, for example the car industry, and for the same reason agricultural land use for many miles round Paris has been greatly changed and intensified (Phlipponeau, 1956). This is as true as the more familiar idea that because there is a prosperous local agriculture there is a shop in the village, a village postman, a schoolmistress, a garage, and so on. One must no longer see tertiary employment as simply the consequence of the presence of the so-called productive industries; it is also an important cause of their presence, and always was, though of less relative importance in the past.

In view of all this, therefore, it makes good sense to begin any discussion of the distribution and density of population and of the economic activities which support it with the population itself. Instead of working from Place through Work to People, it is better to begin with People and proceed through Work to Place, or in some cases from People through Work back to people – that is from the consideration of the presence of sufficient people to constitute an attractive market to the industries which have developed as a result and so back to the size and structure of the population. The treatment this way round is less rigid than the earlier progression in the opposite direction. It leaves it open to anyone attempting to explain population density and distribution to go as deeply into the circumstance of the physical environment as may prove appropriate for the problem in hand. For example in an area like Andean South America where there are still many largely self-sufficient peasant communities it would clearly be important to pay close attention to the characteristics of the local physical environment after having dealt first with the question of the distribution and density of population and the range of material technology at the disposal of the local communities. But it is also possible where the circumstances of the physical environment are unimportant largely to ignore it (as for example in dealing with population and economic activity in London and its vicinity). There is in this procedure no commitment to the priority of the physical environment such as is implied in the 'classical' progression from the treatment of physical environment to the local community, from Place through Work to People. The older procedure was congenial to explanation in terms of cause and effect; the newer to functional analysis and the use of models.

POPULATION AS A CENTRAL FOCUS FOR GEOGRAPHY

The second aspect of an interest in population which may make it especially attractive to geography is that it holds out some promise of being a satisfactory central focus for the subject, a convenient nexus into which all strands can be seen to lead. If it be asked whether there is a common focus for the great mass of published geographical work, ranging from the study of the relationship between wheat yield and precipitation in Kansas to the intensity with which the New Zealand rail network is used or the settlement hierarchy of southern Germany, one possible answer is that they all serve to make clearer either directly or at one remove the circumstances which permit populations of such and such a size, distributed in such and such a fashion, to maintain themselves by such and such activities which afford them a livelihood. Just as in the nineteenth century all the cluster of geographical studies, though very diverse in detail, might be held to point towards the understanding of the ways in which the physical environment influenced social change and function; just as the study of types of farm architecture and of the functional layout of farm buildings or village settlements in the early decades of this century was not only of interest in itself but was related in the minds of the scholars who studied these questions to the more general regional scheme of geography which they had in their mental background (see Chapter 1); so one might suppose that the explanation of the density and distribution of population could serve the same purpose for geography today. This in modern circumstances can be both the starting and the finishing point. A knowledge of the simple facts of population distribution and density is a convenient starting point for the analysis

F

and explanation of these facts, and the course of the argument makes a satisfactory full circle if at its close a fuller understanding of the facts of distribution and density has emerged.

There is nothing very novel or striking about such a suggestion as this. One might say that many of the older types of geography address themselves to the same end, at least in part. The traditional sequence of examination from Place through Work to People, from the physical environment through the material technology of a community to aspects of its social and economic life and organization, did often end with an analysis of population distribution and density among other things. Here the focus lay, at least in the 'classical' treatment of the question, chiefly in the mechanisms by which the physical environment exercised its influence on social change and function, yet there is much common ground between this and the scheme outlined above. Equally the 'regional' view of the subject dominant in the earlier part of this century, with its stress upon the functional interplay between the various elements of landscape, paid much heed to population. Vidal de la Blache, indeed, in the *France de l'Est* (1917) adumbrated a 'population' view of geography in several aspects of his analysis. Any change along the lines suggested can be held to be evolutionary rather than revolutionary; a recombination of some of the traditional elements of geography, the whole to be viewed in a different light; a new focus but much the same set of elements. Not admittedly quite the same set. A line must be drawn somewhere round what can reasonably and fruitfully be included in the subject. It is said that on an average there are only four removes of acquaintance between any two people chosen at random in the population of the United States, between A and B. A knows X who knows Y who knows Z who knows B. In much the same way there is some relationship between all that is taught as geography, between, say, changes in the Cretaceous and the activities of contemporary London, in all the conceptions of geography which have been proposed. There is, after all, connexion between all knowledge, and an endless chain of interconnexion reaching out from any one event to all others. But the temptation to be all-inclusive is one to be resisted. To take *all* that is relevant into account is impossible. It involves too much for the mind to grasp, for any computing programme to handle, for any model to accommodate. One attraction of the view that the chief focus of geographical study should be the attempt to describe and explain the distribution and density of population is that it would enable geographers to be clearer about what should or should not be thought relevant. Things which are relevant only at the second, third, or higher remove from this central question should be scrutinized carefully. A subject develops not only by acquiring new interests but also by dropping old.

The individual scholar is seldom inspired to undertake a piece of research work by his adherence or opposition to a general conception of a subject. For him what is relevant and useful is not determinable in advance. Indeed progress in knowledge has so frequently come by combining interests and techniques from two or more established disciplines that one might be tempted to assert that success comes more often from ignoring disciplinary boundaries than from observing them. Nevertheless, arguments about the general shape and ordering of a subject, about its methodology, are important. They are important because it is natural to ask of any branch of knowledge what its subject matter is and how it is ordered. Upon the answer to this will depend in part the type and quality of people attracted to it. They are important because they help to determine how the subject is taught, how it appears to those beginning their acquaintance with it. And they are important because it is useful constantly to test the compatability and mutual relevance of the components of a field of knowledge, especially in relation to changes in the methodology of cognate subjects. All such arguments form an endless but useful dialogue whose quality both reflects and helps to set the general standard of work and teaching done.

POPULATION AND GEOGRAPHY: AN EXAMPLE

As an illustration of what might be involved in a 'population' view of geography it may be helpful to sketch briefly a study of eastern Australia conceived in this way.

A first look at the density and distribution of population in Australia shows three salient characteristics: that the great bulk of the population is concentrated in the south-east corner of the continent; that very few Australians live more than two hundred miles from the sea; and that the state capitals in four of the five mainland states contain more than half the state population, while there are very few other towns of any size (at the 1961 census Newcastle and Greater Wollongong were the only cities of more than 100,000 people in mainland Australia which were not state capitals). This is the point of departure, and it will also be in a sense the destination. The aim of 'population' geography is to advance the understanding of population distribution and density on the way between these two points.

It is convenient to consider the population as laid down over the surface of Australia in three layers, as it were: a first layer consisting of those engaged in primary industry and their dependants – farmers,

Table 1. *Primary, Secondary and Tertiary Employment* (*as percentages of total employment*)*

A: State
B: Capital city
C: State without capital city

		Primary§	*Secondary*†	*Tertiary*†
New South Wales	A	11·2	29·0	54·4
	B	1·1	36·0	56·4
	C	25·9	19·1	51·4
Victoria	A	11·2	32·7	52·3
	B	1·3	40·1	55·4
	C	29·4	18·9	46·6
Queensland	A	20·4	20·6	54·7
	C	2·3	27·8	66·7
	C	31·7	16·1	47·2

* Calculated from *Census of the Commonwealth of Australia*, 30 June 1954, Part I, Table 6.
§ exc. Mining.
† exc. Electricity, Gas and Water.

pastoralists, miners, and the small number of forest workers and fishermen; a second layer composed of those engaged in secondary manufacturing employments and their families; and a third layer made up of those working in tertiary industries and their families. To consider these three groups in this order is not to imply that any one is more basic than any other. They might indeed just as properly be taken in the reverse order. It is, however, convenient to treat them separately in the first instance since the principles relevant to the understanding of the three layers are substantially different from one another. I shall further simplify this sketch by using illustrations drawn only from the three main eastern states, Victoria, New South Wales and Queensland.

The first layer is numerically much the least important. Primary employment in New South Wales, Victoria and Queensland formed only 11·2, 11·2 and 20·4 per cent respectively of the total work force in 1954. Even if the capital city is excluded from the totals in each of the three states, the figures rise only to 25·9, 29·4 and 31·7 per cent. I shall confine my attention for the sake of brevity to employment in the agricultural and pastoral industries. This is, of course, spread quite differently over the face of eastern Australia from the other main branches of employment, reflecting chiefly the availability of well-watered land and ease of access to market. The general picture is well known. There are huge areas given over to sheep or cattle raising with very slight returns per acre and only a light dusting of population. Such areas figure largely in maps of types of land use but contribute relatively little to the total net value of Australian farm output and afford comparatively little employment. The areas of intensive land use are much smaller but very much more productive. The Riverina-Wimmera, the sugar areas of the Queensland coast, the dairying areas of northern New South Wales and the better mixed farming areas of Victoria do not bulk large on a map but they account for a large part of the farm output of the eastern states. Employment in pastoral and agricultural pursuits naturally reflects the differences between the former and latter types of farming. The importance of the small areas of intensive farming is very marked, though often not sufficiently appreciated. For example, the area of good land inside the arc of the Eastern Highlands, the Riverina-Wimmera, contained 66,666 persons engaged in agricultural and pastoral employment in 1954 out of a total in the three eastern states of 363,899 (18·3 per cent). The importance of this area in agricultural employment alone is even more marked – 43,571 in 163,972 (26·6 per cent). Figures of the net value of agricultural production can be used to underline the same point.[1]

[1] In the first post-war decade the gross value of agricultural output in the Riverina-Wimmera was more than 40 per cent of the total for the three eastern states. For statistical convenience the Riverina-Wimmera is here taken to be the divisions of Riverina and South-Western Slope in New South Wales and the divisions of Wimmera, Northern and Mallee in Victoria.

A consideration of the density and distribution of employment in farming is the quantitative starting point for an examination of those aspects of the material culture and economic structure of the community with which it is associated, and also of the characteristics of the environment which have complemented them in determining the distribution and density of primary employment. In connexion with the Riverina-Wimmera this might involve, for example, the discussion of such things as the structure of agricultural prices within which the Australian farmer operates; the advantages of this part of Australia because of its soil, rainfall and evaporation characteristics; the changes brought about in recent years by the use of subterranean clovers and superphosphates in raising production per acre; the amount and the variability of flow in the Murray and Murrumbidgee rivers compared with those elsewhere in Australia; the significance to the irrigated districts of the progress of the Snowy River Scheme; and so on. A comparison of the trends of agricultural employment and population in the Riverina-Wimmera with those in, say, the Western Districts of New South Wales might also be helpful in bringing to light those factors which in recent decades have tended to lead to a more and more intensive use of the best land rather than the taking in of new land in order to secure an increase of production. The history of the soldier settlement schemes in the Mallee after the First World War makes a very interesting contrast in this connexion with the history of the contemporary settlement in the irrigation districts.

A similar treatment of the sugar growing areas of Queensland would embrace other matters, including, for example, the discussion of the sense in which the policy of the Australian government is the foundation of the sugar economy of the area. Again the treatment of the very scantily populated sheep and cattle grazing areas inland must include the discussion of the peculiar sensitivity of the ecological balance in marginal areas and the great difficulties experienced in securing the provision of social overhead capital for these areas. In dealing with dairying areas of northern New South Wales the significance of access to the Sydney and Brisbane markets needs stressing; and so on.

At the conclusion of the section dealing with the first layer of the population map the reader will have been led as far into related questions as the explanation of the density and distribution of population dependent on primary industry requires. The level at which the discussion is pitched can be varied to suit the needs of the audience. It might not be the same, for example, for those who could be expected to be familiar with writings in the tradition of von Thünen as for those who could not. It is not, in short, geared to a particular technique of analysis but is flexible enough to be used at all levels.

Next in this brief adumbration comes secondary, manufacturing employment. This comprises a much larger share of the total work force than primary employment. Table 1 shows that manufactures in 1954 employed 29·0 per cent of the work force in New South Wales, 32·7 per cent in Victoria, and 20·6 per cent in Queensland. In the capital cities of these three states the figures were 36·0, 40·1, and 27·8 per cent; while in the states without the capital cities the figures were 19·1, 18·9 and 16·1 per cent. The outstanding feature of the distribution of manufacturing employment in Australia is its concentration in a few large cities (more than half the total of persons in secondary employment in the whole of Australia are in Melbourne and Sydney alone). This section might well begin with a discussion of the general circumstances of economic life today which make it more profitable for most manufacturers to establish their plants close to the largest market than at any other point, and also of the conditions in which this generalization does not hold true. The dominance of Sydney within New South Wales may be used as an example of the range of issues which will arise at this stage of the analysis. In 1954 78·0 per cent of the total industrial labour force of New South Wales was to be found in greater Sydney. Table 2 shows the distribution of employment in eight main divisions. In four of the eight divisions more than 70 per cent of employment was concentrated in Sydney. They conform to the rule that the largest market is the best point of manufacture. In the other four divisions the percentages were lower. The lowest, 30·4 per cent, was in Sawmilling and Wood Products where much employment occurs near the source of supply because so much is saved in transport costs by processing the raw material at this point. For this reason many of the workers in this industry were found to be in the Hunter and Manning and North Coast divisions of the state where there are substantial local timber reserves. A similar explanation accounts for the comparatively high percentage of employment outside Sydney in the Food, Drink and Tobacco industry. It is convenient to erect freezing and canning plants, butter and cheese factories, slaughterhouses, etc., in agricultural areas near the point of production because the raw material is perishable or because there is a large loss of weight in the process of manufacture and so a useful saving in transport costs may be had. In the Electricity, Gas and Water group much of the employment is of the service type and is therefore distributed roughly in proportion to the spread of total population with knots of concentrated employment where there are large power stations or gas works.

*Table 2. Industrial Employment in New South Wales**

	Founding, Engineering, etc.	Ships, Vehicles, etc.	Textiles	Clothing, Boots, etc.	Food, Drink & Tobacco	Sawmilling and Wood Products	Paper and Printing	Electricity, Gas & Water	TOTAL
Sydney	82763	38881	13662	38682	31498	5150	25264	18598	317551
Rest of Cumberland	1968	1040	408	920	1735	194	233	500	7984
North Coast	548	846	30	349	2935	3005	384	767	9552
Hunter and Manning	20899	5100	2826	2723	4045	3459	1139	3087	46944
South Coast	13320	626	171	1619	1250	967	310	1194	21933
Tablelands	3418	1393	837	1622	2252	1534	662	1336	15141
Slopes	1228	1464	200	994	2692	1462	563	1013	10654
Plains	136	258	4	73	284	581	64	148	1666
Riverina	242	348	7	155	1211	395	126	236	3093
Western	123	174	2	74	382	197	108	556	1709
Other	36	11	1	18	18	3	14	6	134
Total	124641	50141	18148	47229	48302	16947	28867	27441	436361
Percentage in Sydney	66·4	77·5	75·3	81·9	65·2	30·4	87·5	67·8	72·8

* Calculated from the *Census of the Commonwealth of Australia*, 30 June 1954, Part I, Table 6.

There remain the metal and engineering industries. In many branches of engineering the dominance of Sydney is almost complete as a further breakdown of this group would show, but the manufacture of primary iron and steel takes place not in Sydney but at Newcastle to the north and Greater Wollongong to the south of Sydney where good coking coal outcrops to the sea and the chief raw materials necessary for iron and steel making can be brought together at low cost. Hence the large employment figures in this category in the Hunter and Manning and South Coast divisions. The employment figures make a convenient point of departure for a discussion of the location of iron and steel manufacture. This has been less affected than most other large industries by the tendency to market orientation of the last half century because

Table 3. *Tertiary Employment (expressed as percentages of total employment)**

A: Capital city
B: State without capital city

		Building	Transport	Communication	Finance	Commerce	Public Authority and Professional	Amusement Hotels, etc.
New South Wales	A	7·2	7·5	2·3	3·7	17·7	13·5	6·3
	B	9·8	6·5	2·1	1·6	13·1	10·1	6·0
Victoria	A	7·3	6·2	2·3	3·4	17·2	13·0	6·1
	B	9·9	5·7	2·1	1·5	12·4	9·7	5·4
Queensland	A	9·6	8·1	2·8	3·6	19·7	16·3	6·6
	B	9·4	7·2	1·9	1·7	12·2	9·0	6·0

* Calculated from the *Census of the Commonwealth of Australia*, 30 June 1954, Part I, Table 6.

of the very large weight of raw materials involved and the importance of keeping down the final cost of the product by seeking the point of least cost assembly of the major raw materials involved. As with the food and wood industries it is important in order to understand the distribution of the iron and steel industry to know of the distribution of the raw materials which are used in these industries since the economic and technological circumstances of their production make it profitable to locate manufacture in many cases close to the point at which the raw materials are produced. In other industries where this is not the case there is little point in examining the sources of raw material supply when discussing location. Details of the size and value of production in each industry or industrial area can, of course, be introduced within a framework of analysis of this sort as easily as within a more conventional framework.

Tertiary employment is quite different in its distribution from either primary or secondary. Primary employment is almost entirely outside the great cities. Secondary employment is predominantly within them. Tertiary employment is much more evenly distributed. In 1954, for example, it constituted 56·4 per cent of the employment of Sydney itself, 54·4 per cent in the state as a whole, and 51·4 per cent in the state without its capital city. Table 3 shows the chief differences between Sydney and the rest of the state. In Building employment was relatively greater outside Sydney; in Transport, Communication, and Amusement and Hotels the pattern was similar in the capital city and outside; but in Finance, Commerce and Public Authorities the greater relative importance of tertiary industry in Sydney was marked since these three together employed 34·9 per cent of the total Sydney work force but only 24·8 per cent were similarly employed in the rest of New South Wales. The reason for the difference is very simply understood, at least in outline. As many primary school teachers are needed per thousand children of primary school age in a small country town or in an irrigation district as in Sydney itself; but the employment of men and women to teach in universities can be much more highly concentrated. Branch banks are needed everywhere roughly in proportion to the totals of population but central banking functions can be carried on in one great centre. Hence the distribution of tertiary employment in part reflects the distribution of primary and secondary employment in combination (the primary school teacher or postman type of tertiary employment), but an important element in tertiary employment is not tied in this way but occurs largely in state capitals. Since the existence of this second type of tertiary employment significantly increases the population and purchasing power of the cities in which it occurs it plays an important independent role in stimulating industrial growth. No analysis of the concentration of secondary employment in the Australian state capitals would be complete without reference to the importance of this point. The phenomenon can be observed in its purest form in the case of Canberra where there would be no secondary employment at all were it not for the decision of the Commonwealth in its early days to seek out a new site for a federal capital, but it has also operated powerfully in all the state capitals.

Some considerations necessary to the understanding of the distribution and density of Australian population can be introduced when dealing separately with each layer of the population cake, but some can best be taken after this stage when the population can be treated as a whole. Any treatment of the functional relationship between city and region can then take place. For example, there is scope for an interesting discussion of the differences between Brisbane on the one hand and Sydney or Melbourne on the other within their respective states. Brisbane is very eccentrically placed within Queensland close to the state boundary and contains a much smaller fraction of the population of the state than either Sydney or Melbourne. It is unable because of its position to dominate the industrial, service and administrative activities of Queensland as Sydney and Melbourne do in New South Wales and Victoria. Maryborough, Rockhampton, Mackay, Townsville and other coastal towns are better placed than Brisbane to serve parts of the Queensland hinterland and restrict Brisbane's dominance. The effect of Brisbane's position can be traced in Queensland's economy, transport system and social and political life. The pattern of Queensland road and rail communication is quite different from that of the more southerly states. Jealousy of the southern half of the state has long been a political force to reckon with in the north and gave rise in the past to strong separatist feelings; and so on. An analysis of this sort can be much more illuminating than the bald and misleading assertion that Sydney and Melbourne dominate their respective states to an unhealthy degree. Other general questions of resource utilization, population trends, the water problem in Australia, the transport network, and so on, might also be taken most conveniently at this stage.

CONCLUSIONS

This sketch is too brief to allow an extended discussion, but the prime virtue of the idea of a 'population' geography should be apparent. It is that it provides a nexus in which each line of inquiry can be seen to be anchored; that, to change the metaphor, it makes available a touchstone of relevance which can be applied to any body of material and which will ensure its *connexité* – to use the word to which Brunhes gave currency. If the geography of an area is treated in this way few things which are conventionally present in geographical works need necessarily be excluded, but inclusion would depend upon relevance to the central question of the distribution and density of population. Nor would a 'population' geography imply just a retreat from the periphery to a central citadel. It would involve the acquisition of new interests. This may be seen by comparing it with landscape geography. Whatever the virtues of landscape geography, its weaknesses are very grave. In the circumstances of modern industrial society a great part of the people and the activities by which they make a living escape the net. Unless this is thought an acceptable price something else must replace it. A concept of the subject suitable for a pre-industrial world will not serve for the world today. Geographers interested in modern industrial communities have in recent years materially changed the subject, devising in the process new models and statistical tools to assist them. If it is objected to this that a means should not be confused with an end, that statistical ingenuity needs to be harnessed to some more general conception of what geographers would be at, then perhaps the answer may lie in a 'population' view of the subject.

References

BEAUJEU-GARNIER, J., 1956–8, *Géographie de la population*; 2 vols (Paris).

CLARK, C., 1940, *The Conditions of Economic Progress* (London).

FRIEDRICH, C. J. (ed.), 1929, *Alfred Weber's Theory of the Location of Industry* (Chicago).

GOUBERT, P., 1960, *Beauvais et le Beauvaisis de 1600 à 1730*.

HETTNER, A., 1927, *Die Geographie: Ihre Geschichte, Ihr Wesen, und Ihre Methoden* (Breslau).

ISARD, W. and CUMBERLAND, J. H., 1950, 'New England as a Possible Location for an Integrated Iron and Steel Works', *Econ. Geog.*, **26**, 245–59.

LE PLAY, P. G. F., 1855, *Les Ouvriers européens* (Paris).

MALTHUS, T. R., 1798, *An Essay on the Principle of Population as it affects the Future Improvement of Society* (London).

PHLIPPONEAU, M., 1956, *La Vie rurale de la banlieue parisienne* (Paris).

SAUVY, A., 1956–9, *Théorie génerale de la population*; 2 vols (Paris).

TREWARTHA, G. T., 1953. 'A Case for Population Geography', *Ann. Assn. Amer. Geog.*, **43**, 71–97.

VIDAL DE LA BLACHE, P., 1917, *La France de l'Est* (Paris).

WRIGLEY, E. A., 1961, *Industrial Growth and Population Change* (Cambridge).

ZELINSKY, W., 1962, *A Bibliographic Guide to Population Geography*, Univ. of Chicago, Dept. of Geog., Research Paper 80.

CHAPTER FIVE

Trends in Social Geography

R. E. PAHL

Lecturer in Sociology, University of Kent at Canterbury

The earth's covering of human dwellings is a phenomenon
more geographical, more closely bound to natural conditions, than the earth's covering of human beings itself. . . .
Truly geographical demography is above all the demography of the habitation. [Jean Brunhes, 1920]

No geography can properly be regarded as 'Social' unless it
draws its material from active study of men and women in
their work and homes. [T. W. Freeman, 1961]

Since the idea that 'geographers start from soil, not from society'
(Febvre, 1932, p. 37) was until recently widely held by most geographers, and is indeed still held by some, it is easy to understand
why social geography has been slow to develop. We may define
the field of the subject as the *processes and patterns involved in an
understanding of socially defined populations in their spatial setting.*
The social geographer is thus concerned with 'the lesser divisions of
cities, town and country', which, as Visher noted as early as 1932, are
likely to become more and more sharply differentiated according to
social criteria. Such a definition immediately involves us in a wide
range of problems. By taking socially defined groups and considering
the processes acting upon them the geographer is involved with data
common to other social sciences. However, in the same way that a
geomorphologist has to acquire some knowledge of geology and
physics, so the social geographer has of necessity to have some competence in human ecology and sociology in order to understand the
processes at work in his field of study. Again, the patterns or models
which the social geographer may discern or erect, however closely
based on rigorous quantitative methods, must never be allowed to
dominate thought so that they be made to fit all societies and all places
at all periods of time. The social geographer, as much as the geomorphologist dancing on the coffin of the Davisian cycle, must

constantly test by empirical analysis whatever middle order generalizations may be currently held at a theoretical level.

Social geography must be seen as being more than a loose agglomeration of such sub-divisions as medical geography (Howe, 1963),
religious geography (Boulard, 1960; Zelinsky, 1961), population
geography, linguistic geography (Jones and Griffiths, 1963), and now,
maybe, electoral geography or the geography of education opportunity. This is no more than saying that economic geography is not
limited to the study of the distributions of everything from tea
production to atomic power stations. This is not to say that the
mapping of distributions is not important; much cartographic work
requires great ingenuity in choice of data and construction of indices.
It is simply that distributions, however well presented, are not
enough. Description does not necessarily imply comprehension and
understanding: explanation may involve a wider range of variables
than has hitherto been considered in such work. It is debatable
whether the best map is that which answers or that which provokes
most questions. Just as economic geography is now more concerned
with the theories of the location of economic activity, so social geography has become concerned with the theoretical location of social
groups and social characteristics, often within an urban setting. It is
here that the links within the allied field of social ecology are closest.

THE ORIGINS OF SOCIAL GEOGRAPHY IN BRITAIN

However important it is to emphasize the development of social
geography in an urban setting, before discussing the development of
this branch of the subject from its origins in the Chicago school of
human ecology, it might be useful to develop a parallel theme – the
development of a school of social geography out of the more traditional human geography in Britain. In the inter-war period sociology was still suspect in many British universities, but it was becoming
increasingly accepted that anthropologists, who at the time were
producing some of the great functional studies of primitive communities, had something useful to contribute to human geography.
The reaction by Barrows (1962, p. 6) in the U.S.A. in 1933 against
the determinism of Semple and Huntington – 'How can an inanimate
thing like soil or topography influence man? It would be as foolish to
expect it to send me an invitation to a birthday party.' – was typical

of the school of geographers who approved of the theory of adjustment, so that different human groups adjust in different ways in different places. Unfortunately, attempts 'to deal with the broad features of economic pattern and to consider their relation to physical environment, to social organization and to major factors in the growth of civilization' (Daryll Forde, 1934, p. vi) were entirely concerned with primitive agricultural communities. Whilst it is natural that distance should add enchantment it is curious that only in the last decade or so have local rural communities in Britain been given the same attention that those in Nigeria or Malaya appear to have received. Much of the drive towards an understanding of the social geography of rural communities has come from the University of Wales, where the link between anthropology and geography has been most fruitful, and led, in the 1940's, to the start of the most important work on Llanfihangel Yng Ngwynfa in northern Montgomeryshire (Rees, 1950) which has done so much to influence later work. A selection of further studies of Welsh rural communities, emanating from the same school (Jenkins *et al.*, 1960), helped to deepen the understanding of the cultural basis of community adjustments. A more recent work still, based on a Devon parish (Williams, 1963), emphasizes the spatial relationships of social and economic change and analyses the 'enduring relationship between society and the physical environment' (p. xx). Family farming in this instance is the main manifestation of this relationship. The pattern of the spatial and structural elements of land holding in this area depends partly on the quality of the farmland itself and partly on 'an attitude towards the relationship between family and land' (p. 80). Williams considers that an attempt to create a model of the dynamic family-land relationships is premature, and certainly any model of a stable and simple structure of family-land relationships would be unsatisfactory, when change is part of the system. Further work away from Highland Britain is essential if a meaningful picture of the social geography of rural communities in Britain is to emerge. A study by a social anthropologist of a Cheviot parish (Littlejohn, 1963) gives some indication of the sort of differences that may emerge in different socio-geographic settings, but studies of society-land relationships under the dominance of a major urban centre, say in south-east England, are badly needed.

In a sense the anthropo- or socio-geographic study of a modern rural community was not a great advance on studies of the traditional peasant communities in the 'Man and his Conquest of Nature' school. Certainly kinship analysis may have been substituted for housetype analysis as a meaningful clue to the 'adjustment' of the social group to the land, but in an *urban* environment the relationship with the physical environment may be completely without significance. Initially the study of the origins and morphology of towns provided interest. However, there was a real danger of arid classifications according to origin or present function, and work on the historic structure of towns could easily degenerate into antiquarianism. The importance of the physical environment was implied in much of the early work, with little regard for the social and economic reality of the community, in the manner in which rural communities were studied. So many towns, for example, with an area in the centre designated by urban geographers as the 'medieval core', were held to have much in common on that account, whereas the *use* to which this central area was put might range from slum dwellings, through a tourist centre, to palaces for the rich. Too often towns were discussed solely in terms of their generalized present economic functions – railway towns, spinning towns, political centres and so on with no real attempt to come to grips with the significant sub-divisions based on meaningful criteria.

GEOGRAPHY IN AN URBAN ENVIRONMENT

The tacit determinism in much of human geography, which led workers on the subject to search so readily for the (implicit) influence of the physical environment in patterns of settlement and economic functions of society, held up developments in an urban environment. Once the point had been made that the marshy patch in an eighteenth-century map accounted for a recreation ground in the present city, or that present street patterns followed old field boundaries, there seemed little more to say. In order to find the roots of urban social geography, therefore, we are obliged to go to the United States and consider pioneer work in human ecology, which later proved to be a fertile source of ideas. Thus one of the first serious attempts to formalize the diffuse information on cities was made by R. E. Park whose paper 'The City: Suggestions for the Investigation of Human Behaviour in the Urban Environment' first appeared in 1916.

> There are forces at work within the limits of the urban community – within the limits of any natural area of human habitation, in fact – which tend to bring about an orderly and typical grouping of its population and institutions. [Park, 1952, p. 14]

Park went on to try to discover and explain the regularities which appear in man's adaptation to space in an urban area. In ecological terms a high degree of interdependence and division of labour results in competitive co-operation for space use. As a result *natural areas* of the city emerge. 'They are the products of forces that are constantly at work to effect an orderly distribution of populations and functions within the urban complex' (1952, p. 196). Park has been criticized for seeing in the rapidly developing city of Chicago too much that was not typical of cities in other parts of the world. However, he himself wrote 'The ecological organization of the community becomes a frame of reference only when, like the natural areas of which it is composed, it can be regarded as the product of factors that are general and typical' (1952, pp. 198–9). He even went so far as to claim that he had 'covered more ground, tramping about in cities in different parts of the world, than any other living man' (Park, 1950, p. viii).

The 'forces' which Park mentions are implicitly ecological, and organize at a biotic or sub-social level of society, and it is for this tacit determinism that Park and his followers, the early classical school, have been criticized. For an account of the various models and theories produced by human ecologists the textbooks by Hawley (1950) and Theodorson (1961) are particularly useful. Hawley and others have argued that economic data, being readily available, are often the best indices of social phenomena. Activities will tend towards a central location depending on whether their need for accessibility is such that they can withstand the high cost associated with the land values of central locations. Such central areas – the so-called Central Business Districts (C.B.D.s) – may extend along major lines of communication, which themselves extend by time-cost distance the factor of centrality. Models, whether based on concentric rings round a central point in the C.B.D. or modified by sector development along lines of communication, make physical space, through land values, the final determinant. Now it is quite clear that individuals are not scattered randomly through space, but on the other hand the fact that segregation, according to social and economic criteria, exists does not thus destroy individual will or volition. In this connexion Firey's work (1947) is important, since his empirical analysis of Boston emphasized socio-cultural values as basic to the understanding of 'socially defined populations in their spatial setting'. He argues that *symbolic values* become linked with a spatial area and then social groups seek identification with such an area as an end in itself. There may indeed be a conflict betweeen economic interests and social

G

symbolism as determinants of land use. Firey is concerned to show that 'social values are real and self-sufficient ecological forces' (Firey, 1947, p. 87) and this is demonstrated in his urban study of Boston. Beacon Hill, a district five minutes' walking distance from the city centre, has been an upper class residential district for a century and a half. Three minutes away on the north slope of the same hill is a decayed area of transient roomers, Jewish and Italian immigrants, where prostitution and other such activities flourish. The question is why the South slope has remained fashionable when other such hill sites similarly placed have changed? On the basis of a strictly rational or economic theory this central site should have been developed by the business establishments and exclusive apartment houses, which have repeatedly tried to locate themselves on Beacon Hill.

It is not necessary to describe here the way in which Firey builds up his thesis, based on considerable historical and sociological insight. In short, the relationship of the upper class families, as a social system, to this particular social environment is by no means 'biotic' or subcultural as the space determinists would presumably suggest. Emrys Jones in his more recent work on Belfast again shows that 'human motivation . . . itself tends to conform to a pattern reflecting current social values' (Jones, 1960, p. 268). 'What the geographer must avoid at all costs is the direct and simple correlation between the land and modern urban land use which is suggested by concomitant distributions' (p. 279). In 1840 the Malone ridge, on which the University of Belfast is built, was invested with social values, thus attracting residential building, overcoming the physical differences as between one part of the ridge and another.

Whatever the town or city, it is becoming increasingly accepted that geographers have much to learn from social ecologists and urban sociologists. There is a fashionable trend at the moment to believe that 'only by the complete rejection of uniqueness can geography resolve its contradictions' (Bunge, 1962, p. 13). A certain school of theoretical geographers would agree with Bunge that 'If only we could rake our lawns optimally we could be close to being able to arrange our cities optimally' (p. 27). This is nice to know; but surely the central place theory, which Bunge attempts to illuminate by his metaphor of raking leaves into piles, should be considered in the light of changing technology. There are other ways of clearing lawns than by creating Bunge piles. Unfortunately, so long as geographers sit puzzling over their lawn-raking type problems then the necessary empirical research on urban areas is not going to get done, and workers in other disciplines will feel that geographers can contribute little more than pedantic area delimitations, or the maps and diagrams to illustrate the data gathered by others.

A further point has to be considered. Most of the work on urban social geography has evolved in capitalist societies, where the so-called free play of market forces gives rise to differentially valued areas and the differences in land use related to economic forces. However, the situation is different in a centrally planned economy. How does this affect the theories of urban growth so far propounded? The social geography of Prague gives some interesting indications. The pre-war development of the city showed a concentric differentiation in social areas, which, in the terminology of a sociologist in a country with a centrally planned economy, were due to 'a society which embodied its social differences and injustices in its very territorial structure' (Musil, 1960, p. 237). During the last fifteen years, however, the trend to social area differentiation has halted and Prague is becoming sociologically more homogenous. In this connexion it is interesting to quote from the guide to the Prague development plan.

> The structure of a town has always been reflection of the social order . . . to-day our people are building a classless society . . . without the con-flicts between town and country, manual and clerical workers . . . a town in this sort of society needs a functional structure that differs radically from the one it had under capitalism. The problem of reconstructing a city must be viewed as a complex whole . . . the inequality between the favourable living conditions of people in the newer districts of Prague and the dismal living conditions in the working class suburbs and the city centre must be eliminated . . . it is very important to reconstruct the industrial zone on the same lines as the residential area. [*Rebuilding Prague*, 1962, p. 18]

Now simply because the political or cultural factor changes the situation to be expected under the conditions which give rise to American cities' growth and morphology, this does not mean that segregation ceases to exist. There are strong indications that certain socio-economic groupings segregate themselves in certain districts of Prague, as in the British new towns certain areas appear to attract certain types of people, without such areas being very different in terms of rent or architectural qualities. That segregation continues to persist suggests that the social geographer working in urban areas is likely to become more social in his orientation, as the 'free play of market forces' becomes less free and as the impact of social mobility on the British class structure emphasizes what Goldthorpe and Lockwood describe as the *normative* and *relational* aspects of class rather than the economic aspect alone. A greater concern with total life style, for status differentiation and status enhancement, is likely to lead to the emphasizing of small differences between one residential area and another. Already the indications are strong that new, privately-built estates in the towns and cities of England are fairly homogenous internally, in terms of the social characteristics of their populations, but are socially sharply differentiated from neighbouring estates. Segregation appears to be on the increase.

URBAN GEOGRAPHY IN DEVELOPING COUNTRIES

A most significant and important trend which has developed in the past twenty years is the growing interest in the urban areas of the economically less developed countries.

> The rapid growth of cities, especially of large cities, is an outstanding feature of the modern age. Between 1800 and 1950, the population of the world living in cities with 20,000 or more inhabitants increased from 21·7 to 502·2 million, expanding about 2·6 times in the same period; 2·4 per cent of the world's population lived in urban centres of 20,000 or more in 1800, 20·9 per cent in 1950. [U.N., 1957, p. 113]

Between 1900 and 1950 the population living in cities of 100,000 or more increased by 444 per cent in Asia and 629 per cent in Africa. The phenomenal urban growth of Asia in the first half of the twentieth century has meant that now one-third of the world's large city (100,000 +) population is Asian. Certain countries in Latin America have higher proportions of their populations in localities of 20,000 or more than European countries, such as France or Switzerland. More than one-fourth of all Latin Americans live in cities of 20,000 or more and about one-fifth live in cities of 100,000 or more.

In the same way that a human geographer of an earlier period might visit Latin America or India to do field work on, say, peasant cultivation so, now, social geographers are visiting the booming, choking cities of these areas. Evidence of this interest is given in a recent volume entitled *India's Urban Future*, which is a selection of papers, some of which are contributed by geographers. The separation of residential land use from business or industry is by no means clear-cut in Indian cities. They have not grown in an orderly fashion but rather by addition and agglomeration. Large areas of cities are not very different from villages, so that the life-styles and attitudes of such 'urban' populations are much the same as those of villagers. The analysis and delimitation of the C.B.D. is irrelevant in the Indian situation where 'they can scarcely be said to exist except in the Indo-British seaports' (Turner, 1962, p. 67). Binuclear or polynuclear patterns have been described in various Asian and African cities, being mainly the result of the historical and cultural background of such cities. The social factor is crucial. Indian society is highly differentiated into mutually exclusive groups, and, although castes may lose some of their exclusiveness in urban areas, they continue to exercise a divisive effect in segregating the population.

Singh's study of Banaras (1955), although only very slightly orientated towards social geography, provides a useful case study to compare with Boston and Belfast. The *Inner Zone* of the city is a closely developed labyrinth of lanes connecting the temples, and the main activity is the pursuit of religion. It is mainly a Hindu area with a great concentration both of 'religious minded people' and of the high status homes of the rich. The religious status of the area has led to vertical expansion for residential (not office!) use. Next to the inner zone is the *Middle* or transitional belt of Muslim settlement. Again almost entirely built up, mosques are characteristic and the outer part of this zone contains the poor homes of moslem weavers. The *Outer Zone* is a curious mixture of slums and of the spacious houses of those rich people who were unable to find a house at the centre. In addition, the British administration added the *Civil Lines*, an area of wide, metalled roads and officers' bungalows, and the *Cantonment*, which was designed as a military garrison (Banaras was an important military garrison in the Second World War). There is also the isolated University quarter. In between these areas are formless masses of one- to three-roomed mud houses, typically with a verandah in front, which cannot be described as anything but suburban slums.

Thus Banaras shows considerable social and cultural segregation in its urban geography. There seems to be an almost random scatter of cottage industries throughout the city and the business centres, similarly scattered, provide evidence of what Singh calls 'haphazard' growth, although he does suggest that the concentration of 'feminine commodities' in one lane is on account of the number of women who pass that way to an important temple!

Singh is much less convincing when he attempts to describe the

'Umland' or sphere of influence, of Banaras, taking as his criteria the supply zones of vegetables, milk and grain and agricultural products, and also the circulation zone of newspapers, he derives Umlands of 72, 180, 4,000 and 20,000 square miles respectively. The latter area he claims to be the 'culturally integrated area of the Umland'. Such an approach was thought to be entirely inappropriate by Lambert, who felt that what should be considered was 'not the radiation of influence from cities into rural areas, but rather the presumed radical change in life style which confronts the villager when he moves to the city or the city moves out to encompass the village' (Turner, 1962, p. 131). Urban influences on rural areas appear to depend, in India, not so much on the actual *position* of the village, but on the receptiveness to the absorbtion of new ideas of social groups within the village and the degree of potential flexibility present in the social structure, so that, curiously, 'many of the villages whose unity and isolation have been emphasized lie very close to towns and cities' (p. 126).

Useful teaching material in the social geography of urban areas in economically underdeveloped countries may be found in the three UNESCO reports and also the 1957 U.N. *Report on the World Social Situation*. One might also mention some studies in Africa – Kuper (1958) on Durban, Southall and Gutkind (1957) on Kampala, the Sofers (1955) on Jinja and Marris (1961) on Lagos, which, although not primarily geographical works, provide useful insights into urban ecology and are more easily obtainable than works in French on African towns and in Spanish and Portuguese on some in Latin America. Some further ecological studies are listed in Theodorson's selection of papers (1961, pp. 438–9). The empirical studies need to be related to some sort of theoretical framework; an attempt to provide a typology of urbanization has been recently put forward by Riessman (1964, pp. 198–235) and is likely to be a useful teaching aid.

THE INCREASING DOMINANCE OF SOCIAL FACTORS

We have discussed the study of British towns and rural communities and have made some mention of the social geography of the rapidly growing cities in Asia. Singh's work on the Umland of Banaras was strongly influenced (far too strongly in fact) by some work on urban hinterlands in Britain and the United States carried out some 10–20 years ago. C. G. Galpin's pioneer study, *The*

Social Anatomy of a Rural Community, was published as early as 1915, but it was not until some twenty years later in this country that R. E. Dickinson started to lead the breakaway from the geographers' preoccupation with land use, and in *City, Region and Regionalism* challenged the rigid town/country dichotomy by arguing that 'an area of common living can be designed only in the key trait of that common living, that is, in terms of *social* considerations, not of a particular set of physical factors which condition that pattern of living in part' (Dickinson, 1947, p. 9; my emphasis). However, Dickinson went on to argue that the measurement of the 'service factor' would provide an accurate method of determining the hinterland of urban centres. It was indeed satisfying to see the way the work of Green (1950) or Bracey (1952) could be used to grade service centres, and it was easy to believe that one was bringing order and insight to settlement geography. This was especially so when work based on the analysis of bus time-tables, before the rapid spread of the ownership of motor-cars, could provide a guide to the number of journeys made from a place. However, much of such work became historical geography almost before it was published and it is open to the criticism that it applies a rigid static framework to a developing and dynamic system. It is becoming increasingly understood that the social area of a community may not bear much relationship to the service factor and the detailed studies of rural communities by Rees (1950), Williams (1963) and Littlejohn (1963) are useful correctives here. Further, the superimposition of one pattern over another is becoming accepted and analysed (Pahl, 1965). Farmers may have a different pattern of activities and linkages from farm workers, and middle-class commuters from other villagers. The daily journey to work is increasing in the rural as well as in the urban areas, as Lawton (1963) has shown nationally, and as some interesting work conducted by certain Planning Departments illustrates in detail. More and more research workers in this field are obliged to sub-divide their populations, generally according to social criteria, and to describe the movement of such sub-groups. It is quite inappropriate to consider the population of any settlement, however small, as homogeneous. In many parts of the country an analysis of bus journeys simply reflects the journeys made by those without cars – the working class wives, the old and the poor.

Perhaps the most fundamental trend in the work of social geographers in the last twenty years has been the developing interest in the growth and structure of metropolitan regions. An early description of the growth of the London metropolitan region was given in 1911:

> In many districts, urban and rural, outside the boundary, both the volume of population and its abnormal rate of increase must be partly attributed to their situation with respect to the Metropolis, and although the distance to which the Metropolitan influence extends cannot be defined with accuracy, it can hardly be put at less than thirty miles from the centre. . . . The rates of increase in many seaside towns, which, though outside the thirty-mile limit, are within easy reach of London, have also been far above the average. Men whose daily work is in London often reside at a distance corresponding to a railway journey of not less than an hour's duration. [Report of the London Traffic Branch of the Board of Trade (1911), Cmd. 5972, p. 4]

Parallels could easily be found in America and elsewhere, but certainly the growth of the London Region has been outstanding. The inter-war expansion was described in the Barlow Report (Cmd. 6153, 1940) and Abercrombie later described the 'unbridled rush of building (which) was proceeding in the form of a scamper over the home counties' (1945, p. 2). The post-war trend continued, 788,012 people adding to the region's total from 1951–61, with a redistribution of population such that Central London declined by 11·2 per cent and the inner county ring increased by 46·5 per cent (only half the increase being due to the creation of six New Towns in the area). An American who came to study the situation felt that 'We are more ignorant concerning the physical manifestations of the interlocking character of the Metropolitan community than we are on other seemingly remote questions, such as the internal heat of stars' (Foley, 1961).

For the situation in the United States there is an interesting summary of the social geography of metropolitan regions in the work of Wissink (1962). His argument is that 'the cultural system of a nation – in particular its space-related values, objectives, instruments and institutions – gives rise to a national type of city, a basic theme on which the individual cities are variations' (p. 287). Thus the booming rural-urban fringe round American cities (see also Dobriner, 1963) is paralleled in Britain, with a different cultural and political tradition, by, for example, the planned New Towns of the London Region.

Wissink draws on a broad range of American sociological and ecological literature to provide a valuable summary of the various approaches to the study of the city in relation to its hinterland and also to the history, definition and description of the rural-urban fringe. This book is extremely valuable as a teaching aid and would seem to be an important addition to university textbooks in social geography. In particular it is a useful corrective to the rather arid ecological classifications of Bogue and his associates.

It is estimated that in the United States by 1975, 57 per cent of the total population of metropolitan areas will be in the suburban and fringe area (Dobriner, 1963, pp. 146–7). Over three-fifths of the United States' population is now living within Standard Metropolitan Areas (formally defined in 1950 by the U.S. Bureau of the Census) so that the centrifugal movement out to their peripheries involves a considerable movement of population.

Gottman, in his useful survey of the urbanized north-eastern seaboard of the United States, claimed that all previous patterns of urban regions based on central cities and hierarchies of suburbs and satellite towns were totally inadequate tools with which to analyse his 'megalopolis'. A 'totally new order in the organization of inhabited space' is emerging (Gottman, 1961, p. 9) (thus Dickinson's work published in 1947 is now quite outmoded (ibid., p. 736)). Gottman can find no orderly pattern for this but only a 'nebulous structure'; this structure is perhaps held together by the motor-car.

Returning to this country, it has often been argued that the centrifugal movement can be explained in terms of an anti-urban reaction to the industrial city, with its roots in the nineteenth century and earlier. Ebenezer Howard's vision of 'Garden cities' and its realization at Letchworth and Welwyn Garden City has been maintained in the pressure for more new towns. Vigorous urban renewal or the creation of a new *city* does not seem to meet with the same degree of public support in this country. The green belt idea is a good example of the social basis of land use policy; thus the London green belt was to be 'where organized large-scale games can be played, wide areas of parks and woodlands enjoyed and footpaths used through the farmland' (Abercrombie, 1944, p. 8). The actual use of the green belt has recently been discussed by Thomas (1963), and it is clear that to sterilize land for development, in areas where the economic pressures to build are high, is to put social values before economic ones. In whatever ways the green belt is used in practice and however much it is a 'moat' which commuters have to pay extra to cross, it does help to prevent the continuous spread of urban building, although of course it helps to intensify pressures on its outer side.

It is the interpretation of the national Town and Country Planning legislation by the local authorities which does so much now to determine the distribution of social groups in space. Hence, when a geographer attempts to grapple with the practical problems involved in the megalopolitan London Region, as Peter Hall does in his *London 2000* (1963), this should be welcomed, rather than superciliously rejected as not being 'academic geography'. Clearly there is much in such an unashamedly polemical work which one might want to disagree with or modify. However, as a teaching aid and stimulant, it is first class; would that more present-day geographers were less afraid of their so-called reputations and would follow Hall's lead! The Report on *Traffic in Towns* (H.M.S.O., 1963) estimates an increase of 16½ million vehicles between now and 1980, a trebling of traffic in a little under twenty years. The Registrar General has estimated that the population of England and Wales will rise by 7,318,000 to 54,086,000 between 1962 and 1982. Thus by 1980 there is likely to be in the order of 450 vehicles for every 1,000 of the population. It is clear that, whatever may have been the trends of social geography in the past, if social geography is going to be taught adequately in universities, training colleges and schools, then professional geographers cannot but concern themselves more with the problems that surround them. It may well be that geography will acquire an increasing relevance in future years.

We have moved some way from the changing relationships of farmers to the land at Ashworthy, through the accounts of Boston, Belfast and Banaras, to an analysis of the development of the vast and complex metropolitan regions. Whether one is documenting the characteristics of the rapid urbanization of the economically underdeveloped countries, or the complexity of commuting patterns in the city-regions of advanced countries, it is clear that change is ever-present. The patterns are constantly in a state of flux. Individual families are frequently on the move and whole communities are in the process of changing their characters. It is almost as if there were some sort of cycle at work. At an early stage of the cycle the social factors are important; for example, the Jewish ghettoes are strictly segregated. Then comes the economic stage when the maximum profit determines the use of land and the models of theoretical geographers are useful and illuminating. Thirdly, the social factors reassert themselves and people move to certain areas, not because it is cheaper to do so, but because a form of segregation is part of the way people adapt to their position in society. This is not necessarily an economically determined position. Social geography is the discipline which is fundamentally concerned with the spatial manifestations of social change. In the same way that economic geographers are concerned with the *processes* involved in the study of industrial activity, so too is the social geographer interested in the social processes involved in the spatial location of social groups. To answer questions concerning the distribution of West Indians in London or Catholics in Belfast, the social geographer is obliged to go beyond the simple stage of mapping distributions.

Emrys Jones has expressed some central issues of social geography as follows:

> If the factor being studied has an effect on the residence of people then its distribution will not be a random distribution but irregularities will occur which will exhibit segregation. . . . Human motivation . . . itself tends to conform to a *pattern* reflecting current social values. [Jones, 1960, pp. 199–200 and p. 268; my emphasis]

Hence it is impossible for the social geographer to create a model which would be suited to all societies at all periods of time. Take, for example, the segregation of a country's *élite*. This may congregate round the coast or at a religious centre; it may select a fashionable area of the centre of a town or city or it may choose a particular suburb or area at the periphery. At one period of time the *élite* may be scattered throughout the countryside; at others it may concentrate in a particular place. A particular part of a town may be fashionable at one period, may decay in time, and then become revalued and renovated as fashion changes. Houses built for the artisan at one period may be used by the *élite* of another period, even though these houses may require great expense to make them comfortable and even though they may be economically poorly sited in relation to the workplaces of the earners. Hence it is impossible to limit study to the actual dwelling, since of much greater importance are the socio-economic characteristics of those who live in the dwellings. Under the forces promoting urbanization and in industrially highly developed societies, the field of social geography is concerned less with the relationships of social groups to the physical environment than with the patterns and processes involved in the segregation of social groups and settlements in space. The prime geographical factor is *distance*, whether actual physical distance or economic distance measured in time-cost terms. One might also add that 'social distance' is also of interest to the social geographer. It appears that there is some indication that social mobility and geographical mobility are related and thus, in a society in which certain sections are able to move up a promotional ladder with differential economic rewards, there are strong pressures for these economic differences to be reflected in social differences and hence in the segregated estates mentioned above. The social geographer is thus interested in the broad changes of population structure and distribution as the very necessary first stage in his analysis. From the basic demographic aspect, the analysis might proceed to such aspects as the geography of mortality rates, religion or occupational groups. The binding framework of settlement pattern, whether of villages, towns, cities or metropolitan regions, can be analysed within a particular socio-cultural background. The mobility of socially defined groups is of fundamental interest, as is the changing function and nature of communities as their physical and social space relationships change. Naturally the social geographer is concerned with the *social* aspects of changing space use and is thus likely to have some training in and understanding of sociology. The necessity for some economics to be taught in school geography is already understood, the equal necessity for some sociology to be included is not so readily accepted.

THE TEACHING OF SOCIAL GEOGRAPHY

The preceding part of this chapter has been intentionally discursive. It is an essential part of the argument that social geography does not have a grand structure or theory which may be described and that it would lose much of its point if such a structure or theory were built and then rigidly accepted. The social geographer is obliged to read widely in the field of sociology and the selected essays of Louis Wirth—*On Cities and Social Life* (1964)—provide a useful starting point. A more controversial, but highly stimulating, introduction to sociological insight is *The Sociological Imagination* (1958) by C. Wright Mills. Wirth and Mills would together provide a good introduction for the non-sociologically-trained geographer to a new world of ideas. General textbooks of social geography are useful to provide some general background – for example, those by Pierre George (1961) and R. E. Dickinson (1964). However, in order to teach an imaginative and constructive approach to social geography, it might be helpful to use some of the studies of modern British communities which, fortunately, have become more available in the last fifteen years.

If we take, for example, the English Midlands and we want to introduce a class to certain aspects of social geography, then there are several interesting studies which do much to illuminate the situation, although these are not very appropriate for use by pupils at school. There is an account of Coseley in the Birmingham conurbation by Doris Rich (1953), which relates mobility for leisure time activities to different social and physical factors. About one-third of the leisure time spent outside the home was spent outside Coseley. Using similar methods of investigation, modified according to the local situation, the results of the Coseley survey could be tested and compared with the pattern in the area of the school or college. That is to say, rather than simply measuring what the facilities of a place are, the investigation would be concerned with where people *actually* go in their leisure time. Journeys may be quantified and expressed diagrammatically and an attempt made to account for the pattern. Social factors may then be balanced with economic ones as determinants of mobility patterns.

Studies of Banbury by Margaret Stacey (1960) and of two neighbourhoods in Oxford by Mogey (1956) provide further valuable social insights into the complexities of modern communities. A discussion of these studies in the context of aspects of the modern geography of the area could usefully lead to a consideration of the local area once more and the way it is changing. Quite obviously with an expanding economy, the growth of new industries and the enormous number of houses that need to be built annually, there are few communities that will not be changing in some way in the near future. This emphasis on change presents a challenge to the social geographer to analyse the processes making for the decline of one area and the growth of another, with the consequent development of new patterns of mobility for economic and social reasons. Towns are changing in relation to each other and villages are losing one function to take on another.

It is often possible for schools and colleges to do useful work which may have practical value in the future planning of the communities as well as providing an understanding of the social geography of the area. Local planning authorities may often have basic maps of changes in housing, population and journeys to work and would welcome detailed traffic surveys or mobility patterns of the inhabitants of particular streets or neighbourhoods to provide checks to their more general work. Similarly a class could do practical work on the census material, using either the county volume or more detailed information which could perhaps be made available by the local Planning Department.

Clearly if the study of the local community is taken as the means of gaining an imaginative insight into social geography, there is the danger of assuming that the patterns and processes which emerge and operate in one particular society will necessarily hold true for another. In order to guard against this sort of parochialism there is much to be said for the study in depth, possibly with field work, of another quite different area. The guide produced by Birou, Burdet and Lapraz (1957) to the study of French rural communities provides a mass of information and ideas which could serve as the basis for much useful teaching in social geography. Indeed this is so useful that it is difficult to suggest a better introduction for teachers to practical work in the subject.

It has been the intention of this essay to show that the social geographer, concerned with such things as the commuting patterns of the chief earner, the differential mobility of his family and the new estates on which such people live in the exploding cities of the economically advanced countries, is simply the modern counterpart of the human geographer, interested in, for example, the movement of Swiss peasants up to the upland pastures in spring and down to the vine harvest in the lower valley in the autumn. This is not to say that commuting is more important than seasonal transhumance; it is simply often more relevant to the lives of those in predominantly urban countries. Again, the urban geography of Brisbane is not more or less important than the geography of the Queensland sugar industry; it is simply becoming more relevant to an urban society. The social geographer, armed with sociological insight and conscious of the importance of space, can approach the local community with techniques which may provide a most intellectually satisfying analysis.

References

ABERCROMBIE, P., 1945, *Greater London Plan 1944* (H.M.S.O.).
'BARLOW REPORT', 1940, *The Royal Commission on the Distribution of the Industrial Population, Report* (H.M.S.O.).
BARROWS, H. H., 1962, *Lectures on the Historical Geography of the United States as given in 1933* (Chicago).
BIROU, A., BURDET, R., and LAPRAZ, Y., 1957, *Connaitre une Population Rurale* (Economie et Humanisme, 262 Rue Saint-Honoré, Paris 1er).

BOGUE, D. J., 1950 A, *Metropolitan Decentralisation. A Study of Differential Growth* (Oxford, Ohio).

— 1950 B, *Structure of the Metropolitan Community. A Study of Dominance and Sub-Dominance* (Ann Arbor, Michigan).

— 1955, 'Urbanism in the United States', *American Jour. of Sociology*, 60 (1955), 471–86.

BOULARD, F., 1960, *An Introduction to Religious Sociology* (London).

BRACEY, H. E., 1952, *Social Provision in Rural Wiltshire* (London).

'BUCHANAN REPORT', 1963, *Traffic in Towns* (H.M.S.O.).

BUNGE, W., 1962, *Theoretical Geography* (Lund).

DICKINSON, R. E., 1947, *City, Region and Regionalism* (London).

— 1964, *City and Region* (London).

DOBRINER, W. M., 1963, *Class in Suburbia* (Englewood Cliffs, New Jersey).

FEBVRE, L., 1932, *A Geographical Introduction to History* (London).

FIREY, W., 1947, *Land Use in Central Boston* (Cambridge, Mass.).

FOLEY, D., 1961, 'Some Notes on Planning for Greater London', *Town Planning Review*, 32, 53–65.

FORDE, DARYLL, 1934, *Habitat, Economy and Society* (London).

GEORGE, P., 1961, *Précis de Geographie Urbaine* (Paris).

GOLDTHORPE, J. H. and LOCKWOOD, D., 1963, 'Affluence and the British Class Structure', *Sociological Rev.* (New Series), 11, No. 2, 133–63.

GOTTMAN, J., 1961, *Megalopolis* (New York).

GREEN, F. H. W., 1950, 'Urban Hinterlands in England and Wales: An Analysis of Bus Service', *Geog. Jour.*, 116, 64–81.

HALL, PETER, 1963, *London 2000* (London).

HATT, P. K. and REISS, A. S., 1957, *Cities and Society* (Glencoe, Illinois).

HAWLEY, AMOS H., 1950, *Human Ecology* (New York).

HOWE, G. M., 1963, *National Atlas of Disease Mortality* (London).

JENKINS, D. *et al.*, 1960, *Welsh Rural Communities* (Cardiff).

JONES, E. and GRIFFITHS, I. L., 1963, 'A Linguistic Map of Wales, 1961', *Geog. Jour.*, 129, 192–6.

JONES, EMRYS, 1960, *A Social Geography of Belfast* (London).

KUPER, L. *et al.*, 1958, *Durban, a Study in Racial Ecology* (London).

LAWTON, R., 1963, 'The Journey to Work in England and Wales: Forty Years of Change', *Tijdschrift voor Economische en Sociale Geografie*, 54, 61–69.

LITTLEJOHN, J., 1963, *Westrigg – the Sociology of a Cheviot Parish* (London).

MARRIS, P., 1961, *Family and Social Change in an African City* (London).

MILLS, C. WRIGHT, 1958, *The Sociological Imagination* (New York).

MOGEY, J. M., 1956, *Family and Neighbourhood* (Oxford).

MUSIL, J., 1960, 'The Demographic Structure of Prague' (in Czech), *Demografie* (Prague), **2**, No. 3, 234–48.

PAHL, R. E., 1965, *Urbs in Rure: The Metropolitan Fringe in Hertfordshire* (London School of Economics Monograph).

PARK, R. E., 1950, *Race and Culture* (Glencoe, Illinois).

— 1952, *Human Communities – The City and Human Ecology* (Glencoe, Illinois).

REBUILDING PRAGUE, 1962, *K. Problemum Vystavhy Prahy* (Prague).

REES, A., 1950, *Life in Welsh Countryside* (Cardiff).

REISSMAN, L. 1964 *The Urban Process – Cities in Industrial Societies* (Glencoe, Illinois).

RICH, D., 1953, 'Spare Time in the Black Country', *Living in Towns*, Leo Kuper *et al.* (London).

SINGH, R. L., 1955, *Banaras: A Study in Urban Geography* (Banaras).

SOFER, C. and R., 1955, *Jinja Transformed* (Kampala, Uganda).

SOUTHALL, A. W. and GUTKIND, P. C. W., 1957, *Townsmen in the Making* (Kampala, Uganda).

STACEY, M., 1960, *Tradition and Change* (Oxford).

THEODORSON, G. A., 1961, *Studies in Human Ecology* (Evanston, Illinois).

THOMAS, D., 1963, 'London's Green Belt: The Evolution of an Idea', *Geog. Jour.*, **129**, 14–24.

TURNER, ROY (Ed.), 1962, *India's Urban Future* (Berkeley and Los Angeles). (See: Brush, J. E., *The Morphology of India's Cities*, pp. 57–70. Ellefson, R. A., *City-Hinterland Relationships in India*, pp. 94–116. Lambert, R. D., *The Impact of Urban Society Upon Village Life*, pp. 117–140. Hoselitz, B. F., *A Survey of the Literature of Urbanisation in India*, pp. 425–43).

UNESCO, 1956, *Social Implications of Industrialisation and Urbanisation in Africa South of the Sahara* (Paris).

— 1957, *Urbanisation in Asia and the Far East* (Calcutta).

— 1961, *Urbanisation in Latin America* (Paris).

UNITED NATIONS, 1957, *Report on the World Social Situation* (New York).

VISHER, S. S., 1932, 'Social Geography', *Social Forces*, **10**, 351–4.

WILLIAMS, W. M., 1963, *A West Country Village* (London).

WIRTH, L., 1964, *On Cities and Social Life* (Chicago and London).

WISSINK, G. A., 1962, *American Cities in Perspective* (Assen, The Netherlands).

ZELINSKY, W., 1961, 'Religious Geography of the United States', *Ann. Assn. Amer. Geog.*, **51**, 139–93.

CHAPTER SIX

Changing Concepts in Economic Geography

P. HAGGETT

Professor of Geography, University of Bristol

It has been a recurring criticism of our military command that it was well prepared to fight the Boer War by 1914 and adept at trench war-fare by 1939. How far such criticism may be unjust of military preparedness there were disturbing signs in the academic world that geographers were in the post-war period girding their loins to fight pre-war battles. The long debate over possibilism and determinism was carried wearily on long after it had been dismissed in other subjects as '. . . simply a misunderstanding of history' (Bronowski, 1960, p. 93). Broadsides continued to be launched at quantitative methods which had been tested, tried and assimilated decades before in subjects no better suited to their use. Textbooks in regional geography continued to be fashioned around the same conceptual constructs which, as Wrigley (Chapter 1) points out, Vidal de la Blache had seen collapsing about him two generations before.

Ackerman (1963) has seen the origins of this frustration in the extreme separatism of geographers as a group; a separatism traced in Hartshorne's *Nature of Geography* (1939). He suggests that in neglecting the course of science as a whole, we neglected the axiom that the general progress of science determines the progress of its parts. How far Ackerman's case may be true, it is certain that the last decade has seen strenuous efforts to close the gap. More external concepts, notably from the fields of *mathematical statistics* and *systems analysis*, have been introduced into geography in the last decade than in any comparable period. These imports have been of critical importance in the new geography, but since they are wholly general in application rather than specific to economic geography I have not treated them at length here. Burton (1963) has provided an extensive review of quantification in geography and Chorley (1962)

H

has demonstrated the impressive potential of systems analysis for geographical research.

I propose then to discuss some of the other aspects of contemporary change in economic geography and leave the quantitative and systems-analysis revolution as a self-evident truth. I have in mind three lines of more distinctive development: convergence of economic geography with other branches of human geography; the extension of models in teaching and research; and the growing interest in chance or stochastic processes. This is not an exhaustive list of possible topics, but it does include three that I regard as important in my own approach to economic geography.

THE IDEA OF CONVERGENCE

Convergence with Other Parts of Human Geography

In a penetrating study of the relations between history and geography, Darby (1954) has suggested that contacts between disciplines can be both *active* and *passive*. Using this analogy we can think of an active relationship between geography and economics in which the one is actively influencing the other or a passive juxtaposition. McCarty's *Geographic Basis of American Economic Life* (1940) is concerned with the active role of regional geography in shaping economic activity; conversely Chisholm's *Rural Settlement and Land Use* (1962) is concerned with the active role of the economics of transport in structuring familiar features of regional geography. Other contributions, like that of Ginsburg's *Atlas of Economic Development* (1961), are harder to classify and may well represent a more passive juxtaposition of economics and geography in equal partnership. Certainly the swing of the pendulum has now set in strongly towards the second sort of active relationship with a growing emphasis on the role of economics in determining the form of geographic patterns. To be sure this emphasis is currently on a rather specialized part of economics, transport economics, with studies of highway economics and urban change in the United States to the fore (Garrison, Berry, Marble, Nystuen and Morrill, 1959; Berry, 1959), but there are signs that this is being extended to the wider range of economic analysis.

It is of course as convenient as it is misleading to regard economics as a fixed point and plot the course of geography in relation to it. Economics itself has been undergoing developments at least as fundamental (Kendall, 1960). Robbins (1935, p. 17) in his *Nature and Significance of Economic Science* drew attention thirty years ago to the swing in economics from a classificatory to an analytical position: 'We do not (now) say that the production of potatoes is economic activity and the production of philosophy is not . . . in so far as either kind of activity involves the relinquishment of other desired alternatives, it has its economic aspect.' But this move away from a classificatory view of economics had little apparent effect on textbooks of classical 'economic' geography (Smith, Phillips and Smith, 1955; Jones and Darkenwald, 1954) where rubber gathering and steel production continued to be regarded as the quintessence of economic activity. In so far as this remains true, economic geographers may be guilty of teaching a view of economics that has dwindling acceptance among economists.

Even if geographers were unfamiliar with the changing functions of economics they were unable to ignore the changes in the landscape they described. And here they saw, as Vidal de la Blache had seen, the fundamental trends towards an urban world; a world where the number of large cities with over one million inhabitants had doubled since 1930 to over a hundred. Although accurate measurement has been blurred by the problems of city boundaries and the varying thresholds adopted for 'urban population' (from 250 in Denmark to 10,000 in Spain (Alexander, 1963, p. 528)) the growing dominance of the metropolis in the organization of world economic activity has become self-evident.

This trend created both problems and opportunities in economic geography. It meant a newer and more complex world in which the traditional stand-bys of climate and soil provided a less sure guide to its puzzling land-use patterns; it meant that traditional locational theories like those of Weber (1909) lost their immediate relevance in a welter of 'footloose' industrial expansion; it meant that the study of processes, processes of urban growth, processes of diffusion, processes of system adjustment, emerged as one way of making sense of a rapidly changing world.

It was this fundamental concern with the centres of organization, the cities and the city hierarchy, that underscores the idea of convergence. For in concentrating on the cities, on viewing the economic landscape as an urban-centred system, the economic geographers came back towards those workers in settlement geography, historical geography, and social geography who had been continuing to work on their own separate aspects of urbanization. Indeed so complete has been the fusion that it is often difficult to distinguish old divisions in new studies. Historians, such as Russell (1964), have adopted concepts from economic geography like basic–non-basic activities (Isard, 1960) and applied them to Domesday population. Workers in the field of settlement geography (Berry and Pred, 1961) are then welding an integrated 'human', 'social' or 'behavioural' geography in which economics is playing an important but not overbearing role. In the agricultural field Spencer and Horvath (1963) have reminded us that at least five main groups of factors need to be considered before the regions such as the 'Corn belt' of the United States can begin to be satisfactorily explained. They identify sets of psychological factors (e.g. farming attitudes), political factors (e.g. farm subsidies), historical factors (e.g. 'lags' in the spread of technological knowledge), and agronomic factors (e.g. improvement of hybrid corns) besides the conventional economic factors.

Convergence with Regional Geography

Regional studies have been booming in North America since World War II. A recent survey (Perloff, 1957) reported about 140 U.S. universities had established programmes in regional studies, while two new institutions, the Regional Science Association and Resources for the Future, have polarized regional research on a new scale. In Britain the Hailey and Parry Committees on Afro-Asian and Latin American studies in British universities have seen the founding of new regional research centres; at Cambridge, a new South Asian Studies Centre has been created with a geographer, B. H. Farmer, as its first director.

While such regional studies tend to deal with many features and involve the use of several academic disciplines, the strongest development has come from economics or, more specifically, from econometrics. Thus it is that the first major textbook on regional science, Isard's *Methods of Regional Analysis* (1960) is essentially concerned with economic regions. The problems he sees as paramount is the economic 'performance' of a region (p. 413); what industries does it need to smooth out employment irregularities?; how can it optimize the use of its often niggardly resource endowment? Questions of this kind throw the weight of interest solidly towards economic development and Fisher (1955, p. 6) has summed up this view: '. . . the most helpful region . . . is what might be called the *economic development region*'.

The approach of economists to regions has been strongly mathematical. Two of the most important tools used have been those of *input–output analysis* and *linear programming*. In input–output analysis an attempt is made to trace the flow of goods between industrial sectors and express these as a matrix of input–output coefficients. The difficulty lies in the lack of data on many movements and in the sheer size of the matrix; indeed Meyer (1963, p. 33) has shown that if we are interested in five regions, each of which has fifty industrial sectors, then the matrix would contain 62,500 coefficients. Leontief (1963) in a very readable summary has shown, however, that when such matrices are complete they form a very powerful tool both in showing how regions work and pay their way, and equally important, in diagnosing defects in regional structures. The second major technique, linear programming (Isard, 1960, p. 413–92), is used where the problem demands that some function is maximized or minimized. A typical case might be to relocate hospital, school, or commercial 'collecting areas' so as to minimize the cost of transportation and optimize the size of units; such a study was carried out by Yeates (1962) for schools in Grant county in Wisconsin and is being used at least by one rural English county, Somerset. Perhaps the most ambitious study using mathematical programming in locational programmes is the Penn-Jersey study which relates housing development to a very complex and complete set of environmental conditions (Herbert and Stevens, 1960).

Whether the present interests of economists in regions is a major departure or whether in the future '. . . regional economics may increasingly be indistinguishable from the rest of economics' as argued by Meyer (1963, p. 48) remains to be seen. Whatever the long run importance for economics the impact on geography has been catalytic. Both economic geographers and regional geographers have been either exposed to the literature or drawn in to participate on inter-disciplinary regional research of a rigorously high standard. As Garrison's review (1959–60) indicates, the boundary line work has been immensely productive both of new ideas and new techniques and this is already being translated into action in a few geography schools. No one who follows through the research theses published by the University of Chicago's geography department since 1948 can be unaware of both the nature and pace of the revolution.

Convergence with Physical Geography

This trend is more speculative. Hartshorne (1959) has summarized the views of orthodox geographers in regarding the divisions between physical and human geography as a fundamental dichotomy. Whether this was regarded as a source of division and weakness or as a fundamental educational advantage in a pedagogic world of arts and science, the existence of the difference was never in dispute. It is one of the curious by-products of an apparently disrupting influence, quantification, that it brought geographers to a realization of certain common problems of relating form and process. Bunge in his *Theoretical Geography* (1962, p. 196) has memorably found many things in common:

> Davis's streams move the earth material to the sea and leave the earth etched with valleys; Thünen's agricultural products are moved to the market and leave their mark on the earth with rings of agriculture; . . . farmers scattered on plains move to their hamlets and form Christaller's hexagonal network on their landscape; . . . agricultural innovations creep across Europe, as do glacial fronts, to yield Hägerstrand's regions of agricultural progress and terminal moraines.

Whether Bunge's refreshingly unified view of a discipline organized around the duals of 'geometry and movement' is a pipe-dream or prophecy, the immediate situation is that there is now more co-operation and borrowing across apparently immutable boundaries than for many decades. Geography has, in Ackerman's idiom, dropped both its internal and external separatism.

THE IDEA OF MODELS

Types of Models

In everyday language the term model has at least three different usages. As a noun, model implies a representation; as an adjective, model implies ideal; as a verb, to model means to demonstrate. We are aware that when we refer to a model railway or a model husband we use the term in different senses. In scientific usage Ackoff (Ackoff, Gupta and Minas, 1962) has suggested that we incorporate part of all three meanings; in model building we create an idealized representation of reality in order to demonstrate certain of its properties.

Models are made necessary by the complexity of reality. They are a conceptual prop to our understanding and as such provide for the teacher a simplified and apparently rational picture for the classroom, and for the researcher a source of working hypotheses to test against reality. They convey not the whole truth but a useful and apparently comprehensible part of it.

It is clear then that classifications are very simple models. How vitally important such classifications are in the development of understanding may be seen from the order which Linnaeus and others brought to botany, an order which was to be the starting point for models of evolution and genetics. Why the Linnaean classification was to be so important was difficult to foresee. Other classifications had been tried before and we may find the answer simply in the fact that it worked. Our own working classifications need to be extended and continued despite early failures for I can see no *a priori* grounds for thinking our own periodic tables or fundamental indices are not waiting to be found.

There are indeed several typologies of models, of which that by Chorley (1964) is the most fully developed for the earth sciences. A simple three-stage breakdown has been suggested by Ackoff (Ackoff *et al.*, 1962) into *iconic, analogue* and *symbolic* models, in which each stage represents a higher degree of abstraction than the last. Iconic models represent properties at a different scale; analogue models represent one property by another; symbolic models represent properties by symbols. A very simple analogy is with the road system of a region where air photographs might represent the first stage of abstraction (iconic); maps, with roads on the ground represented by lines of different width and colour on the map, represent the second stage of abstraction (analogue); a mathematical expression, road density (Taaffe, Morrill and Gould, 1963), represents the third stage of abstraction (symbolic). At each stage information is lost and the model becomes more abstract but more general.

It will be clear that the examples cited so far are 'static' models, little more than rarefied descriptions. Most of the models used in economic geography to date tend to be of this kind – Christaller (1933) and Lösch (1954) describe the pattern of settlement with its basic hexagonal structure as a static economic landscape, Auerbach (1913) describes the structure of cities by a rank-size rule, Von Thünen (1826) describes the structure of land use as a rent-distance function, Weber (1909) describes the location of industry by a weight-loss rule, Zipf (1949) describes movement between centres in terms of Newtonian physics, and so on. Each represents a rule-of-thumb model of how part of the economic landscape behaves at a given point in time rather than providing a predictive model into which changes and forecasts can be built.

This concentration on static rather than dynamic models is rather a serious shortcoming of economic geography at this stage. Berman (1961, p. 300; cited by Meyer, 1963, p. 39) has put it: 'It may be argued that dynamic models are harder to construct than static, or that we cannot begin to fashion dynamic models until we have a static model of some believability. But for practical purposes . . . a crude dynamic model may be better than a highly tooled, multi-jeweled static creation.' Economic geography lacks the (dynamic) iconic models of the geomorphologist (e.g. the wave tanks of the coastal geomorphologist) or dynamic analogue models (e.g. the hydraulic models of sector flow of the economist). The contrast between static and dynamic models is nowhere more clearly seen than in biology: Charles Darwin's great contribution was to take a well-known static model, evolution, and inject it with a dynamic mechanism, the mechanism of natural selection. A dynamic infusion for geography's static models is being pursued through the study of stochastic processes, the subject of the final section.

Approaches to Model Building

In economic geography, model building has proceeded along two distinct and complementary paths. In the first, the builder has 'sneaked up' on a problem by beginning with very simple postulates and gradually introducing more complexity, all the time getting recognizably nearer to real life. This was the approach of Von Thünen in his first 1826 model of land use in his *Isolierte Staat* (Chisholm, 1962). In this 'isolated state' he begins by assuming a single city, a flat uniform plan, a single transport medium, and like simplicities and in this simple situation is able to derive simple rent gradients which yield a satisfying alteration of land-use 'rings'. But Von Thünen then disturbs this picture by reintroducing the very things that he originally assumed inert and brings back soil differences, alternative markets and different transport media. With their introduction the annular symmetry of the original pattern gives way to an irregular mosaic far more like the pattern we observe in our land-use surveys. Nevertheless Von Thünen's model has served its point; in Ackoff's terminology it has 'demonstrated certain properties' of the economic landscape.

The second method is to 'move down' from reality by making a series of simplifying generalizations. This is the approach of Taaffe (Taaffe *et al.*, 1963) in his model of route development. The study begins with a detailed empirical account of the development of routes in Ghana over the period of colonial exploitation. From the Ghanaian pattern a series of successive stages is recognized. In the first, a scatter of unconnected coastal trading posts; in the last, an inter-connected phase with both high-priority and general links established. This Ghanaian sequence is finally formalized as a four-stage sequence common to other developing countries like Nigeria, East Africa, Malaya and Brazil.

Not all such models have developed inductively from observations within geography. Some of the most successful have come from borrowing ideas from related fields, especially the field of physics. Thus Zipf (1949) attempted to extend Newton's 'divine elastic' of gravitation to social phenomena and his $P_i \, P_j/d_{ij}$ formula for the interaction between two cities of 'mass' P_i and P_j at a distance d_{ij} is a direct extension of Newtonian physics. When modified by Isard's refined concept of distance (Isard, 1960) and Stouffer's addition of intervening opportunity (Stouffer, 1962) it has proved a very powerful predictive tool in the study of traffic generation between points. A less widely known borrowing was used by Lösch (1954, p. 184). He related the 'bending' of transport routes across landscapes of varying resistance and profitability to the sine formula for the refraction of light and sound. While such borrowing may have its dangers, it is a most fruitful source of hypotheses that can be sobelry tested for their relevance to the problems of economic geography. A book like D'Arcy Thompson's *On Growth and Form* (1917) illustrates how many subjects find common ground in the study of morphology; there is inspiration still to find in his treatment of crystal structures or honeycomb formation, as Bunge (1964) has illustrated.

Perhaps the biggest barrier that model builders in economic geography will have to face in the immediate future is an emotional one. It is difficult to accept without some justifiable scepticism that the complexities of a mobile, infinitely variable landscape system will ever be reduced to the most sophisticated model, but still more difficult to accept that as individuals we suffer the indignity of following mathematical patterns in our behaviour.

THE IDEA OF CHANCE PROCESSES

Indeterminacy and Game-theory

In the spirit of optimism that seized science after Newton's triumphant demonstration of his laws of gravitation there was much nonsense dreamed about scientific prediction. It was the French mathematician Laplace who suggested it was conceptually possible to forecast the fate of every atom of the universe both forwards and backwards through time. Although all doubted that the technical possibility lay remotely far in the future it served as a grail towards which science might slowly progress. The break-up of such physical determinism with the rise of quantum physics and the enunciation of Heisenberg's 'uncertainty principle' in 1927 is a part of scientific history of unrecognized importance to geography.

The realization that even physical laws were statistical approximations of very high probability based on immense uniform populations seeped rather slowly through to the social sciences. Economics itself recovered slowly from what Bronowski (1960, p. 67) has called '. . . the fatal reasonableness of Adam Smith's *Wealth of Nations*' and remained largely wedded to a causal system of growing complexity. This attempt to reduce economics to a set of principles has come under increasing attack from both within and without economics departments. Kendall regarded the system '. . . as mistaken as the attempts of the early physicists to explain everything in terms of four elements, or of the early physicians to explain temperament in terms of four humours' (Kendall, 1960, p. 7). The breakthrough here waited until 1944 with Von Neumann and Morgenstern's *Theory of Games and Economic Behaviour* (1944). In this the uncertainty principle was introduced through a mathematical treatment of games, substituting the formalism of demand, supply and perfect assumptions for the half guesses and probabilities of an uncertain market. This is the world which we know intuitively as individuals making our own economic decisions; a world which is neither wholly rational nor wholly chaotic but a probabilistic amalgam of choice, calculation and chance.

Gould (1963) has made direct use of Von Neumann and Morgenstern's ideas in a study of land-use patterns. He illustrates the problem of farmers in a small west Ghanaian village who have five major crops which yield very differently in 'wet years' and 'dry years'. Yams for example yield nearly eight times as heavily in wet as in dry years, while another crop, millet, has much lower yields but is little affected by weather fluctuations. In this situation Gould uses game-theory to derive the crop combination that optimizes yield under these unpredictable conditions. His answer, to specialize in maize (77 per cent of the area), and rice (23 per cent of the area) accords roughly with the land use actually adopted in the area. Since the 'natural' solution was evolved only through trial and error, error which meant starvation, the practical implications of such problem-solving techniques is clear.

Origins of Indeterminacy

How does such randomness arise? Morrill (1963) in his study of town location in Sweden suggests that randomness enters the locational process in three ways. First, randomness arises from the imperfectness of human decisions. We are not always able to distinguish between equally good choices and we cannot always recognize optimum locations even should these exist. There are, Morrill contends, basic uncertainties in the pattern of human behaviour that we cannot wish away. Secondly, randomness arises from the multiplicity of equal choices. There are far more potential routeways than there are routes, more town sites than towns. Thirdly, randomness arises from our inability to take into account the effect of many small sources in any reasonably comprehensible view of reality. Following Newton's view that an alighting butterfly disturbs the earth it is clear that each locational decision stems from an infinity of cause and counter-cause which leads back only into an unending labyrinth. The net effect of all these small causes may be considered as random even though each may be, in the last analysis, rational; we can in practice only hope to disentangle some of the major threads, the rest we can regard as 'noise', a background Brownian motion.

These problems were recognized by medieval scholars like Aquinas with his search for the First Cause, and are reflected in contemporary geography by Meinig's (1962) study of the routes chosen by the railways in the Pacific north-west. He argues that not single routes but sets of possible routes were normally available to the developing company from the purely engineering viewpoint. To explain the location of the route actually selected would need detailed archival research into boardroom decisions and '... leave one stranded in the thickets of the decision-making process' (p. 413). Meinig argues that what graft, ignorance, whimsy or good sense leads to a given locational decision we may never know.

Development of Stochastic Models

One of the ways in which such indeterminacy is being built into a number of locational models is through the use of probability matrices. Fig. 6.1 shows a hypothetical example of part of such a matrix where it is postulated that eight centres are to be established within an area. Clearly there are more possible locations than centres and we need some method of assigning them to the area. If the area is graded into three different levels of attractiveness, I, II, III (Fig. 6.1–A), so that the probability of each area being chosen is respectively 3:2:1 (Fig. 6.1–B), then we can assign a sequence of numbers between zero and ninety-nine to the areas (Fig. 6.1–C) which represent the 'chances' of each cell in the area receiving a centre. The actual assignment process is random in that eight numbers are drawn between zero and ninety-nine from a random numbers table to determine the location of the centres. With numbers 05, 42, 59, 61, 67, 78, and 80 the pattern is as shown on Fig. 6.1–D. The case is of course a trivial one and actual matrices used (e.g. by Morrill, 1962) contain far more centres over a far wider area with additional growth and spacing constraints 'built in' to make the model a better approximation of reality. Nevertheless the basic principles of simulation by setting determinate behaviour in a probability framework are apparent even in this simple case; we clearly cannot predict the outcome of *individual* events, but we can simulate the *general* locational pattern.

Using techniques of this kind with rules related to distance from original points of growth, migration proportional to size of settlement, the availability of the transport net, and so on, Morrill (1962) was able to build up the pattern of settlement around hypothetical centres and check this with the evolution of settlement in a specific area, the Värnamo area, of southern Sweden (Morrill, 1963). The advantages of running this type of model is that where data is available, it is possible to adjust the 'rules of the game' to make the model simulate what we know actually to have taken place. Morrill was able to do this for twenty-year periods between 1860 and 1960. It is tempting to allow the model to run on into the future; not to predict what *will* happen but to predict what could, *might* happen if conditions remained unchanged. Such a model might have very important consequences for testing alternative legislative schemes for land-use development in an area of rapid growth, say the south-east of England or the north-eastern seaboard of the United States. One disadvantage of the method is the very great amount of computation needed. In Morrill's small area the migration phase of the model alone required the computation of probabilities for nearly 150,000 migration paths.

Neyman and Scott (1957) have carried this idea a stage further with a general stochastic theory which they suggest may be applicable to phenomena as unlike as star galaxies and animal populations. Basically their scheme hinges on chance, distribution of population centres, chance variations in population increase, chance mechanisms

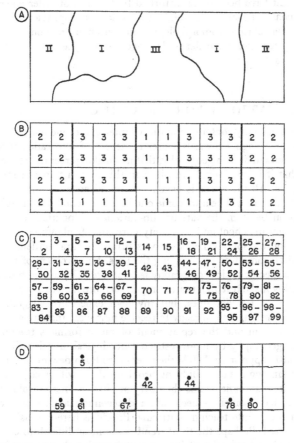

FIG. 6.1. *Hypothetical probability matrix for the allocation for eight settlements in relation to three environmental classes, I, II, III.*

of dispersal, and chance mechanisms of survival. The curious result of these random processes is to build up regular hierarchies and patterns not unlike those of the Christaller–Lösch landscape. The idea has recently been carried much further by a geographer (Curry, 1964) and there is every indication that if Lösch had survived he would have taken up this idea with enthusiasm. He wrote in 1940: 'I doubt that the fundamental principles of zoological, botanical, and economic location theory differ very greatly' (Lösch, 1954, p. 185). He might have been more disturbed by the idea of order emerging from chance processes, certainly a disturbing metaphysical concept, but evidence from geomorphology points to similar random processes creating recognizable order in drainage systems (Leopold and Langbein, 1962).

IMPLICATIONS FOR TEACHING

It has been my contention in this chapter that economic geography has been changing towards a greater degree of integration with its neighbours in human geography, that it is making freer use of models, and that it is increasingly involved in thinking in terms of random processes. All three trends have some significance for our teaching. The first trend will be a welcome one in that it reduces the inevitable confusion between the various sub-varieties of 'political' or 'social' geography in school and university curricula. It allows clearer concentration on specific fields such as urban or rural settlement and should go some way to clarifying the persistent problem of whether or not a specific topic is geography *sensu strictu*. The second trend will inevitably make teaching more complex as more models are produced and replace familiar ones, but my own experience suggests that students are often more ready to receive new ideas than we are ready to teach them and the replacement of much formless teaching of human geography by more restricted but logical models will surely meet little opposition. While the attempt to learn new models before understanding the old will bring problems these will be considerably less acute than in fields like chemistry or physics where revolutionary changes have somehow been met and accommodated, more or less satisfactorily, into teaching curricula. Problems here are easily overemphasized. Perhaps the third trend is the most difficult to comprehend and to teach in that it represents a major change in thinking patterns. The concept of stochastic processes is not an immediately assimilable one and our best hope here may lie in the spread of elementary statistical teaching into school mathematics with a basic concern with probability and chance. The encouraging work by the Association of Teachers of Mathematics promises revolutionary change in the 'numeracy' of the children who will come to read geography in the next decade; an increasing number of schools' VIth-form time-tables are being freed to allow mathematics and geography to be taken together.

My main impression as one who teaches as well as researches in this field is that human geography is entering a most exciting phase. Problems that puzzled Vidal de la Blache and Ellsworth Huntington are beginning to yield to an attack that owes as much to changed thinking as to the hardware of the statistical revolution. It is of vital importance in this field as in others that the thread linking research, university teaching and school teaching, a thread already pulled taut, should not be allowed to part.

References

ACKERMAN, E. A., 1963, 'Where is a Research Frontier?', *Ann. Assn. Amer. Geog.*, **53**, 429–40.

ACKOFF, R. L., GUPTA, S. K. and MINAS, J. S., 1962, *Scientific Method: Optimising Applied Research Decisions* (New York).

ALEXANDER, J. W., 1963, *Economic Geography* (New York).

AUERBACH, F., 1913, 'Das Gesetz der Bevölkerungskonzentration', *Petermann's Mitteilungen*, **59**, 74–76.

BERRY, B. J. L., 1959, 'Recent Studies Concerning the Role of Transportation in the Space Economy', *Ann. Assn. Amer. Geog.*, **49**, 328–42.

BERRY, B. J. L. and PRED, A., 1961, 'Central Place Studies: a Bibliography of Theory and Applications', *Regional Science Research Institute Bibliographic Series*, **1**, 1–153.

BRONOWSKI, J., 1960, *The Common Sense of Science* (London).

BUNGE, W., 1962, *Theoretical Geography* (Lund).

— 1964, 'Patterns of Location', *Michigan Inter-University Community of Mathematical Geographers, Discussion Papers*, **3**, 1–39.

BURTON, I., 1963, 'The Quantitative Revolution and Theoretical Geography', *Canadian Geographer*, **7**, 151–62.

CHISHOLM, M. D. I., 1962, *Rural Settlement and Land Use: an Essay in Location* (London).

CHORLEY, R. J., 1962, 'Geomorphology and General Systems Theory', *U.S. Geol. Survey, Prof. Paper*, 500–B, 10 pp.

CHORLEY, R. J., 1964, 'Geography and Analogue Theory', *Ann. Assn. Amer. Geog.*, **54**, 127–37.

CHRISTALLER, W., 1933, *Die zentralen Orte in Süddeutschland* (Jena).

CURRY, L., 1964, 'The Random Spatial Economy: an Exploration in Settlement Theory', *Ann. Assn. Amer. Geog.*, **54**, 138–46.

DARBY, H. C., 1954, 'On the Relations of Geography and History', *Inst. Brit. Geog. Pub.*, **19**, 1–11.

FISHER, J. L., 1955, 'Concepts in Regional Economic Development Programmes', *Regional Science Association, Papers*, **1**, W1–W20.

GARRISON, W. L., 1959–60, 'Spatial Structure of the Economy', *Ann. Assn. Amer. Geog.*, **49**, 232–39 and 471–8; **50**, 357–73.

GARRISON, W. L., BERRY, B. J. L., MARBLE, D. F., NYSTUEN, J. D. and MORRILL, R. L., 1959, *Studies of Highway Development and Geographic Change* (Seattle).

GINSBURG, N., 1961, *Atlas of Economic Development* (Chicago).

GOULD, P. R., 1963, 'Man against His Environment: a Game Theoretic Framework', *Ann. Assn. Amer. Geog.*, **53**, 290–97.

HARTSHORNE, R., 1939, *The Nature of Geography* (Lancaster, Pa.).

— 1959, *Perspective on the Nature of Geography* (London).

HERBERT, J. D. and STEVENS, B. H., 1960, 'A Model for the Distribution of Residential Activity in Urban Areas', *Journal of Regional Science*, **2**, 21–36.

ISARD, W., 1960, *Methods of Regional Analysis* (New York).

JONES, C. F. and DARKENWALD, G. C., 1954, *Economic Geography* (New York).

KENDALL, M. G., 1960, 'New Prospects in Economic Analysis', *Stamp Memorial Lecture*.

LEONTIEF, W., 1963, 'The Structure of Development', *Scientific American*, **209**, 148–66.

LEOPOLD, L. B. and LANGBEIN, W. B., 1962, 'The Concept of Entropy in Landscape Evolution', *U.S., Geol. Survey, Prof. Paper, 500–A*, 20 pp.

LÖSCH, A., 1954, *The Economics of Location* (New Haven).

MᶜCARTY, H. H., 1940, *The Geographic Basis of American Economic Life* (New York).

MEINIG, D. W., 1962, 'A Comparative Historical Geography of Two Railnets: Columbia Basin and South Australia', *Ann. Assn. Amer. Geog.*, **52**, 394–413.

MEYER, J., 1963, 'Regional Economics: a Survey', *Amer. Economic Rev.*, **53**, 19–54.

MORRILL, R. L., 1962, 'Simulation of Central Place Patterns Over Time', *Lund Studies in Geography, Series B, Human Geography*, **24**, 109–20.

— 1963, 'The Development and Spatial Distribution of Towns in Sweden', *Ann. Assn. Amer. Geog.*, **53**, 1–14.

NEUMANN, J. VON and MORGENSTERN, O., 1944, *Theory of Games and Economic Behaviour* (New York).

NEYMAN, J. and SCOTT, E. L., 1957, 'On a Mathematical Theory of Population conceived as a Conglomeration of Clusters', *Cold Spring Harbor Symposia on Qualitative Biology*, **22**, 109–20.

PERLOFF, H. S., 1957, *Regional Studies at U.S. Universities* (Washington).

ROBBINS, L., 1935, *An Essay on the Nature and Significance of Economic Science* (London).

RUSSELL, J. C., 1964, 'A Quantitative Approach to Medieval Population Change', *Jour. Econ. Hist.*, **24**, 1–21.

SMITH, J. R., PHILLIPS, M. O. and SMITH, T. R., 1955, *Industrial and Commercial Geography* (New York).

SPENCER, J. E. and HORVATH, R. J., 1953, 'How Does an Agricultural Region Originate?', *Ann. Assn. Amer. Geog.*, **53**, 74–92.

STOUFFER, S. A., 1962, *Social Research to Test Ideas* (New York).

TAAFE, E. J., MORRILL, R. L. and GOULD, P. R., 1963, 'Transport Expansion in Underdeveloped Countries: a Comparative Analysis', *Geog. Review*, **53**, 503–29.

THOMPSON, D'ARCY, 1917, *On Growth and Form* (Cambridge).

THÜNEN, J. H. VON, 1826, *Der Isolierte Staat in Beziehung auf Landwirtschaft und Nationalökonomie* (Hamburg).

WEBER, A., 1909, *Ueber der Standort der Industrien* (Tübingen).

YEATES, M., 1962, 'The "Transportation Problem" in Geographical Research', *Northwestern Univ., Dept. Geogr. Disc. Paper*, 2: 1–9.

ZIPF, G. K., 1949, *Human Behaviour and the Principle of Least Effort* (Cambridge).

CHAPTER SEVEN

Historical Geography: Current Trends and Prospects

C. T. SMITH

Lecturer in Geography, University of Cambridge

As long as geographers are concerned with the study of places and what they are like, how they differ from each other, and how their parts are interrelated, they will want to know how these places came to be what they are, and what they were like in the past. It is obvious that such studies may be pursued with varying relevance to the geography of the present day, and with varying interest and importance to geographers, according to their predilections. But historical studies have a contribution to make which lies very much closer to the heart of the subject, for they are frequently essential to the understanding of why things are where they are. This problem of location is, indeed, seen by some as the central theme of the subject of geography as a whole.

Now it is apparent that the problem of the location of phenomena of interest to the geographer may be treated either in a genetic way or a functional way or both. Thus, the location of an industry, a town, or a political frontier, for example, may be discussed in terms of the ways in which it is related to the conditions of its present environment and how it fits into or functions in its present context. For some types of study this may be enough. Studies of crop distribution, for example, may sometimes satisfactorily concentrate on the context of a physical, social, economic and political environment in which little reference need be directly made to the past. The same is true, in general, of studies of urban spheres of influence, migrations of populations, the movement of trade and traffic, or the economic operation of an industry on a given site. But it is equally obvious that many geographical features require historical study for a satisfactory explanation of how they come to be where they are. The location of farms, villages, towns, industries and communications, for example, can only be understood in terms of a sustained consideration of the conditions under which they were founded, grew, and survived to the present.

Experience of examining the work of candidates for Advanced Level and other examinations reveals many misconceptions and very many quite appalling gaps in the knowledge of even good candidates about the importance of the historical past to an understanding of the facts of human and regional geography. Towns and industries seem to appear fully grown overnight as an anonymous 'response' to bridging points, heads of navigation, proximity of raw materials, fuel or water. It is rare indeed for any candidate to give an indication of a real historical perspective or to hint at the differences which separate economies and societies of the past from those of the present. It is, on the other hand, far too common for considerations widely different in their chronological relevance to be lumped together in a cavalier manner. Some of the commonly used textbooks are, unfortunately, greatly deficient in explaining the circumstances under which, for example, cities, industrial regions or even certain zones of highly specialized agriculture have developed and survived. The over-simplifications and distortions which have been so frequently perpetrated may in part be the result of undue emphasis on 'the influences of the physical environment' and a misguided neglect of 'non-geographical' factors, but they also arise from the failure of historical geographers to fill the many gaps which occur between what is available in historical writings and what is needed by the modern geographer.

It is clearly necessary to be much more precise and rigorous about the problems and topics which emerge out of these general and preliminary comments. First, the aims and purpose of historical geography should be clarified, and in particular, what function it should play in relation to the mainstream of geography. Very different answers have been given to this question in the past, as will be seen. Secondly, it may be profitable to sum up briefly the current trends which have been shown by research in the subject and to attempt to chart the broad directions of future activities in the light of new developments in mainstream geography. Finally, presentation often produces difficult literary problems of organization. Place, time and topic present three major headings under which a study must be organized and it is the ordering and relative placing of these, together with their integration into a modern geographical study, which create a challenge to the ingenuity of authors. But this is an issue about which little need be said here in view of Darby's excellent presentation of the problem and discussion of the various means which have been taken solve it (Darby, 1963).

THE AIMS AND METHODS OF HISTORICAL GEOGRAPHY

Every practitioner may have his own views about the meaning of historical geography, and these will almost certainly be very strongly coloured by the nature of his own research work in the field, but one could, perhaps, reduce to some six or seven the definitions which have been given to the subject in the past. Of these, three or four are archaic and should be of no more than historical importance. Historical geography is a term which has been used as a synonym for the history of geography as a discipline, and although Sauer seems to have included it within the boundaries of historical geography in 1941 (Sauer, 1941), it is not normally now so used. In the nineteenth century it was commonly used to mean the history of exploration and discovery and also the history of the mapping of the earth (Baker, 1936; Gilbert, 1932). Keith Johnston's study of the development of knowledge about the surface of the earth was called *A Sketch of Historical Geography* in 1872, and the term still survives in continental literature in these senses, as in *Petermanns Geographische Mitteilungen* or the *Bibliographie Géographique Internationale*. As long as geography could be seen as primarily concerned with the description and naming of geographical features and with survey and discovery, historical geography could clearly and legitimately retain these meanings, but they are now also archaic. Similarly, the preoccupation of historians with political and national history was reflected in the emergence of an historical geography which dealt with the history of changes in the boundaries of political units, with or without lists of the battles, conquests, marriages confiscations and the like by which new pieces of territory were won or lost. Freeman's *Historical Geography of Europe*, was of this type. Mirot's *Géographie Historique de la France* (1929) is also in this genre, and Kretschmer (1904) wrote a German counterpart, dealing with the historical geography of Central Europe. Yet it should be noted that both Freeman and Kretschmer held much wider views about the nature of historical geography than appears in the contents of these two books. Kretschmer in particular put much greater stress on a more modern concept of historical geography as the reconstruction of past human and regional geographies.

Other views about the nature of historical geography cannot be so rapidly dismissed, however. They may be listed briefly as: the operation of the geographical factor in history, the evolution of the cultural landscape, the reconstruction of past geographies, and the study of geographical change through time. As will be seen, all are subject to a number of limitations, and discussion about them must be qualified by a few remarks. Firstly, the value of all of these portmanteau definitions is limited by the difficulty of comprehending in a single slogan attitudes or points of view which have many subtle shades of meaning. As will be seen, each of the four concepts quoted above shade one into the other; many studies would be difficult to classify under no more than one of the headings given above; yet the advantages of classifying them in this way are probably greater than the danger of creating that kind of spurious opposition between imaginary viewpoints which is not uncommon in methodological writings.

Secondly, it is also apparent that each of these headings is closely associated with a particular view about the nature of geography as a whole. Each of them appears to stand in a symmetrical and orderly relationship to geography as a whole in a way which is logical, but which rarely fits at all exactly the somewhat rambling and haphazard way in which knowledge in fact grows. The tendency for profitable research to be done at the margins between disciplines is well known and almost axiomatic, but the new fields of study which are thus created may tend to gravitate towards the discipline from which most of the research has been conducted even when it may logically belong elsewhere. Whether particular topics have been approached from history or geography may be largely a matter of accident, personality, the structure and traditions of institutions concerned with teaching and research, or it may even be partly a result of the geography and history of the area concerned. A single example must suffice.

Until fairly recently much of the work on field systems, agrarian structures and the history of rural landscapes has been done in England by historians, but by geographers in France. It may be hazarded that a number of factors have contributed to this state of affairs: the structures of open-field agriculture are still an integral part of the rural geography of France and therefore a suitable topic for geographical study, whereas in England the enclosure movements have removed the visible evidence of open fields from the landscape. In

France the close association of history and geography, coupled with an early preoccupation from the time of Brunhes with the facts of the 'cultural landscape', may also have worked towards a greater geographical interest in this direction, whereas in England a more precocious development of economic and social history and the lack of any obvious and outstanding relevance of the problem for geography tended, perhaps, to swing the topic another way. And finally the interests of men such as Bloch, Dion, Seebohm, Gray, Hoskins and Darby have clearly had important consequences.

THE HISTORICAL FACTOR IN GEOGRAPHY

In the 1870's Wimmer considered that the primary functions of historical geography were two-fold: it should be concerned with the operation of the geographical factor in history and also with the interrelationship of phenomena in space at a particular period, or the geography of past periods (Wimmer, 1885). It is the first of these views that must be examined at this stage. Many of the writings on this theme of the geographical factor in history, and there is quite a considerable bulk, use the term geographical in an ambiguous way. Where, as in the hands of more recent writers such as Whittlesey or East, the term 'geographical' in this phrase is given its full value, the study of the operation of the geographical factor in history may in practice involve a study not very different from reconstructing the geography of a past period as a part of the necessary context within which the flow of historical events may be the better understood. But where, as is often the case, the adjective 'geographical' indicates simply the facts of physical geography, this study of the geographical factor in history is seen as an old friend in a very thin disguise, for it may simply be rephrased as 'the control/effect/influence of the physical environment on man's activity in the past, or on history and historical events'. To some extent it carries with it the dusty haze which surrounds the acrid discussions of determinism and possibilism; and it suggests a view of causation by which it was thought that phenomena could best be understood by studying, listing and classifying the operation of particular groups of factors and their effects, each group producing its own particular pattern of determinism: social, economic, technological determinisms thus take their place side by side with the geographical determinism.

Both historians and geographers converged on this theme from different directions, however. In the last quarter of the nineteenth century growing interest in economic and social history meant a new concentration on the everyday activities of past societies. Historians were therefore becoming interested in many of the features which also interested geographers: agriculture, settlement, field systems, towns, industries, trade and communications; and they also became interested therefore in the simple facts of physical geography to which these features were in some degree related. Thus, J. R. Green in the *Making of England* (1882) used the evidence of geology, relief, soils and vegetation together with knowledge of navigable river systems in order to throw a flood of new light on the history of England in the Dark Ages. G. A. Smith's *Historical Geography of the Holy Land* sought to discover what geography had to contribute to questions of biblical criticism, venturing boldly into speculative realms on the effects of environment on religious thought. It is interesting and revealing that H. B. George in his *Relations Between History and Geography*, 1901, should still find it worth making the point, now so much a part of the historian's stock in trade, that 'geographical knowledge affords much valuable data for solving historical problems'.

As long as geography was seen as essentially concerned with the relationship of man with his environment geographers could clearly search for material in the records of the past just as logically as in those of the present for examples of 'responses' to physical environment which could be catalogued and classified as a stage *en route* to the formation of geographical laws.

The work of E. C. Semple is full of this type of selection from history, sometimes cavalier in wrenching information from a proper and necessary historical context (e.g. Semple, 1903, 1932). Possibilists, on the other hand, have used the past as a source from which to collect information about the ways in which man has used his physical environment through many generations and are concerned to demonstrate the varied uses to which it might be put. From here it is but a very short step to a kind of study in which the same place is studied at different times and a kind of comparative regional geography built up dealing with the same place at different times instead of similar places in their modern setting. Conclusions might thus be expected to emerge about the use made of position and resources under very different social, economic and technological environments. Something like this has indeed been the task of many writers in the French regional school, and it is an approach which has yielded much of interest and significance. But by the time the study of the geographical factor in history has reached this level of interpretation, particularly with the emphasis on social, economic and political matters that is the logical consequence of the possibilists' position, there is very little difference at all from a viewpoint which is defined as a reconstruction of past geographies.

Two charges in particular were levelled against this view of historical geography by Hartshorne (1939). It seems right to object that the elucidation of the geographical factor in history is a task for the historian rather than the geographer, since he may be the better equipped to set the geographical factor in its full context and to put the operation of the geographical, as opposed to other factors, into its proper context, but it is worth noting that this is a criticism which flows from a changed attitude to causation and explanation, for it implies a modern attitude towards the study of complex inter-relationships rather than the now largely outdated attempt to study causes. The other serious objection to this type of historical geography, that it sets out to clarify history, not geography, seems to be largely valid. And one might also concur with Sauer that at its worst, this type of historical geography may be nothing more impressive than 'adding the missing environmental notations to the work of historians' (Sauer, 1941).

THE CHANGING 'CULTURAL LANDSCAPE'

Whatever its methodological inadequacies, the idea of geography as the study of the landscape, with its corollary that historical geography should be the study of the changing cultural landscape, has been extremely productive of good work and stimulating ideas. This is not the place to explore the roots of the concept in the ambiguities of German *Landschaft* or French *paysage*, the irritating discussions over what may be called problems of visibility, or the attraction of an approach to the subject which rids it of the duality and question-begging of the man–environment approach, yet which offers the geographer a moderately reliable touchstone whereby to distinguish what is relevant to him and what is not. Historical geography as the study of the changing landscape is a very obvious extension of this idea, and one which was followed even before 1914 by Brunhes, in for example the first volume of the Hanotaux's *History of the French Nation* which is a general introduction to the human geography of

France, or by Kretschmer in his historical geography of Central Europe or by Wimmer long before him. Many French and German geographers have followed this line of attack, which has been taken up in England notably by R. E. Dickinson and Darby in geography and by some historians, chiefly Hoskins (1955) and Beresford (1957). Whittlesey has sometimes followed this approach in the U.S.A. and Sauer's emphasis on culture-history has been very closely associated with studies which concentrate upon the changes wrought by man upon the land.

Studies of the changing landscape lead naturally to the extension that historical geography should be concerned particularly with the transformation of natural landscapes by man, and this theme is given form and substance by Darby's systematization of the major items: the clearing of the woodland, the drainage of marsh, the reclamation of heath, the changing arable, the landscape garden and towns and industry. These are indeed the headings under which 'the changing landscape' is considered in a paper of 1951 which parallels a discussion by S. W. Wooldridge of the physical landscape of Britain (Darby, 1951). With a few additional headings, which might include irrigation, soil erosion and conservation, perhaps, or a greater emphasis on the features associated with population and settlement this approach can be seen to yield a set of more or less standardized categories under which historical geography may be written. And what is more important, these are categories which are so clearly relevant to and peculiar to geography that there is little possibility of that kind of confusion with economic history that has so bedevilled the subject in the past. Yet in some respects it is a system which is fore-shadowed already in the writings of Brunhes and his insistence on the classification of landscape features as the essential facts of geography. It may be noted, however, that part of the price which is paid for this rediscovery of unity in geography by way of landscape studies is the danger that man may be deposed from a central position in the study of geography and regarded as no more than an anonymous automaton whose task it is to produce the visible features of the cultural landscape, as impersonally as the processes responsible for soil creep. By this token, man plays the role of geomorphological agent, and while this is a casting which has been very productive of new ideas in human geography and in geomorphology, it should surely be regarded as only one of several roles.

One of the most appealing virtues of the approach to historical geography through the changing cultural landscape is the apparent symmetry with genetic geomorphology. Both are seen to be concerned with the evolution of landscape features; both, in the words of H. C. Darby (1953), are concerned with laying the foundations of geography, though one must stress, as he does, that these are no more than foundations, on which better things are ultimately to be built by social and economic geographers. Consideration of this analogy between historical geography and geomorphology may draw attention, in particular, to some of the problems of evidence and method which beset both of these subjects. Genetic geomorphology sets out to understand landscape features by arranging them according to the manner and chronology of their development. Past circumstances which relate to the development of landforms must be reconstructed in the light of knowledge about the processes which act to produce the observed landforms themselves. Evidence external to the landscape features themselves comes only from the composition of deposits of various kinds. Now it is also clear that in historical geography progress may be made in reconstructing past geographies by a careful analysis of the manner and chronology by which features of the landscape have taken the form they now possess. Emphasis on the study of such features and their interpretation is undoubtedly the greatest single contribution of this school of thought. Studies of the morphology of towns, rural settlement patterns, agrarian structures and field systems arise directly out of this approach, and have leaned very heavily on reconstructions of chronological sequences out of the pattern of landscape features (the arrangement of field systems, for example). Indeed this method of approach has been dignified by such terms as 'the morphogenesis of the cultural landscape', and the analogy with genetic geomorphology is even carried to extremes by attempts to characterize the development of settlement structures in terms of 'structure, process and stage' and by attempts to identify 'cycles' of development, as for example, the attempt to characterize a cycle of development in French *bocage* landscapes.

In breaking the type of circular argument which results from interpreting landscape elements in terms of themselves, historical geography has always so obviously had recourse to documentary evidence (analogous perhaps to the evidence of the deposits for the geomorphologist) that his normal task has more frequently been seen as the reconstruction of past geographies than the understanding of landscape features which derive from the past. Indeed, as techniques of dating and interpreting the ecology of relatively recent deposits are improved and refined in precision, and as knowledge of geomorphological processes becomes more elaborate, so attention in geomorphology has itself begun to shift beyond an emphasis on the reconstruction of the narrative of events towards a much more informed reconstruction of the circumstances of particular periods of the past in the light of more elaborate methods of analysis.

The next stage in the discussion should clearly be concerned with the approach to historical geography as the reconstruction of past geographies, but before proceeding to it, it seems worth making two further points about the study of landscape. Firstly, it sometimes appears that disproportionate attention is given to landscape elements which are either minor in themselves or which have a significance which is marginal to what is normally considered to be part of geography. Roof forms, house types, the distribution of marl pits, for example, may seem to fall into the first category and the study of landscape gardening, so obviously relevant in itself to landscape studies, seems to fall in the second category in so far as it seems to lead us away from what is normally comprehended in geography towards social history and even towards the history of the fine arts. The same may be said to be true, for example, of the study of baroque town plans and it is also quite obviously true of the study of architectural forms.

A second question is thus raised. To identify and put into chronological order the elements of a landscape is an interesting exercise, and may well be a difficult one to carry out (Yates, 1960), but though Yates succeeded admirably in his selected area, it is obvious that the understanding of the distributions of relict forms depends very heavily on knowledge of the historical circumstances of their origin and survival. Dating alone is not enough. It is perhaps more frequently profitable to regard the landscape elements as a source of invaluable evidence for the reconstruction of past geographies than as the phenomena which are to be explained by historical study.

The magnificence of medieval churches may thus be seen as possible evidence for the distribution of medieval prosperity or piety; the distributions of roof and house types fall into place as items, together with field systems, settlement patterns and documentary evidence, which throw light on, for example, cultural contrasts between northern and southern France, as in the excellent study of Limagne (Derrau, 1949). Marl pits in Norfolk are revealed as a hitherto neglected source of evidence for the distribution of activity by eighteenth-century improvers. The distribution of 'ridge and furrow' is important as a potential source of information about early farming activities in the open fields. Study of the origin of the Norfolk Broads began as an exercise in the origin of a landscape feature first thought to be of natural origin and only at a much later date considered to be of historical origin. But in the historical part of the study, the most interesting problems which emerged were those which had to do with the economic, social and physical circumstances of the period when the turf pits were abandoned and were subsequently flooded. It is helpful and interesting to consider the landscape or the topographical map as a palimpsest, but it is not enough to identify and put into chronological order the fragmentary inscriptions which are legible. The ultimate aim of the study should surely be to read and interpret the inscriptions themselves.

THE RECONSTRUCTION OF PAST GEOGRAPHIES

By far the most orthodox and, indeed, unexceptionable view of historical geography is that it should be concerned with the reconstruction of the geographies of past times. Now it is clear that these can be as varied as the adjectival geographies of modern times. There can be not only regional geographies of the past but there can also be urban and rural or agricultural and industrial geographies. Many of the most important historical geographies to have been published employ this method. A series of cross-sections is presented in H. C. Darby's classic work on the historical geography of England before 1800, and this is the method which is used so systematically and ingeniously by Ralph Brown in 'Mirror for Americans' – a reconstruction of the geography of the Eastern Seaboard as it was in 1811, using contemporary sources. It is the method used by Fernand Braudel in his equally classic work on the history and geography of the Mediterranean world in the second half of the sixteenth century, though he calls it *geohistory* rather than historical geography. These are but fairly recent examples of a theme which goes back at least as far as the last quarter of the nineteenth century. Kretschmer considered that the task of historical geography was to discover the changing relationships of land and people at particular periods according to their causal interdependence, and like Whittlesey at a much later date advised that 'It is necessary to study periods of peace and stability before or after great changes.

It must not be concerned with processes of development. It describes and explains the geographical interrelationships for a fixed period' (Whittlesey, 1929). Kretschmer's became the standard view, with a few variations in emphasis. Thus, Mackinder wrote that historical geography is a study of the historical present: 'the geographer has to try and put himself back into the present that existed, let us say one thousand or two thousand years ago; he has got to try and restore it'. And finally the view that historical geography was concerned with the reconstruction of the geography of a past period or periods was the only view of the subject which Hartshorne would allow in 1939 though it is important to recognize that his opinions seem to have changed since then (Hartshorne, 1959).

There appear to be two reasons why consideration of the geography of past periods should be of significance to the geographer. First, if successive cross-sections are made of a given area, a kind of comparative geography may be built up in which it may be possible to view over a whole period of time the way in which such factors as resource and position, soils and climate have been used under varying conditions of technology, social structure, population trends and so on. Useful conclusions may be drawn about the relationships of man and environment from such comparative studies, and since both physical position and physical environment are substantially the same, the historical method should make it possible, at least in terms of theory, to isolate some of the extremely complex variables of any geographical situation. It was probably this which Wimmer had in mind when he wrote as early as 1885, 'The individual aim of historical geography is to compare the geographies of different periods in the same area.' East (1935) has written in the same vein that: 'The significance of historical geography is the more readily grasped when it is possible to review side by side the geography of a whole series of historical periods.'

Secondly, the reconstruction of the geography of past periods is necessary before the relationship between past and present geography can be fully understood. J. B. Mitchell (1954) has written in this context that the value of the work of the historical geographer, *qua* geographer, 'lies in the fact that some elements of the geographical design that develop in response to passing conditions are extremely stable in their form or long lasting in their effects, and the understanding of the present demands the study of the geography of the period of their establishment and development'. Little further comment seems necessary, however, since this is clearly very similar to the view already expressed above in connexion with the study of relict forms in a cultural landscape. Such forms can only be properly understood when placed in the general geographical or cultural context of their origin or even, in certain cases, the context of the conditions which allowed them to persist. Thus, like the case of the dog that did not bark in the night, the survival of so much Georgian and Regency architecture in the market towns of East Anglia is in itself a negative comment on the development of these areas in the nineteenth century.

The concept of historical geography as a reconstruction of past periods is orthodox and it is also clear and distinctive. But this clarity and distinctiveness need further examination, for they rest on assumptions about the nature of explanation in geography and history which were never wholly valid. To put the matter very briefly indeed, there is general agreement that geography is essentially concerned with the functional interrelationship of the phenomena it studies. Nineteenth-century German writers stressing *causal interdependence* and the interrelatedness of the *zusammenhang* of phenomena, French writers such as Brunhes stressing the idea of *connexité*, Mackinder, Hettner and subsequent authors, notably Hartshorne, are all agreed on this point. The area of disagreement had been largely about whether geography should be concerned with genetic studies, that is, studies of growth and development, or the processes which have operated to produce change. Wimmer, Kretschmer, Mackinder, and Hartshorne in 1939 felt that genetic studies belonged properly to historical method and should be excluded from geography as much as possible. It is from this position, in fact, that historical geography can be most clearly seen as a series of period pictures or cross-sections, using its own methods to produce a geography which would be methodologically quite distinct from any kind of history of the same place during the same period. Geography should be concerned with functional interrelationships; genetic studies were essentially historical.

Unfortunately, it is quite clear that this position leads to the empirically untenable conclusion that although geographers may legitimately consider how an industry or town may function at the present time, it would not fall within their task to study how that industry or town came to be where it is. Most geographers have rejected this extreme position, including Hartshorne in his more recent methodological work.

There is, however, another and more strictly logical reason for rejecting this excessively simple view. The opposition between functional and genetic studies is more apparent than real. Genetic studies may, in fact, involve one of two different components. Many so-called genetic studies are, indeed, no more than an attempt at reconstructing the functional relationships of some period in the past. Thus, a study of industrial location frequently proceeds by showing how successive patterns of industrial location were adjusted to the physical, social and economic circumstances of particular periods. The second component is, however, rather different in character and concerned with the evolution of situations continuously through time, and may be conveniently labelled the dialectic method. This has involved '. . . a pre-occupation with linked historical sequences. A dialectic explanation of an historical situation will demonstrate how it arose from the situation which preceded it; a dialectic prognosis will show how a certain future is fashioned by forces operating in the past. Past, present and future are but steps in the "ascending ladder of necessity": a ladder in which every step is so fully supported by the step below and so fully supports the step above that its relative position between the two will tell us all we need to know about itself as well as about the ladder as a whole' (Postan, 1962, p. 399). Dialectical materialism is one example in the historical field of a theoretical approach using this type of method; and in the geographical field it would seem that the idea of *stage* in the Davisian trinity of structure, process and stage owes a great deal to this kind of dialectical thought.

Now it is apparent that many historians have retreated from the ambitious claims of a dialectic method such as this; so much so that Postan quotes a historian's fears 'lest by playing down historical changes we remove history itself from the work of historians' (Postan, 1962, p. 399). Instead, historians have increasingly pursued a more limited aim of understanding the relationships within societies in a much more precise and detailed way than was thought possible by nineteenth-century historians. Studies of change through time of a single element or of a complex of related elements offer the possibility of evaluating the part which it may play in a society, but this is by no means the same idea as the study of changing phenomena in their totality through a continuous flow of events. From the publication of Namier's classic study historians, indeed, have tended more and more to concentrate on the functional interrelationships of phenomena at a given period, and although the emphasis is different from that of the historical geographer, it is quite obvious that neither history nor geography can claim exclusive rights to methods of analysis.

It is thus practically undesirable and methodologically unnecessary to exclude genetic study from geography. But if genetic studies of particular elements in the complex situations of modern geography are desirable, this must also be logically true of geographies of past times. And so the conceptual clarity of a series of relatively static period pictures is greatly obscured. It does, indeed, then become highly desirable to link successive period pictures by studies of intervening changes and the social and economic determinants of change (Broek, 1932; Darby, 1960).

There are, moreover, other difficulties involved in restricting historical geography to the large scale period picture, and the fact that these are practical difficulties and can be quite briefly discussed does not make them any less important than the methodological problems which have been raised at great length above. Whittlesey and others have followed Kretschmer in considering that periods of stability should be chosen as suitable for the reconstruction of period pictures and it is true that the geography of certain periods may be particularly significant in contributing to a geographical under-standing of areas such as the Great Plains. But stability is by no means contemporaneous over a whole area or over a whole sector of an economy or society. Moreover, in a series of geographical period pictures even the simple elimination of repetition by omitting those sectors in which change has not taken place will clearly result in a concentration of some sort of change rather than stability. The axiom itself that periods of stability should be chosen needs careful examina-tion, for in studies of settlement, for example, or of clearing and reclamation, agriculture, town growth and communications, it is precisely on the moment of change and on the process of change that most geographers concentrate their attention, for it is frequently this which is the most interesting and profitable study. The kind of questions which are currently being asked often demand consideration of change as well as static relationships – how did settlement take place? Why did a town foundation succeed here and fail there? In what circumstances and with what result did the clearing of forest take place in this or that area?

GEOGRAPHICAL CHANGE THROUGH TIME

Indeed, in many cases the position of the historical geographer may be quite indistinguishable from that of the historian in so far as he concentrates on what has been described above as 'studies in the changes of a single element or of a complex of related elements in order to evaluate the part it plays' – but, of course, in a geographical situation related to place rather than an historical situation related to society. A. H. Clark stresses this point also in asking for a greater emphasis on the description of the processes by which 'selected elements . . . that are believed to contribute largely to regional character have changed through time' (Clark, 1960). There is much to be said for an approach to historical geography which would see it as a study in 'geographical change through time' as Clark suggests (Clark, 1954). To accept this view is, in a sense, to do no more than recognize what historical geographers are already doing in their studies of settlement, field systems, changing industrial locations, urban growth and so on. There are few methodological stumbling blocks, and although it begs questions about the nature of geography like any other of the viewpoints mentioned above, it is admirably loose and permissive rather than precise and restrictive. But it offers only a partial solution and extends the field without adequately defining the whole. For it is unsatisfactory to *insist* on the study of change. Changing situations may enable a clearer picture to emerge about the function of position or resource (for example in the successive uses of river, sea and land routes in the history of the Low Countries), but it is clearly a prerequisite to examine, say, the trading structure of seventeenth-century Holland in a relatively static context. And it is still basic for geographers to study how the things which interest them about periods and places are interrelated.

What conclusions, then, emerge from this analysis of viewpoints that have been taken in the past about historical geography? The first and most important comment to make is that no doctrinaire solution to the problems of relevance which beset historical geographers is possible, and that there is no formula which will decide for them what is history and what is geography. The problem itself is often artificial and arid, since what matters is the contribution being made to knowledge, but it is frequently helpful to bear in mind the rule of thumb that geographers are essentially concerned with places and what they are like, whether in the past or the present. No precision instrument can be fashioned for this purpose out of the most carefully constructed definition. Secondly, each of the attitudes which have held the field at one time or another have served a useful function in connexion with prevailing views about the nature of geography and also about the nature of explanation. Thirdly, each of

K

these viewpoints has helped to contribute to, or has facilitated, or has (perhaps most often) simply accepted and recognized a real contribution to knowledge which was being made to geography or history or both.

Finally, the abandonment of doctrinaire attitudes towards the nature and content of historical geography opens the way for a much more flexible approach to the organization of work and its presentation. This is a theme which has quite recently been discussed at some length by Darby (1962). There is little that needs here to be added except that the extent of variation in the methods of presenting historical elements in geography is not only a measure of the difficulty of the problem, but it is also a measure of the degree to which historical geography has felt itself emancipated from the restrictions of outworn methodologies. It is obvious that the nature of the material itself may dictate the form of organization, particularly in research topics. The aim and purpose of an author may perhaps lead him to adopt a retrospective approach, or to adopt a topical approach by which changes in settlement or agriculture, for example, are followed through time. This has now become so common as to be orthodox.

CURRENT TRENDS AND PROBLEMS IN HISTORICAL GEOGRAPHY

Trends in the literature of historical geography in recent decades illustrate these themes. The relationship of man to his environment was from the nineteenth century onwards a theme which stimulated and was often associated with studies of primitive societies, among which adjustment to physical environment played a part much more obvious and direct than among Western industrialized societies. Prehistory was the corresponding field in which appreciation of position, relief, soils, drainage and water supply could throw new light on human geography and could also contribute something to archaeology. But there has been a shift away, in recent years, from the kind of study in prehistory made by Fleure, Daryll Forde, Cyril Fox or Wooldridge and Linton on Anglo-Saxon settlement. Prehistorians have learnt their geographical lessons, but it is also true that the links which connect prehistory (in Europe, but not in America and Asia) with modern geography tend to be long and tenuous, debatable and often of little apparent importance in the modern world. They have left relict field patterns and traces of settlement patterns, and these now form a starting point for studies in rural history. They have affected distributions of race and language, but geographical interest in these topics has waned with the decline of political geography, having reached its peak, perhaps, in the period before 1939.

The study of period pictures has long been standard in the repertoire of historical geographers and will undoubtedly continue to be so, for much remains to be done that is profitable and that still fills a gap between the parochialism of much local history and the larger interests of the economic historian, who still occasionally tends to ignore regional differences from one area to another and who is not usually *primarily* interested in areal differentiation. The success of *Historical Geography before 1800* (Darby, 1936) and the value of the Domesday Geography make it clear that there are considerable possibilities and that there would be great value in the further extension of this principle by detailed examination and mapping, wherever possible, of the data relating to population, tax returns and land use which are available in many forms from 1250 to the sixteenth- and seventeenth-century population censuses and hearth taxes. Much has already been done on a local scale and there is a wide field of opportunity still for pilot studies of small areas. But the evaluation of the data, the compilation of maps and the interpretation and comparison from one period to another of conclusions drawn from such studies on a larger scale represents a major task for the future. Topographical accounts, travellers' writings and early maps are other sources of peculiar interest to historical geographers of the early modern period, and these too are being exploited at greater depth than was possible only a few years ago. And only a beginning has been made on the vast resources of the nineteenth century for the reconstruction of the geography of an industrializing country. This is not the place to list such sources in any detail, and a broad outline is available in Darby's paper on 'Historical Geography, Twenty Years After' (Darby, 1960). But two points are worth making. The first is that the analysis and mapping of such quantitative data as is available for the early periods seems to be one of the fields in which there are considerable possibilities for judicious use of elementary statistical techniques to compare distributions at successive periods, to help in throwing light on the factors involved in particular distributions, and perhaps occasionally to help in checking the validity of samples. The second is that it may be possible to provide a regional framework for specific purposes which could profitably replace unsatisfactory political units or a modern regional framework which is often equally unsatisfactory. Such regions, sometimes provided ready-made like the agricultural regions identified by authors of the *General View of Agriculture* are more often the product of detailed labours on, for example, land-use statistics (Mitchell's land-use divisions of medieval Suffolk based on studies of the *Inquisitiones Nonarum* of 1341 is a case in point).

Studies in the genesis of landscape features and in geographical change through time have both led to considerably more emphasis on tracing the evolution of agriculture, settlement, trade and industry in what have been called 'vertical' or topical studies through time. It would be tedious to make a survey of the contributions which have been made to the historical aspects of systematic studies in agricultural and industrial geography, etc., but it may be helpful to pick out the areas of considerable growth in recent years.

Changes in physical landscapes in historical time have been studied more intensively with new standards of criticism and new sources of material drawn from local and national archives. Studies of climatic change before the appearance of instrumentation have been made with the help of meticulous attention to year-by-year weather conditions and cropping in monastic accounts, or by systematic study of the dates of vine harvests in France, and these add new criteria to the very many others which have been used to throw light on climatic conditions in historical time and their relationship to economic conditions. The view of man as an agent shaping the face of the land is apparent in the title of the work edited by Thomas on *Man's Role in Changing the Face of the Earth*, and it is in this light that one might also view a considerable recent output on the clearing of woodland, reclamation of heath, drainage and irrigation. Clearing, soil erosion and sedimentation have been treated from a historical point of view in work on the French Alps and by Haggett in work on Brazil. Many studies of this kind have combined techniques of research in physical and historical geography in a new and profitable way which seems often to demonstrate the absence of any great gulf between historical and 'scientific' methods.

The study of the Norfolk Broads (Lambert *et al.*, 1960) led to the accumulation of evidence both from stratigraphy and historical documents in order to explain how they originated. And in the Low Countries the history of reclamation and flooding have been greatly informed by the examination of recent sedimentation, and here, as in Western Germany and France, steps have been taken to establish the medieval history of clearing and even the major outlines of the development of land use by studies of pollen and of other plant remains.

Studies of rural settlement, field systems and agrarian structures have undoubtedly been given a great stimulus by the emphasis which has been placed on the understanding of elements in the landscape. In Western Europe, pioneer works in settlement and field systems were written between the end of the nineteenth century and the late 1920's. But after the publication of such standard works as that of Meitzen in Germany, Marc Bloch in France, the works of Seebohm and Gray in England, and after the publication of the proceedings of the Commission on Rural Habitat by the International Union in 1928, work in this direction appeared to dry up. During and since the Second World War there has been a great expansion of such studies on a more local and often on a more intensive scale than the pioneer works of early writers. In France, the Low Countries, Germany and Sweden a great deal of progress has been accomplished, not only towards the understanding of present rural landscapes and their problems, but also towards a much more profound knowledge of the historical circumstances under which field systems and settlement patterns developed. In England historians, notably Hoskins, Beresford and Allison, have continued the traditions established by social historians or by the long line of local topographers, but geographers too have made important contributions, particularly in recent years. Jones, Sylvester, Vollans, Thorpe and Bowen may be mentioned, but there are many others who are or have been researching in this field with considerable profit and mutual benefit for historian and geographer. This type of study has been one of the growing points of geography as a whole, and it has become a mainstay of agrarian history. Articles on this kind of theme have indeed served to fill the pages of new or relatively new periodicals such as *Études rurales* and the *Agricultural History Review*.

The organized study of settlement, both urban and rural, and the study of field systems clearly fills one of the gaps that was left between history and geography in the late nineteenth century, and which was only partially filled by the pioneer studies of that time and later. In the inter-war years it was largely neglected, and its revival has come mainly after the Second World War. One is inclined to speculate why this should be so and to list some of the qualities about this type of study which seem to have made it peculiarly susceptible to attack at this period and by people equipped with geographical training.

Firstly, there is usually a relationship of settlement and field system to the characteristics of the physical environment. It was indeed this fact which attracted geographical interest in the early phase of settlement studies. Relief, soils and water supply were the factors quoted by geographers to counter the ethnic views of Meitzen and his followers. This interest in the facets of physical environment is of enduring interest. Field systems and rural settlement patterns are prominent landscape features in areas of continental Europe where enclosure movements have not taken place, and the relevance of medieval and early modern rural history is all the more apparent. Indeed, one is inclined to wonder how far the early reticence of English geographers on this theme is associated with the extent to which enclosures obscured earlier arrangements of fields. Thirdly, the sources of evidence are heavily weighted towards the kind of sources geographers may feel most competent to tackle – early maps and plans, air photographs, and, of course, field observation and measurement.

Fourthly, in the absence of the abundance of documentation such as one often has in English studies, the most appropriate methods of study have often been those of genetic geomorphology. It is in connexion with studies of rural settlement and field systems that the ugly phrase 'morphogenesis of the cultural landscape' has been forged (Vadstena Symposium, 1960, 1961). In more general terms, it might be added that this is also a field in which interest is frequently centred on social and technological factors rather than strictly economic matters, and its revival thus coincides with renewed interest in social aspects of geography.

Rather similar comments apply also to urban studies. For here, too, there was a wave of interest in the late nineteenth century which was concerned with questions about the origin of towns and the nature of their institutions. This wave of interest also receded before 1939 and it is only since the war that there has been a revival of interest in urban history by historians, geographers, archaeologists, architects and town-planners. Many of the qualities which made rural settlement an attractive field of settlement for geographers apply also to the study of urban history, particularly when it is centred on problems connected with physical growth, economic function and with the changing functions of particular areas within the town. But it seems unnecessary to elaborate on this theme, essentially similar to that of rural settlement in certain ways.

It may be useful, though hazardous, to conclude by suggesting the ways in which historical geography may develop in the near future.

The application of statistical methods is a development which geography shares with economic history. It may be in the more refined analysis of the quantitative data relating to distributions of prosperity, population and land use that statistical methods have most to contribute. If so, the study of distributions at particular periods may be provided with more reliable tools for the comparison of successive distributions and the checking of data. Studies such as those of Buckatzsch (1950) provided a pointer in this direction, perhaps. Secondly, one may expect historical studies of vegetation and sedimentation to make much greater progress as it becomes possible to set conclusions drawn from pollen analysis and the organic content of recent sedimentary deposits dated by radio carbon methods against the fragmentary record of the documents or place-names. The clearing of forest, soil erosion, the reclamation of heath and the drainage of marsh may be the better understood, and so also may the history of land use in some areas.

In the study of rural settlement there is also need for more systematic co-operation in the use of evidence from different disciplines: local or architectural historian, geographer, soil scientist, and particularly archaeologist may profitably work together, as in the village survey project of Norfolk. A beginning has already been made towards the elaboration of a theoretical model for the study of the expansion of settlement through time making certain assumptions about the nature of new colonization (Bylund, 1960). But this is no more than a beginning and the application of models of this type seems to be greatly limited by the difficulty and sometimes the impossibility of gauging the approximation of real systems to the theoretical. Nevertheless, in matters of rural settlement, both in the sense of settlement and colonization and in the sense of the interpretation of settlement forms and patterns there is much scope, it would seem, for a more systematic approach through generally acceptable classifications and terminologies.

It is perhaps in urban geography and in theories of location that many of the recent advances have been made in the subject as a whole. This is no place to enter into a discussion of the utility and validity of the theoretical concepts which these involve: but it is not impossible that new approaches to problems of location, 'spatial analysis' or the *Standortsproblem* may provide a conceptual framework of reference which may greatly help in the organization and interpretation of past geographies.

An illustration may clarify this approach. In modern highly

industrialized countries the multiplication of consumer industries and the emancipation of industry from the coalfields, the universal availability of power, and the market-orientation of many 'footloose' industries have all tended to mean that an industrial geography based on the study of individual commodites has become increasingly impracticable, or at best, divorced from the realities of modern industrial structures. Geographers have been faced for some time by the need to examine industrial development either in the general terms of the economist or in terms of the study of industrial regions or of individual cities and conurbations. This must necessarily complement the study of a few selected industries which are, for one reason or another, considered suitable for analysis: iron and steel, heavy chemicals, textiles. The concentration of industry on coalfields or on other favoured sites, e.g., ports and capitals, which formerly gave coherence to industrial geography is now lost.

The situation which has emerged with emancipation from the coalfields has something in common with the 'industrial' geography of pre-industrial society. Before the rise of coal as a dominant locative factor, there were relatively few industries – iron, cloth, ship-building, etc. – which warranted detailed studies of location. Then, as in recent years, perhaps the greater part of industry took place in very scattered locations in the small workshops and in the homes of both rural and urban populations. Market forces seem often to have dominated the localization of crafts and even of rural industry to some degree. It is perhaps relevant therefore for the historical geographer of pre-industrial societies to use and modify the ideas and principles of organization developed in the literature on central place theories, urban hierarchies and spheres of urban influence. It seems likely, for example, that ideas such as these may help to make coherent and intelligible the varied occupational structure of small towns and regional centres, and they may help to interpret some aspects of the distribution of rural domestic industry round the towns which supplied the raw material and in which the finished product was marketed, e.g. in the framework knitting industry of the East Midlands.

It is also evident that new trends will transgress the boundaries of the various viewpoints about the nature of historical geography which have been outlined above and it is right that they should. Topical studies in historical branches of systematic geography seem to offer possibilities of fundamental development, particularly in terms of a better understanding of the processes by which geographical change through time may take place. Some of the developments in urban geography, notably those concerned with central place theory and spheres of influence, statistical techniques of regional analysis and the further development of distributional studies would seem to give a new and considerable stimulus to studies of the geography of particular periods. So also will the transference to historical cases of geographical studies of 'underdevelopment' and progress therefrom. It is perhaps not too ambitious to hope that these may all be fields in which the geographer has something to contribute to historical understanding as well as to the study of place, whether in the past, present or future.

What have recent trends in historical geography to offer to the teaching of geography in schools, and in what ways may they help to correct some of the shortcomings pointed out in the beginning of this chapter? Perhaps one of the most important opportunities it opens up is towards a much deeper understanding of local environments through an ability to follow up the wider context of writings on urban and rural settlement, field systems or enclosure history. There is still abundant opportunity for field studies and practical work which may overlap with local history (an opportunity for co-operation and mutual profit rather than an occasion for competition), but which can usually avoid the barrenness of many land-use studies. The ideas which come with the study of local urban and rural settlement may help students to make a profitable link between textbook abstractions about other and larger regions, and the landscapes of these areas as they are seen in film or filmstrip, or increasingly in the modern affluent world, on holidays abroad.

It is also clear that recent emphasis in historical geography towards clarifying the ways in which the past has contributed to present geographies is of much greater relevance to the teacher than academic exercises in the geography of past periods. Textbooks are rightly introducing more and more material by which the importance of a cultural heritage or of the early settlement of an area may be gauged by the student. Study of Latin America, for example, gains immeasurably by knowledge of the cultural contribution of the Indian cultures, the Spanish colonial period and nineteenth-century independence. Appreciation of the use of position, routes and resources gains by some knowledge of their use in the past. Use of the Rhine can as well be linked with the growth and vicissitudes of the towns and ports in its neighbourhood as with the movement of oil and coal. And it is also clear that crudely deterministic 'explanations' of the location of industry or the siting and growth of cities can be judiciously avoided by the introduction of a historical summary of how they came to be where they are. But it is for the historical geographer no less than the historian to provide the textbook writer with the data, to summarize as accurately as possible, and to provide methods by which problems and local studies may be the more readily understood.

References

BAKER, J. N. L., 1936, 'The Last Hundred Years of Historical Geography', *History*, New Series, **21**, 193–207.

BERESFORD, M. W., 1957, *History on the Ground* (London).

BROEK, J. O. M., 1932, *The Santa Clara Valley, California* (Utrecht).

— 1943, 'Relations between History and Geography', *Pacific Historical Rev.*, **10**, 321–5.

BUCKATZSCH, E. J., 1950, 'The Geographical Distribution of Wealth in England, 1086–1843', *Econ. Hist. Rev.*, 2nd series, **3**, 180–202.

BYLUND, E., 1960, 'Theoretical Consideration regarding the Distribution of Settlement in Inner N. Sweden', *Geog. Annaler*, **42**, 225–32.

CLARK, A. H., 1954, 'Historical Geography', Chapter 3 in *American Geography: Inventory and Prospect*, ed. P. E. James and C. F. Jones (Syracuse).

— 1960, 'Geographical Change as a Theme for Economic History', *Jour. Econ. Hist.*, **20**, 607–17.

DARBY, H. C. (ed.), 1936, *An Historical Geography of England Before 1800* (Cambridge).

— 1951, 'The Changing English Landscape', *Geog. Jour.*, **117**, 377–98.

— 1953, 'On the Relations of Geography and History', *Trans. Inst. Brit. Geog.*, *Pub. No.* **19**, 1–13.

— 1960, 'Historical Geography, Twenty Years After', *Geog. Jour.*, **126**, 147–59.

— 1962, 'The Problem of Geographical Description', *Trans. Inst. Brit. Geog.*, *Pub. No.* **30**, 1–13.

DERRUAU, M., 1949, *La Grande Limagne* (Clermont Ferrand).

EAST, W. G., 1935, *Historical Geography of Europe* (London).

FREEMAN, E. A., 1881, *Historical Geography of Europe* (London).

GILBERT, E. W., 1932, 'What is Historical Geography?', *Scot. Geog. Mag.*, **48**, 129–36.

— 1951, 'The Seven Lamps of Geography: an Appreciation of the Work of Sir Halford Mackinder', *Geog.*, **36**, 21–40.

HARTSHORNE, R., 1939, *The Nature of Geography* (Lancaster, Pa.).

— 1959, *Perspective on the Nature of Geography* (Chicago).

HOSKINS, W. G., 1955, *The Making of the English Landscape* (London).

KRETSCHMER, K., 1904, *Historische Geographie von Mitteleuropa* (Munich).

LAMBERT, J. M. *et al.*, 1960, 'The Making of the Broads', *Roy. Geog. Soc.*, Research Series No. 3.

MIROT, A., 1929, *Manuel de la Géographie Historique de la France*, 2 vols (Paris).

MITCHELL, J. B., 1954, *Historical Geography* (London).

POSTAN, M., 1962, 'Function and Dialetic in History', *Econ. Hist. Rev.*, 2nd series, **14,** 397–407.

SAUER, C. O., 1941, 'Foreword to Historical Geography', *Ann. Assn. Amer. Geog.*, **31,** 1–20.

SEMPLE, E. C., 1903, *American History and Its Geographic Conditions* (Boston).

— 1932, *The Geography of the Mediterranean Region: Its Relation to Ancient History* (New York).

SMITH, D. M., 1962, 'The British Hosiery Industry at the Middle of the Nineteenth Century', *Trans. Inst. Brit. Geog.*, *Pub. No.* **32,** 125–42.

VADSTENA SYMPOSIUM, 1961, 'Morphogenesis of the Agrarian Cultural Landscape', *Vadstena Symposium*, 1960, *I.G.U. Congress*, *Geog. Annaler*, **43,** 328 pp.

WHITTLESEY, D., 1929, 'Sequent Occupance', *Ann. Assn. Amer. Geog.*, **19,** 162–5.

WIMMER, J., 1885, *Historische Landschaftskunde* (Innsbruck).

YATES, E. M., 1960, 'History in a Map', *Geog. Jour.*, **126,** 32–51.

PART TWO

TECHNIQUES

CHAPTER EIGHT

The Application of Quantitative Methods to Geomorphology

R. J. CHORLEY

Lecturer in Geography, University of Cambridge

It would be wrong to assume that the changes which have taken place in geomorphology during the past quarter of a century or so have resulted from the application of radically new techniques to the study of landforms. The new approach to the subject drives its roots much deeper than this, and involves a series of responses to questions which are fundamentally different from those posed either by Davis or by the denudation chronologists. In fact these questions were being asked before Davis formulated his cyclical approach to the subject, but it is only recently that even a small body of geomorphologists has thought it worth while to try to answer them. Partly this attempt has been promoted negatively, in that the shortcomings of both the cyclic and denudation chronology approaches to the subject had by the Second World War so restrictively formalized geomorphology as to reduce the virile and proliferating study of the late nineteenth century to a series of narrow scholastic exercises. There was, however, a much more positive reason for the recent change of geomorphic emphasis, in that modern quantitative methods both of the study of geomorphic processes and forms (the latter largely through improved mapping and aerial photography) began to yield data which could be processed by simple statistical techniques. It is the purpose of this chapter to start from the 'ecological niche' created in the western world by the decline of the Davisian system and denudation chronology, to balance the factors which have impeded and promoted quantification in geomorphology, and to proceed to a general outline of the quantitative treatment of geomorphic data.

THE WEAKENING OF DENUDATION
CHRONOLOGY

The decline of the Davisian geomorphic system was treated in Chapter 2 and there is no need to return to this matter here. However, although some indication has been given regarding the failure of denudation chronology to form an adequate geomorphic basis for *geographical* work, it is necessary here to show in what ways its former dominance over *geomorphic* work in Western Europe and the United States has diminished. It should be made clear at the outset that studies of the sequential development of landforms having changes of baselevel as a fundamental focus of interest still form a respected and fruitful branch of geomorphology, and that many of the criticisms which follow apply most strikingly to later, more derivative, work in localities poorly suited both to the aims and methods of denudation chronology. The point which I wish to make in this respect is that much of the former dominance claimed for denudation chronology in the field of geomorphology has now vanished, and that both a cause and an effect of this is the development of what has been termed, rather inaccurately, quantitative geomorphology.

One of the major limitations on the study of denudation chronology is that it cannot exist satisfactorily on a purely morphological plane. Many of the most impressive studies of this type rely heavily on the known origin and date of associated *deposits* (commonly terrace deposits or those resting on erosion surfaces), and it is characteristic that the higher the elevation or the longer the uninterrupted period of erosional history the more difficult it is to produce an unambiguous denudation chronology for the area. For a few classic areas significant historical reconstructions have been made, but for most regions attempts at a denudation chronology commonly end with the presentation of a cleverly-integrated body of ambiguous circumstantial evidence the interpretation of which is strongly coloured by the previous findings of similar workers in other areas. The ambiguity of purely *morphological* evidence relating to valley features was pointed out in Chapter 2, and Rich (1938) has given a summary of many of the arguments against purely morphological denudation chronology. It is instructive to take the best-known work of the latter sort, Johnson's book on the Appalachians (1931), and read it in the light of more modern work (e.g. Flint, 1963) which is stressing lithological control over elevation, rather than that of baselevel. Denudation chronology did provide, however, the only important pre-Second World War stimulus for quantitative work in morphometry (i.e. the study of the *geometry* of landscape), which mainly took the form of altitude/frequency analysis (e.g. Hollingworth, 1938). Such studies were, of course, based upon certain articles of faith, chiefly that areas of low slope are probably erosional surfaces related to appropriate baselevels, and that elevation above sea-level is a measure of relative age for such features. Irrespective of such commitments, however, this quantitative data was collected in such a subjective manner and treated by means of such coarse and uncritical techniques as to destroy the real significance of quantification and to reduce these studies to a quasi-scientific veneer embellishing essentially qualitative work.

Another aspect of denudation chronology which has tended to bring it into disrespect is the very prevalent use of '*ad hoc* postulates'. The abandoned eustatic swings of sea-level or the cavalier casting of Cretaceous covers which are required by some ambitious reconstructions are examples of such *ad hoc* postulates which are employed to 'explain' certain observed features without any thought as to the other concomitant effects of such occurrences. Long ago Chamberlin (1897) formalized the method of 'multiple working hypotheses' for use in the earth sciences where evidence is ambiguous and sparse. The inability of some workers in denudation chronology to follow up the other associated effects which might be logically deduced from their hypotheses has done little to create respect either for their aims or their methods.

It is, however, on a much more fundamental plane that the study of denudation chronology is being shown to provide an inadequate *general* basis for the study of geomorphology. The historical bias inherent in the denudation chronology approach, together with the assumption that rates of change involving landforms are commonly very slow, has meant that landforms have been viewed in much the same manner as the light from a distant star, in which what is perceived is merely a reflection of happenings in past history. Denudation chronologists have therefore almost universally directed their attention to those (often minute) elements of landscape having supposed evolutionary significance, in such a manner as to exclude from geomorphology the study of the basic dynamics which are controlling the continuous development of the major part of all landscapes. This dichotomy between the so-called *historical hangover* and *dynamic equilibrium* approaches has been presented elsewhere (Chorley, 1962 and 1964) as highlighting the fundamentally different

L

methodological approaches to geomorphology of W. M. Davis and G. K. Gilbert (1877). Gilbert was primarily concerned with the manner in which equilibrium landforms become adjusted to geomorphic processes, and an interest in the progress towards such adjustment and the changes to which such adjustment is susceptible through time replaced for him a simple cyclical basis such as that which preoccupied Davis. Obviously, however, most present landscapes possess in highly variable proportions evidence of past history and of the reasonably contemporaneous processes which are gaining ascendency, blurring and destroying them at very different rates. These rates are related to the rates of operation of geomorphic processes which may be very slow (as in the case of some ancient pediment surfaces in Africa) or practically instantaneous (as for the features of hydraulic geometry). Most landform assemblages lie somewhere between these extremes, and one of the most pressing needs, the existence of which has done so much to give modern work its distinctive character, is for this fact to be recognized and for the relative importance of past and present processes to be evaluated. The key to such investigations lies obviously in the proper understanding of the current rates of operation of geomorphic processes, and it is precisely in this aspect of geomorphology that most ignorance exists. If James Hutton has anything to teach the modern geomorphologist it is that we should not assume the inadequacy of present 'causes' without establishing the surest quantitative grounds for so doing. We can, after all, directly observe and measure present conditions (which is more than we can for those in the past!) and before we sweep many of our geomorphological problems under the mat of historical speculation we should try to learn in what ways they can be treated by attention to presently-observable conditions.

THE DISTRUST OF QUANTIFICATION

The influences which have tended to inhibit quantification in geomorphology stem largely from two sources – the past historical and geographical affinities of the subject. Many of the historical influences follow naturally from the previous discussion of denudation chronology. Chief among these is the blanket belief that most topographic forms are essentially 'fossil' and relate to antique processes operative in climatic, tectonic or historical circumstances different from those of the present. From this point of view studies of the relationships between present forms and processes are irrelevant. As has been shown above, the truth of the above assertion varies widely with the environmental circumstances, and a recent example indicates the dangers of its uncritical application. In a work on cliff forms in the Colorado Plateau Ahnert (1960) identified rounded cliff tops with processes occurring during past pluvial conditions, and from this assumption went on to imply a fundamental disharmony between these conditions and those obtaining at present. Shortly afterwards Bradley (1963) demonstrated that the existence of such rounding was widely due to the occurrence of pressure-release joints in the sandstone – from which it may be assumed that the existence of cliff rounding is a structural matter and not necessarily indicative of the nature of past processes. Most purely *morphological* evidence is so ambiguous that theory feeds readily on preconception.

Closely associated with these historical inhibitions are those relating to the measurement of form and process and their affinities. Precise measurement and quantitative expression of the geometry of landforms were not only irrelevant to the denudation chronologist but absolutely unnecessary for the Davisian synthesis, and, for example, Strahler (1950 A) has noted the lack of precisely surveyed slope profiles in Davis' work. Davis did, in passing, point to the parabolic form of the typical longitudinal stream profile, but his reasoning was that of the intuitive artist rather than that of the quantitative scientist. It is also commonly believed that most landforms are too complex to treat satisfactorily in a quantitative manner, although one of the important features of recent work is that significant and diagnostic aspects of landscape geometry have been isolated and so treated. Thus, for example, Horton (1945) analysed areal properties of the erosional drainage basin, Strahler (1950 B) the straight middle segments of valley side slopes, and the same author (1952) the hypsometric aspects of drainage basins (in which a 3-dimensional problem was reduced to a 2-dimensional one).

Allied with the above belief in the irrelevancy and complexity of form is the traditional attitude to geomorphic processes, which have commonly been regarded as of little consequence in the analysis of existing landforms; . . . 'I regard it as quite fundamental that Geomorphology is primarily concerned with the interpretation of forms, not the study of processes' (Wooldridge, 1958, p. 31). Although this extreme attitude is less fashionable now than hitherto, there still remains the view that process (even more than form) is usually of such a complex character that significant measurement is well-nigh impossible. Processes commonly seem to operate either too rapidly, too slowly, too infrequently, too capriciously, or too variably to make for ease of observation and measurement. In part these objections have been met in a similar way to those relating to the complexity of form, in that significant *aspects* of process have been isolated and treated – e.g. *bankfull* discharge (Leopold and Wolman, 1957; Dury, 1961), soil moisture *changes* on slopes (Young, 1960; Kirkby, 1963), the '*significant* wave' (Shepard, 1963, pp. 54–55), and the '*weighted resultant* wind' (Bagnold, 1951, pp. 80–82; Rosenan, 1953). Such considerations have naturally given rise to a recent interest in the magnitude and frequency of geomorphic processes (Wolman and Miller, 1960). However, the 'traditional' view of such processes commonly held by geomorphologists in the humid temperate regions is that rates of operation are too slow to be susceptible of useful measurement. Even excepting the increased erosional rates observed in other environments (see Langbein and Schumm, 1958, in Chapter 2), the few measurements which have yet been made of humid temperate processes (e.g. Young, 1960; Kirkby, 1963) by no means support this view. Likewise, measurable rates of diastrophic movement have been reported (Gilluly, 1949; Chorley, 1963; Schumm, 1963). In short, we can no longer proceed on the assumption that current rates of operation of earth processes can be ignored in an assessment of existing landforms. Probably in some instances they ultimately can, but even such elimination must only be made on the basis of reliable measurement.

The second source of inhibition to quantification in geomorphology derives from its many geographical affinities. Bunge (1962) has pointed to the geographer's traditional preoccupation with the unique and distrust of generalization (i.e. 'idiographic' attitude) during the present century, and there is no doubt that the attraction which geographers have commonly felt towards the intuitive and artistic aspects of natural and social science has militated against the quantification of geography in general and of geomorphology in particular. It is characteristic for such scholars to hold the erroneous belief that 'you can prove anything with statistics'; for them to dwell with satisfaction on the fact that on Charles Darwin's death his rulers were found to be of inexact length and his conversion tables incorrect; and for them to believe that the important, interesting and worthwhile aspects of natural phenomena are somehow bound up with departures from predictable regularity (Chorley, in press). However, even in geography quantification is proving an increasingly valuable research stimulus (Ackerman, 1963), so much so that it can be legitimately held that the quantitative 'revolution' is already upon us and that it is profitless to pursue methodological discussions which are based upon a disregard of this fact (Burton, 1963). Such a view is even more applicable to geomorphology.

THE PROMOTION OF QUANTIFICATION

It is less easy to isolate the factors which have prompted quantification specifically in geomorphology from those operative in the fields of natural and historical science as a whole. Indeed, in common with students of these subjects, geomorphologists are becoming increasingly impressed with the weight of quantitative arguments and are developing a healthy distrust of their purely visual sense and of the basic preconceptions which seem to have restricted their science in the past.

The employment of reasoning based on measurement is not new in geomorphology (e.g. Geikie, 1868), but since the 1930's the increasing opportunity for such work in terms of facility of measurement, availability of data, and the imaginative and technical advances in treatment has given post-war studies much of their distinctive flavour. The desire to relate form to process, inspired by the pioneer work of Gilbert, has given rise to the important fluvial studies of the hydrologist R. E. Horton (1945), as well as the associated work of Strahler (1950 B) and his students; to the work on hydraulic geometry mainly associated with the name of L. B. Leopold (with Maddock, 1953; with Wolman, 1957; with Wolman and Miller, 1964); to research on dune formation (e.g. Bagnold, 1941) and periglacial forms (Peltier, 1950; Jahn, 1961); and to investigations into beach forms and dynamics (e.g. Krumbein, 1944). It is one of the features of this work that much of its stimulus has come from the investigation of practical engineering problems and from recent improvements in topographical maps and air photographs. The quantitative analysis of landscape geometry (i.e. 'morphometry') has been especially revivified by the latter improvements.

The developments of quantification in geomorphology can be viewed as indicating a general progress of the discipline towards a more secure scientific footing through its concern with measurement and the analysis of data (Strahler, 1954; Melton, 1957), through its attempt to break out of the narrow scholasticism which characterized much of its pre-war development, and through its increasing association with kindred sciences. It is becoming recognized that sensible quantification of those aspects of the subject which call for it translates many geomorphic questions into a common language which enables them to be attacked in the light of the experience of the other sciences, such that geomorphology can both draw on and contribute to the wealth of scientific experience which is the most distinctive intellectual achievement of our age. There is no remedy for bad or irrelevant measurement, or excuse for unnecessarily elaborate treatment and analysis, but once it is recognized that useful properties and associations can be expressed in quantitative terms then one is *automatically* committed to a programme (no matter how simple in conception and execution) of operational definitions, choice of scales of measurement, sampling and collection of data, descriptive statistical methods and analytical methods. It is with a brief treatment of the features of such a programme that the remainder of this chapter is concerned.

ASPECTS OF QUANTIFICATION

The appropriateness of quantification in geomorphology depends basically upon the character of the science which one wishes to exploit. The methodological features of an historical natural science have been recently set forth in some detail (Albritton, 1963), and, for those workers concerned with geomorphic phenomena as part of the latter phases of historical geology, the main preoccupations continue to be the elucidation of *what has occurred* and *in what order*. Quantification, of course, often plays an important part in both these aims, but it is not indispensable in setting up a sequence of recognizable events. One has only to recall Davis' reticence on the subject of the precise dating of cyclic events to be impressed with the fact that, even in an historically-oriented science, quantification of *time* is not immediately basic to the main thesis, which is one of relativity. Where a conflict of views on the value of quantification in geomorphology occurs, it commonly resolves itself into a dispute between those who view the whole subject as part of historical geology and those who do not.

Those who find the quantitative approach to the study of landforms valuable can recognize four major classes into which geomorphic parameters may be grouped. Those relating to *force* or *energy*, of which discharge and wave energy are examples; to *strength* or *resistance*, by far the most neglected of the groups, which includes infiltration capacity and shearing resistance; to *time*, involving absolute dating of events, rates of erosion, etc; and to *form*. This latter class, including slope angles, drainage density, relief and height, is the one which has obviously been most subjected to quantification in the past, and measurements of stream profiles and hypsometric attributes have even become part of the standard techniques in denudation chronology.

Whichever of these classes is the subject of interest, it is necessary for the geomorphologist to adopt a formalized procedure for the collection of his quantitative data. It is necessary at an early stage, having decided on the nature of the problem and the type of questions to be asked about it (i.e. the *design of the experiment*; Krumbein, 1955), to lay down some clear *operational definition* involving a precise statement of the attribute concerned, such that there shall be no confusion regarding what parameter has actually been measured. Thus 'valley side slope angle' or 'stream discharge' are manifestly inadequate as operational definitions. It is surprising how the establishment of a meaningful operational definition assists in clarifying the character of the investigation in the mind of the investigator. Next one has to decide on an appropriate *scale of measurement* (Stevens, 1946) with reference to which the above parameter may be stated and the units in which it shall be expressed. Space does not permit an extended treatment of the theory of scales of measurement here, and it must suffice to point to the ratio scale as the most versatile, and to suggest that the expression of much data in geomorphology (but much more so in geography as a whole) falls short of this ideal. Another group of highly versatile numbers are those which are derived in such a manner as to make them 'dimensionless' (e.g. Strahler's (1952) 'hypsometric integral'). *Errors of measurement* must be a prime consideration at this stage and a conscious attempt made to reduce 'operator variation' in measurement – i.e. involving important differences in the recognition and measurement of identical phenomena either by the same 'operator' or by different operators, should more than one be involved in the investigation. Many of these errors can be eliminated by a meaningful operational definition and a logical experimental design. Finally, before the actual collection of data can take place a *sampling design* must be established. This topic, too, is a very large one, for obviously geomorphic sampling can be carried on both in space and time. However, two concise guides to areal sampling have been given by Strahler (1954) and Krumbein (1960), although it is largely a matter of trial and error at this stage in determining the size of sample necessary to give an adequate representation of a given phenomenon.

The first step in the treatment of the data collected as the result of the procedure outlined above is to organize and describe it in some convenient manner. Such *descriptive statistical methods* are described fully in standard texts (e.g. Croxton and Cowden, 1939; Croxton, 1959), as well as for geographical (Gregory, 1963), geological (Miller and Kahn, 1962) and geomorphic data (Strahler, 1954). These commonly involve the plotting of frequency diagrams (e.g. histograms), the recognition of the 'characteristics of the population' (i.e. whether it is normally distributed, skewed, etc.), the definition of the 'measures of central tendency' (the mean, mode or median), and the definition of the 'dispersion' (e.g. the standard deviation) of the data. Descriptive statistical methods involve both the description of the sample data and the drawing of inferences regarding the general characteristics of the total 'population' from which the sample was drawn. This latter inference enables one to introduce the concept of *probability* upon which the whole of statistical analysis rests.

The most simple analytical techniques which have proved useful in geomorphic research involve the testing of the significance of difference between data grouped into classes. The *'chi-square'* test (Croxton and Cowden, 1939, pp. 282–7; Strahler, 1954, pp. 9–10) is used to investigate the significance of difference between frequencies of 'variates' (i.e. occurrences) within classes, and in this respect is useful in testing the normality of distributions. The *'t' test* (Croxton and Cowden, 1939, Chapter 12; Strahler, 1954, pp. 12–14) enables one to test the significance of difference between the arithmetic mean values of two samples. This type of analysis has proved most important in quantitative geomorphic reasoning, and Strahler (1950 B) demonstrated a significant difference between maximum angles of valley-side slopes which were being basally corraded and slopes where the basal accumulation of debris was taking place (Figure 8.1A). This significance of difference between sample means cannot always be objectively analysed by visual inspection, and in figure 8.1C, for example, there is no reason to assume (at a probability of 5 per cent) that the difference in mean valley-side slope angles shown by the two samples from the Athens sandstone and the Pennington sandstone and shale in part of the folded Appalachians reflects a real difference between mean angles on the two formations (i.e. the observed sample

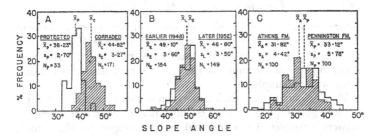

FIG. 8.1. *Paired histograms showing:*

A. *Maximum slope angles of valley-sides in the Verdugo Hills, California, the bases of which are being protected and corraded, respectively (after Strahler, 1950 B).*

B. *Maximum slope angles of valley-sides in badlands at Perth Amboy, New Jersey, measured at an interval of four years (after Schumm, from Strahler, 1954).*

C. *Valley-side slope angles measured on the Athens sandstone and the Pennington sandstone and shale of western Virginia (after Miller, from Strahler, 1954).*

(In all instances $\bar{X}$ = the arithmetic mean of the sample, s = the standard deviation and N = the number of variates in the sample.)

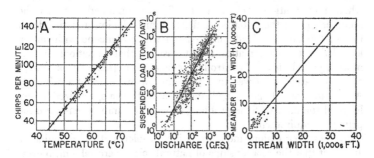

FIG. 8.2. A. *Linear relationship between the number of cricket chirps per minute and the temperature (after Croxton and Cowden, 1939).*

B. *Non-linear (logarithmic) relationship between the amount of suspended load and the discharge of the Powder River at Arvada, Wyoming (after Leopold and Maddock, 1953).*

C. *Linear relationship between meander belt width and stream width for some Wisconsin rivers (after Bates).*

difference could be explained by chance sampling alone). It is one of the functions of the versatile *analysis of variance* (Croxton and Cowden, 1939, pp. 351–9; Strahler, 1954, pp. 16–17; Krumbein, 1955) to perform the same comparison of arithmetic means between three or more samples.

More important than the examination of the significance of differences in geomorphology is the establishment of general relationships between variables (i.e. *regressions*) and the degree to which these relationships express the reality presented by observed data (i.e. *correlation*). Some regression lines are shown in figure 8.2, with a high degree of correlation indicated in figure 8.2A and poorer correlation in figure 8.2B. It will be readily appreciated that reasoning from such associations commonly involves the assumption that one variable (the *independent*) is exercising some control over the other (the *dependent*), but especial care must be taken in this respect. Sometimes a causal relationship can be readily supported (as in figures 8.2A and 8.2B), but on other occasions the observed relationships may result from the yoked behaviour of two dependent variables responding in harmony to variations in some undisclosed independent variable or variables – in figure 8.2C, meander belt and stream widths are responding to differences in other variables, notably stream discharge and the calibre of bed debris. In other instances a high degree of correlation may result from purely *chance* co-variation, as when a correlation coefficient of 0·87 was obtained between the membership of the International Machinists Union and the death rate of the State of Hyderabad – from which someone with only a knowledge of mathematics might assume that variations in the mortality of Hyderabad controlled 75·69 per cent (i.e. 0·87^2 × 100%) of the variations observed in the number of members of a primarily American Trade Union! In all statistical matters a specialized knowledge of the nature of the problem, as well as common sense, are the primary requirements.

In most regression and correlation problems it is immediately apparent that the relationships of natural science are seldom, if ever, in terms of a simple one cause–one effect framework. Even in figure 8.2A, where there is a very high degree of correlation, such deviations which occur from the general regression trend indicate that temperature, although by far the most important, is not the only factor controlling the rate of cricket chirping (others might be the age of the cricket, the absolute humidity, etc.). The poorer correlation exhibited by figure 8.2B indicates similarly that discharge is not the only factor controlling the rate of suspended sediment transport in a river, and that other factors are involved (i.e. calibre of the sediment, specific gravity of the sediment, temperature of the water, etc.). The recognition of this multi-factor or *multivariate* character of most problems in the earth sciences has long existed, but until comparatively recently it was possible to do little more than to pay lip service to the complexity of most problems by the use of such phrases as 'X seems to be the most important control' and 'other things being equal, Y is a function of X'. This difficulty often resulted in the past in the assumption of gross oversimplifications in cause and effect, and, in the most extreme instances, in an extremely narrow determinism. Refined statistical techniques, facilitated in recent years by the application of electronic computers, now enable one to face up more realistically to the challenge of a multivariate reality, such that the following kinds of questions can be asked – and largely answered:

1. How many significant controlling factors are involved in determining the effect with which we are concerned?

2. Are these factors interrelated in some manner?

3. Can these controls be ranked in order of importance, both relatively and absolutely?

4. Do the factors retain this importance under all circumstances? – i.e. Does the intervention of one factor reinforce or damp-down the effect of another?

Of these questions, the first can only be approached through experience and intuition, although the relevance of a choice made in this manner can be later checked statistically, and the last three are amenable to solution by the techniques of multivariate statistics. These techniques can be divided into two broad groups – eliminative investigations and more sophisticated multivariate techniques.

Eliminative investigations are those involving the careful selection of a number of instances (either situations simulated and controlled in the laboratory, or 'real world' situations) such that the assumed controlling variables are successively eliminated or included in various associations. This has been the standard method of testing the factors which, for example, control crop yields by the use of experimental plots. Examples in geomorphology include the eliminative laboratory investigations into the factors controlling tractive capacity (i.e. the mass movement of stream bed load) (Gilbert, 1914) and the angle of repose of fragmented material (Van Burkalow, 1945). The number of factors involved in most natural phenomena, together with the restricted number of locations in which their effects can be examined in detail, has tended to inhibit eliminative field work in geomorphology, but an example of this technique has been provided in the examination of factors influencing the geometry of erosional landscapes (Chorley, 1957).

An obvious limitation facing eliminative investigators is that questions 2 and 4 (above), involving the mutual effects of independent variables, cannot be readily answered. This is because one can never realistically hold one set of factors constant while the changing effect of others is examined (i.e. other things are *never* equal!) for, as Sir Ronald Fisher pointed out, nature 'will best respond to a logical and carefully thought out questionnaire; indeed, if we ask her a single question she will often refuse to answer until some other topic has been discussed'. Such a 'questionnaire', in which all the assumed independent variables together with the dependent variable are examined and evaluated simultaneously in a large number of different situations, combinations and magnitudes, forms the basis of *multivariate analysis*. The development of statistical methods assisted by the use of high-speed electronic computers now enables the data from such a questionnaire to be processed and evaluated in such a manner that all four of the above questions can be answered. Two such, closely-allied, methods are those of *multiple correlation* and *multiple regression*. Melton (1957) examined by means of the first method the control exercised by five factors over drainage density (i.e. the total length of drainage lines per unit area) and found that all five operating together could account for more than 93 per cent of the observed variation in drainage density between different localities. Using a multiple regression technique Krumbein (1959) evaluated the influence of four factors in controlling beach firmness, concluding that together they contributed over 76 per cent of the observed firmness variations.

Quantitative techniques, supported by statistical analysis, provide a standardized, rigorous, conservative and objective framework for the investigation of many of the problems of earth science – although those of an historical character are less obviously susceptible to such treatment at present. However, these techniques and analyses are only an adjunct to, and not a substitute for, the initial qualitative stage of any investigation. This stage is entirely a matter for the exercise of experience, controlled intuition, imagination and creativity, in which quantitative methods are of no help – although they may subsequently be used to test the efficiency of this qualitative design. Quantitative methods and statistical analyses are merely tools, but tools with which one may sharpen the imagination and, like Galileo's telescope, which enable this imagination to operate on higher planes than ever before.

References

ACKERMAN, E. A., 1963, 'Where is a Research Frontier?', *Ann. Assn. Amer. Geog.*, **53**, 429–40.

AHNERT, F., 1960, 'The Influence of Pleistocene Climates upon the Morphology of Cuesta Scarps on the Colorado Plateau', *Ann. Assn. Amer. Geog.*, **50**, 139–56.

ALBRITTON, C. C., 1963, *The Fabric of Geology* (Reading, Mass.), 372 pp.

BAGNOLD, R. A., 1941, *The Physics of Blown Sand and Desert Dunes* (London), 265 pp.

— 1951, 'Sand Formations in Southern Arabia', *Geog. Jour.*, **117**, 78–85.

BRADLEY, W. C., 1963, 'Large Scale Exfoliation in Massive Sandstones of the Colorado Plateau', *Bull. Geol. Soc. Amer.*, **74**, 519–28.

BUNGE, W., 1962, *Theoretical Geography* (Lund), 210 pp.

BURTON, I., 1963, 'The Quantitative Revolution and Theoretical Geography', *Canadian Geog.*, **7**, 151–62.

CHAMBERLIN, T. C., 1897, 'The Method of Multiple Working Hypotheses', *Jour. Geol.*, **5**, 837–48.

CHORLEY, R. J., 1957, 'Climate and Morphometry', *Jour. Geol.*, **65**, 628–38.

— 1962, 'Geomorphology and General Systems Theory', *U.S. Geol. Survey, Prof. Paper 500-B*, 10 pp.

— 1963, 'Diastrophic Background to Twentieth-century Geomorphological Thought', *Bull. Geol. Soc. Amer.*, **74**, 953–70.

— 1964, 'The Nodal Position and Anomalous Character of Slope Studies in Geomorphological Research', *Geog. Jour.*, **130**, 70–73.

— (In press), 'The Application of Statistical Methods to Geomorphology', in *Essays in Geomorphology*, ed. by G. H. Dury (London).

CROXTON, F. E., 1959, *Elementary Statistics with Applications in the Medical and Biological Sciences* (Dover Paperbacks), 376 pp.

CROXTON, F. E. and COWDEN, R., 1939, *Applied General Statistics* (New York), 944 pp.

DUNCAN, O. D., CUZZORT, R. and DUNCAN, B., 1961, *Statistical Geography: Problems in Analysing Areal Data* (The Free Press of Glencoe, Illinois), 191 pp.

DURY, G. H., 1961, 'Bankfull Discharge: an Example of Its Statistical Relations', *Int. Assn. Scientific Hydrology*, 6th Year, No. 3, 48–55.

FLINT, R. F., 1963, 'Altitude, Lithology, and the Fall Zone in Connecticut', *Jour. Geol.*, **71**, 683–97.

GEIKIE, A., 1868, 'On Denudation now in Progress', *Geol. Mag.*, 5, 249–54.

GILBERT, G. K., 1877, *The Geology of the Henry Mountains*, U.S. Dept. of the Interior (Washington) (Chapter 5, 'Land sculpture').

— 1914, 'The Transportation of Debris by Running Water', *U.S. Geol. Survey, Prof. Paper 86*, 263 pp.

GILLULY, J., 1949, 'Distribution of Mountain Building in Geologic Time', *Bull. Geol. Soc. Amer.*, 60, 561–90.

GREGORY, S., 1962, *Statistical Methods and the Geographer* (London), 240 pp.

HOLLINGWORTH, S. E., 1938, 'The Recognition and Correlation of High-level Erosion Surfaces in Britain: a Statistical Study', *Quart. Jour. Geol. Soc.*, 94, 55–84.

HORTON, R. E., 1945, 'Erosional Development of Streams and Their Drainage Basins: Hydrophysical Approach to Quantitative Morphology', *Bull. Geol. Soc. Amer.*, 56, 275–370.

JAHN, A., 1961, *Quantitative Analysis of Some Periglacial Processes in Spitzbergen* (University of Warsaw, Poland), 54 pp.

JOHNSON, D. W., 1931, *Stream Sculpture on the Atlantic Slope* (New York), 142 pp.

KIRKBY, M. J., 1963, 'A Study of the Rates of Erosion and Mass Movement on Slopes with Special Reference to Galloway' (*Unpublished Ph.D. Thesis, Cambridge University*).

KRUMBEIN, W. C., 1944, 'Shore Processes and Beach Characteristics', *U.S. Beach Erosion Board, Tech. Memo.* 3.

— 1955, 'Experimental Design in the Earth Sciences', *Trans. Amer. Geophys. Union*, 36, 1–11.

— 1959, 'The "Sorting Out" of Geological Variables illustrated by Regression Analysis of factors Controlling Beach Firmness', *Jour. Sedimentary Petrology*, 29, 575–87.

— 1960, 'The "Geological Population" as a framework for Analysing Numerical Data in Geology', *Liv. and Man. Geol. Jour.*, 2, 341–68.

LEOPOLD, L. B. and MADDOCK, T., 1953, 'The Hydraulic Geometry of Stream Channels and Some Physiographic Implications', *U.S. Geol. Survey, Prof. Paper 252*, 57 pp.

LEOPOLD, L. B. and WOLMAN, M. G., 1957, 'River Channel Patterns: Braided, Meandering and Straight', *U.S. Geol. Survey, Prof. Paper 282-B*, 39–85.

LEOPOLD, L. B., WOLMAN, M. G. and MILLER, J. P., 1964, *Fluvial Processes in Geomorphology* (San Francisco), 522 pp.

MELTON, M. A., 1957, 'An Analysis of the Relations among Elements of Climate, Surface Properties and Geomorphology', *Office of Naval Research Project NR 389–042*, Tech. Rept. 11, Dept. of Geol., Columbia Univ., New York, 102 pp.

MILLER, R. L. and KAHN, S. J., 1962, *Statistical Analysis in the Geological Sciences* (New York), 357 pp.

PELTIER, L. C., 1950, 'The Geographic Cycle in Periglacial Regions as it is related to Climatic Geomorphology', *Ann. Assn. Amer. Geog.*, **40**, 214–36.

RICH, J. L., 1938, 'Recognition and Significance of Multiple Erosion Surfaces', *Bull. Geol. Soc. Amer.*, **49**, 1695–722.

ROSENAN, E., 1953, 'Comments on the Paper by R. A. Bagnold; In "Desert Research"', *Research Council of Israel, Spec. Pub. 2* (Jerusalem), 94.

SCHUMM, S. A., 1963, 'The Disparity between Present Rates of Denudation and Orogeny', *U.S. Geol. Survey, Prof. Paper 454-H*, 13 pp.

SHEPARD, F. P., 1963, *Submarine Geology*, 2nd Edn. (New York), 557 pp.

STEVENS, S. S., 1946, 'On the Theory of the Scales of Measurement', *Science*, **103**, 677–80.

STRAHLER, A. N., 1950 A, 'Davis' Concepts of Slope Development Viewed in the Light of Recent Quantitative Investigations', *Ann. Assn. Amer. Geog.*, **40**, 209–13.

— 1950 B, 'Equilibrium Theory of Erosional Slopes, approached by Frequency Distribution Analysis', *Amer. Jour. Sci.*, **248**, 673–96 and 800–14.

— 1952, 'Hypsometric (area-altitude) Analysis of Erosional Topography', *Bull. Geol. Soc. Amer.*, **63**, 1117–42.

— 1954, 'Statistical Analysis in Geomorphic Research', *Jour. Geol.*, **62**, 1–25.

VAN BURKALOW, A., 1945, 'Angle of Repose and Angle of Sliding Friction; an Experimental Study', *Bull. Geol. Soc. Amer.*, **56**, 669–708.

WOLMAN, M. G. and MILLER, J. P., 1960, 'Magnitude and Frequency of Forces in Geomorphic Processes', *Jour. Geol.*, **68**, 54–74.

WOOLDRIDGE, S. W., 1958, 'The Trend of Geomorphology', *Trans. Inst. Brit. Geog.*, **25**, 29–35.

YOUNG, A., 1960, 'Soil Movement by Denudational Processes on Slopes', *Nature*, **188**, 120–22.

CHAPTER NINE

Scale Components in Geographical Problems

P. HAGGETT

Professor of Geography, University of Bristol

One of the characteristic features of geographical research is its concern with a particular scale of reality. If we conceive this scale as a continuum running from the reality of the electron microscope, up to the one-to-one reality of our everyday life, through to the astrophysical reality of the galaxies, and on finally to the dimensionless reality of mathematics, then geographical research occupies a rather well-defined position in this succession (Haggett, Chorley and Stoddart, 1965). It ranges from highly localized studies of individual villages or river basins at magnitudes of 10^{-1} square miles through to world wide studies of the order of 10^7 square miles. Unlike the microscopic sciences where results have to be brought up to a one-to-one world for our understanding, geography is macroscopic in that it has to shrink reality to make it comprehensible.

The concern of this chapter is to draw attention to this shrinkage problem. It suggests that scale obtrudes into geographical research in three main ways: in the problem of covering the earth's surface; in the problem of linking results obtained at one scale to those obtained at another; and in standardizing information that is available only on a mixed series of scales. For convenience these are simply referred to here as the 'scale-coverage problem', the 'scale-linkage problem', and the 'scale-standardization problem'.

THE SCALE COVERAGE PROBLEM

Nature of the Problem

The scale coverage problem is simple and immediate. The earth's surface is so staggeringly large that, even if we omit the sea-covered areas, each of the profession's 3,000 nominal practitioners (Meynen,

1960) has an area of about 5,000 square miles to account for! These gross ratios caricature rather than characterize the problem. But if we agree with Hartshorne that the purpose of geography is '. . . to provide accurate, orderly, and rational description and interpretation of the variable character of the earth surface' (Hartshorne, 1959, p. 21) or follow Sauer in regarding it as a 'focused curiosity' (1952, p. 1) then we need to be aware of the magnitude of the task we set ourselves, or alternately the size of the object we are trying to get into focus.

This can hardly be regarded as a new problem. From at least the time of Eratosthenes the size of the problem has been apparent and it may well be that our predecessors were more keenly aware of its importance. Many a doubtful isopleth now strays self-importantly across areas that our more honest forbears might have filled with heraldic doodles or labelled 'Terra Incognita'. There are more reassuring signs, however, that the scale-coverage problem is partly being solved at this time by changes both inside and outside the discipline.

Internal Solutions: Sampling

Sample studies have long been used in both research and teaching. Platt (1942, 1959) was acutely aware of the '. . . old and stubborn dilemma of trying to comprehend large regions while seeing at once only a small area' (1942, p. 3) and he skilfully used sample field studies to build up an outstandingly clear series of pictures of the regions of Latin America. Similarly Highsmith (Highsmith, Heintzelman, Jensen, Rudd and Tschirley, 1961) has used a world-wide selection of sample studies as the basis for an extremely useful teaching manual in economic geography.

There is an important difference, however, between these attempts to use sampling to circumvent the scale problem, and the way in which sampling is now being used in research. This essential difference is between *purposive* and *probability* sampling. While it is outside the scope of this chapter to discuss sampling theory [there are excellent general summaries by Cochran (1953) and Yates (1960)] it is important to note that in probability sampling it is possible to estimate how accurate the survey is likely to be from the information actually collected during the sample survey. This means that given a limited budget the accuracy of any sample survey can be determined; or, vice versa, given a fixed limit of accuracy, the necessary size of sample and time-cost estimates can be made. The simplest case of

M

Table 4. Model of Sampling Systems

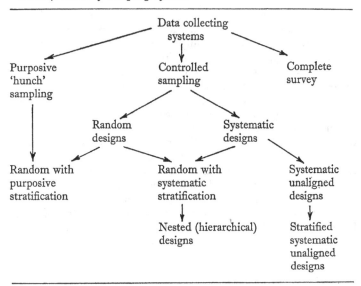

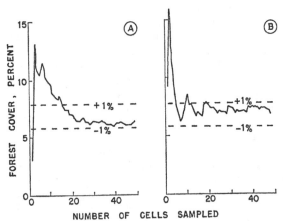

FIG. 9.1. *Changes in sampling error with size of sample (source: Haggett, 1963).*

this type of relationship is in the simple random design where the random sampling error (accuracy) is proportional to the square root of the number of observations (effort). An empirical illustration of this relationship is shown in Figure 9.1 where as sample size is steadily increased the values settle down around the known true value.

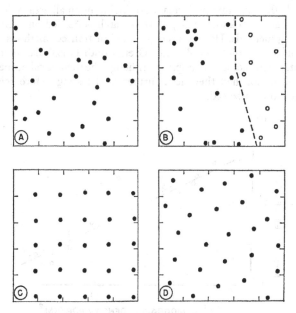

FIG. 9.2. *Alternative types of sampling design (source: Krumbein, 1960; Berry, 1962).*

Various types of sampling design have been used in recent years for specific problems (Table 4). Four of these designs are shown in Figure 9.2. Where the problem is exploratory and little is known about the characteristics of the 'population' being studied then a simple random design may be adopted, *A*. Wood (1955) introduced stratification, *B*, into the random design to allow certain parts of his study areas, eastern Wisconsin, to be more heavily sampled than others. The disadvantage of random designs is that they are more difficult to use as control points in mapping the results and systematic samples, *C*, may be substituted in their place where mapping is a prime consideration. Berry (1962) in an extensive study of the application of sampling design to land-use surveys of flood plains has found that a compromise design, the 'stratified systematic unaligned sample', *D*, gave the most accurate results with the additional advantages of facilitating both punched card storage and machine mapping.

While the sampling performance of 'point' or small area collecting systems are well known now, more research is needed on their line sampling methods. Haggett (1963) found line transect methods more accurate than point samples and Greig-Smith (1964) has reported advantages in ecological sampling. Although the theoretical properties are well established they need further field testing before relative merits can be assessed.

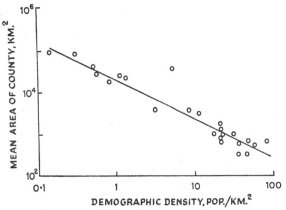

FIG. 9.3. *Relationship of the area of Brazilian county divisions to state population densities.*

External Solutions: Improving Sources

More information is available about the earth's surface today than at any previous time. The trickle of maps and census reports from government agencies a century ago has now risen to a torrent and there are indications that it is increasing logarithmically over time. This vast increase has not, however, been evenly spread so that the information contrasts between one part of the earth's surface and another are rather acute. Commonly the level of information is related to the development of the area. Figure 9.3 shows the relationship between the fineness of the data-collecting grid and the population density for one country: Brazil. Similarly, Berry (Ginsburg, 1961, p. 110) has pointed to the inverse relationship between the economic development of an area and the amount of information available about that development.

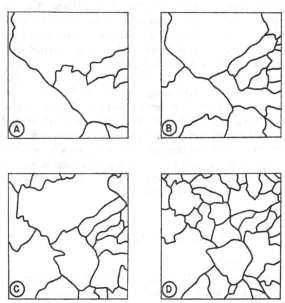

FIG. 9.4. *Increasing dissection of administrative areas in an area of rapid population increase; Santa Catarina state, Brazil, 1870–1960.*

Comparisons over time also run into difficulties. The very fact of improving information may make comparisons with earlier periods invalid. Figure 9.4 shows the successive subdivision of an area of rapid population increase over successive thirty-year time periods. Although far more is known in detail about the area in the final period, 1950, than in the initial period, 1870, the degree of comparable detail is controlled by the coarsest grid or the largest areal denominator. Dickinson (1963) has illustrated similar problems of subdivision and boundary change in England and Wales, while Hall (1963) has noted the problems in tracing the industrial growth of London from census data. For map coverage, Langbein and Hoyt (1959) have shown that for the United States there are some curious lacunae in both coverage and age with the poorly mapped areas being revised rather less frequently. Again the gap in coverage is tending to grow.

A vitally important supplement to such 'archival' data in maps and censuses is the growth of airphoto coverage. Although this has a history going back to at least 1858, the effect of World War II and the 'cold war' that followed has been virtually to complete and/or revise the airphoto coverage of the whole earth's surface. Rapid improvements have been made in both lens and camera, in vehicles (through to U2's and satellites) (Colwell, 1960), in mapping with electronic plotters, and in interpretation with electronic scanners (Latham, 1963). More revolutionary changes are foreshadowed in completely automated terrain sensing systems in which information about the earth's surface is recorded by satellite, relayed back to base, and made available on magnetic tape (Lopik, 1962). This threatens to cut out the airphoto as such and replace it by continuous information on a co-ordinate system. It seems possible then that continuous recording of certain simple terrain information may replace discontinuous mapping within the forseeable future.

THE SCALE LINKAGE PROBLEM

Nature of the Problem

A direct consequence of the scale problem discussed above has been to restrict field work to rather small areas. This leads in its turn to the second problem of researchers, that of '. . . seeing the link between their own local field work and the standard regional courses on the continents of the world' (Bird, 1956, p. 25). This scale linkage problem was brought home forcibly to the writer in comparing the inferences on forest distribution based on a small 100 square kilometre survey area made in 1959 with a later survey of the same features over a wider surrounding area (Haggett, 1964). These dangers have been very neatly summarized by McCarty: 'In geographic investigation it is apparent that conclusions derived from studies made at one scale should not be expected to apply to problems whose data are expressed at other scales. Every change in scale will bring about the statement of a new problem, and there is no basis for assuming that associations existing at one scale will also exist at another' (McCarty, Hook and Knos, 1956, p. 16). Such difficulties are not of course confined to geographical research. Duncan, Cuzzort and Duncan (1961) have pointed out the difficulty of linking individual with mass economic behaviour, while Bronowski (1960, p. 93) has raised the whole problem of individual action in the 'stream of history'.

For the more particular problems of scale linkage two sets of solutions are discussed here. First the qualitative attempt to accommodate scale in regional systems and secondly, the quantitative attempt to isolate and measure the impact of scale at series of levels.

Qualitative Solutions: Scale in Regional Systems

The fact that scale realizations have long troubled geographers is rather plainly shown in the series of attempts that have been made to define regions on scale terms. With formal regions, the early system applied by Fennemann (1916) to the landform divisions of the United States with his recognition of major divisions, provinces, and sections had a major effect on other writers (Table 5). Unstead (1933)

Table 5. *Comparative Scales and Terminology of Regional Systems*

Approx. size (sq. mls.)	Fennemann (1916)	Unstead (1933)	Linton (1949)	Whittlesey (1954)	Map scales for study*
10^{-1}			Site		
10		Stow	Stow	Locality	1/10,000
10^2	District	Tract	Tract		1/50,000
				District	
10^3	Section		Section		
		Sub-Regn			
				Province	
10^4	Province		Province		1/1,000,000
		Minor Regn			
10^5	Major Divn		Major Divn	Realm	1/5,000,000
10^6		Major Regn	Continent		

* Whittlesey, 1954.

in an interesting paper on 'systems of regions' put forward the scheme which filled in at the smaller levels the system Fennemann has begun at the larger. Linton (1949) integrated both preceding systems in a seven-stage system which ran through the whole range from the smallest unit, the site, to the largest, the continent. More recently Whittlesey (In James, Jones and Wright, 1954, pp. 47–51) presented a 'hierarchy for compages' with details of the appropriate map scales for study and presentation and followed this with a model study on Southern Rhodesia to illustrate his method (Whittlesey, 1956). The decade since the Whittlesey scheme was put into operation and the call was made to '. . . fill this lacuna in geographic thinking' (James *et al.*, 1954, p. 47) has not seen any rush to adopt it. Of the few significant papers published in this field, only one, that by Bird (1956) subjected Whittlesey's scheme to field testing. Bird's two-scale comparison of the western peninsulas of Brittany and Cornwall suggested that, while a general (or small-scale) approach showed the two areas to be similar, the intensive (or large-scale) study showed that the two peninsulas were quite dissimilar in most details. Bird's deft illustration of a fundamental and very common geographic problem passed scarcely without comment.

The second major move in the period since Whittlesey's papers came from Philbrick (1957) who published a very full scheme based on the concept of a sevenfold hierarchy of functions. Corresponding to each function is a nodal point with its functional region. Here scale is introduced through the 'nesting' concept with each order of the hierarchy fitting within the next highest order. As a theoretical model Philbrick illustrates the case where each central place of a given order is defined to include four central places of the next lower order. This gives a succession for a seventh-order region of 4 sixth-order places, 16 fifth-order places, and so on down to the final level of 4,096 first-order places. His attempt to apply this scheme to the eastern United States with New York and Chicago in the role of seventh and sixth-order centres was only partly successful but the attempt to introduce a scale component into a system of nodal regions has given an important lead.

Quantitative Solutions

Quantitative attempts to isolate and measure scale components in geographical patterns have not been widely attempted and there is, to the writer's knowledge, no extensive literature in this field. Hence the sections which follow two possible methods of attack – those of filter mapping and of nested sampling – are illustrated from work in hand at Cambridge on patterns of forest distribution in central Portugal.

(1) *Filter Mapping:* The basic ideas of filter mapping can be seen from a fairly simple example. Figure 9.5 shows the stages by which a given distribution may be broken down into regional and local components. Map *A* shows the original distribution in a section of central Portugal. For statistical purposes this pattern can either be expressed as a ratio of (forested/non-forested area) or as a fraction

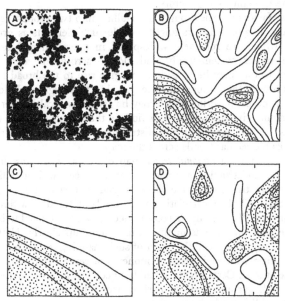

FIG. 9.5. *Mapping of regional and local components in a land-use pattern: Tagus–Sado basin, central Portugal. Shaded areas show densities above average.*

(percentage forest in total land area), i.e. either 0·352 or 26·30. By covering the area with a rectangular grid these ratio values can be collected for small areas (in this case square cells of forty square kilometres) and contoured. The resulting map, *B*, completely describes the area in two-dimensional form. Like contour values for terrain it could be converted to a three-dimensional plaster model but in any case can be regarded conceptually as a three-dimensional trend surface.

This surface may be thought of statistically as a *response surface* (Box, 1954). That is the height (i.e. degree of forest cover) at any one point may be regarded as a response to the operation of that complex of '. . . geology, topography, climatic peculiarities, natural composition, economic disparities, and local and regional history' (Köstler, 1956, p. 82) which together determine forest distribution. Variation in the form of the surface may be regarded as responses to corresponding areal variations in the strength and balance of these hypothetical controlling factors.

These factors may be thought of as falling into two groups, *regional* and *local*. Regional factors might include such elements as growing season which are relatively widespread in operation and tend to change rather systematically and slowly across the area. Such regional factors may be considered to give rise to the broad larger-scale trends in the response surface. Local factors might include such items as soil composition which may be relatively local in operation. Such factors give rise to local variations in the response surface which are unsystematic and spotty in distribution and do not give rise to recognizable secular trends across the map.

Map *C* shows a *regional trend* map of the area. It was derived simply by constructing a circle around each cell with a radius of 28·20 kilometres so as to include a ground area of 2,500 square kilometres and then calculating the woodland fraction within this circular unit. Plotting this '2,500' surface caused local detail to be lost but the main lineaments of the pattern show up clearly. Nettleton (1954, p. 10) has likened the effect of such mapping to that of '. . . an electric filter which will pass components of certain frequencies and exclude others'. Certainly the detail has been lost in a predictable and controllable manner and comparison with other maps based on a similar 'grid' is made more reliable.

Separation of the local anomalies can be very rapidly derived from the regional map. For each cell the values of the original 40 square kilometre cell are subtracted from those of the 2,500 square kilometre cell. Positive values (i.e. where local values exceed regional values) are shaded and negative values (i.e. regional values exceed local values) are unshaded. Map *D* therefore shows the operation of local factors as a pattern of positive and negative residuals.

Clearly an infinite number of trend-surface maps can be drawn and the nature of the resultant maps will vary with the grid interval chosen. To this extent the trend map is a quantitative expression of a qualitative choice. However, by including details of the generating grid with the map (in the same way that scale and orientation are conventionally included on a map) and by standardizing mapping around multiples of conventional levels – the 100 square kilometre unit would seem a useful basis for both aggregation and subdivision – this disadvantage can be nullified.

An alternative approach to filter mapping which is not dependent on a regular grid has been proposed by Oldham and Sutherland (1955). By using orthogonal polynomial equations they were able to compute a 'best fit' quadratic surface. This approach which has been further developed by Krumbein (1959) is taken up again in another context later in the chapter, but it is worth noting at this stage that more rigorous bases for determining trend surfaces are being explored and applied. These alternatives are discussed at length by Chorley and Haggett (In Press).

(2) *Nested Sampling:* One approach to the problem of local and regional variation that cuts out the need for complete information on all the area considered is that of nested sampling. (Olson and Potter, 1954; Krumbein, 1960). It is particularly valuable in exploratory studies where there is a need to cover as large a region as possible but at the same time pay attention to local variations. The basic idea of the nested approach (also termed the 'multilevel' or 'hierarchical' approach) is to divide the region into a few major areas of equal size. Several of these major regions are then chosen at random and broken down into a number of smaller sub-regions. Several of these sub-regions are chosen at random and broken down again, the process being continued until the smallest meaningful unit is reached or data cease to be available. Figure 9.6 illustrates this process by breaking down a 150 by 100 kilometre quadrangle into six 'regional units' each 50 by 50 kilometres square, and then subdividing each square a further four times until the smallest units, squares 3·125 by 3·125 kilometres, are reached. By selecting randomly at each level only two of the four available squares, only 96 of these smaller units are selected for study out of a possible total of over 1,500 such units within the original area, i.e. a sampling fraction of $\frac{1}{16}$. Their location is shown in Map *B*. Sampling on this hierarchical framework ensures not only that every part of the region is represented but that the field work time in visiting each point is reduced well below that of a simple $\frac{1}{16}$ random sample.

The main value of collecting data in this frame comes at the analysis stage. Here any 'local' value (X) can be regarded as being generated by the sum of independent deviations at each level of variability; i.e.:

X – Overall mean value (150 × 100 km)

    + Region (50 × 50 km) deviation from overall mean value.

    + Sub-region (25 × 25 km) deviation from region mean.

    + District (12·5 × 12·5 km) deviation from sub-regional mean.

    + Sub-District (6·25 × 6·25 km) deviation from district mean.

    + Locality (3·125 × 3·125 km) deviation from sub-district mean.

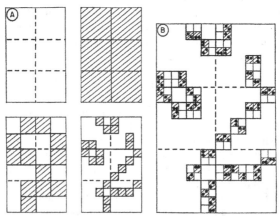

FIG. 9.6. *Five-stage nested sampling procedure for extracting scale components: Tagus–Sado basin, central Portugal.*

Values for each level can be determined using an appropriate type of variance analysis so that it is possible to specify the contribution each level makes to the total variability of the pattern. Table 6 shows the

Table 6. *Tagus-Sado Basin, Central Portugal: Contribution of Five Areal Levels to Variability in a Forest Pattern*

Level	Areal unit (with area in square kilometres)	Variance component	Percentage contribution
I	Region (2,500)	208	32
II	Sub-region (625)	0	0
III	District (156)	258	41
IV	Sub-district (39)	51	7
V	Locality (10)	124	19

results of an analysis made by the writer into the pattern of variability shown by the forest distribution in central Portugal using this approach. The original pattern is shown in Figure 9.5A. It indicates the great contrast between the increment in variability at the third level, the district, compared to the negligible effect of the next higher level. In this case it proved possible to link the high variability at levels I, III and V with the operation of specific factors at those levels.

It is clear that either by mapping techniques applied over the whole area (filter mapping) or by carefully selected sampling designs (nested sampling) the known variability of areal patterns can be broken down and examined. Techniques developing outside geography, notably in geophysics and in plant ecology (Greig-Smith, 1964) are being increasingly applied and the final forging of traditional cartography and applied statistics is proving a most powerful tool in dissecting areal problems.

THE SCALE STANDARDIZATION PROBLEM

Nature of the Problem

While the two previous scale problems apply with equal force to work in all sides of geographical research the third scale problem applies with particular force to work in human geography. Here two almost intractable problems are faced. Firstly, much of the data is released for areas rather than for points; secondly, these areas vary wildly in size and shape both between countries and within countries. This variation had little more than nuisance value when the limit of sophistication was the choropleth map but as more refined statistical analysis is carried out the problem has grown in importance. Weighting for area as proposed by Robinson (1956) can overcome some of the problems but there remains a range for error and misinterpretation so great that Duncan found it necessary to devote the greater part of his pioneer book on statistical geography to just such problems in analysing areal data (Duncan, Cuzzort and Duncan, 1961).

The problem can be seen at its simplest in city comparisons. If we ask an apparently simple question: 'Is City X bigger in population than City Y?' then the answer will often hinge on our areal definition of the city. Dickinson (1963, p. 68) has shown that Liverpool may be either smaller or larger than Manchester depending on how each city is demarcated. Similarly, Duncan has shown that Chicago may be more densely or less densely populated than Detroit depending on which of the available definitions – that of the 'city', the 'urbanized area', or the 'standard metropolitan area' – is taken (Duncan *et al.*, 1961, pp. 35–36). Table 7 shows how this problem extends to inter-

Table 7. Belgium and Netherlands, 1947: Comparison of Apparent Commuting Differences

Region	Out-commuters	Mean size of sub-divisions*
Belgium	40·0%	1,880 hectares
Netherlands	15·2%	6,670 hectares

* Weighted according to population resident in each sub-division.
Source: Chisholm, 1960, p. 187.

national comparisons of commuting movements. Chisholm (1960) has demonstrated how the apparent differences in commuting described by Dickinson (1957) between two neighbouring states probably owes more to differences in the size of administrative sub-divisions than to any inherent differences in social behaviour. This arises from the definition of a commuter as 'moving to work outside his residence area': clearly *ceteris paribus* the smaller the residence areas the greater the chance of a worker being recorded as a commuter.

Unless great care is taken such 'mirage effects' are likely to become more common as more refined indices are derived. Kendall as a statistician has warned that with certain coefficients of geographical association we can get any coefficient we choose by juggling with the collecting boundaries! (Florence, 1944, p. 113). It is an open question whether detailed medical maps (e.g. Murray, 1962) in which mortality indices are most carefully standardized for age and sex should not equally well be standardized for the areas for which they are collected. Certainly we need to be reassured that the apparently unhealthy small pockets of disease in Lancashire and Yorkshire owe nothing to the fragmented system of local government areas.

These problems are central and critical and in suggesting a series of part-solutions below, the writer is keenly aware that these are really ways of making the best of a bad job. The eventual answer lies in (*a*) collecting more field data ourselves on a controlled sampling basis; and (*b*) persuading local and central government to follow the lead of the Swedish census in collecting and publishing data for specific *x, y* co-ordinates rather than for irregular, inconvenient and anachronistic administrative areas.

Aggregation Solutions

Where the collecting units are many and irregular a fairly simple counter-measure is to group them into fewer but more regular areas. Such a technique was adopted by Coppock (1960) in a study of parish records in the Chilterns. Here not only are parishes irregular in shape and size and running orthogonally across the major geological boundaries but the farms on which the data was collected themselves had land outside the parishes for which their acreages were recorded. Grouped parishes in this case allowed both more regular units and reduced the farm 'overlap' problem since with larger units the farm area outside the combined parish boundaries was proportionally a rather insignificant part of the larger total area.

It is important to remember in using the aggregation method that it is possible to throw the baby out with the bathwater in the sense of throwing away detail to gain uniformity. Haggett (1964) has suggested that the coefficient of variation may be used as an indication of both loss of detail and of gains in uniformity, and that only when the latter exceeds the former is the detail loss justified. Table 8 shows a

Table 8. *South-east Brazil, 1950. Comparison of Original and Grouped Counties*

Characteristics	County (*município*)	Super-county
Number	126	24
Mean area	133 square miles	699 square miles
Coefficient of variation	74·2	7·91

Source: Haggett, 1964, p. 371.

case in point in a regression analysis of county-data in south-eastern Brazil. Here the original 126 counties were grouped into only 24 'super counties' but the 82 per cent loss in detail was less than the percentage gain in uniformity as measured by comparisons in the coefficient of variation. Where units are in any case very regular, as in the American Middle West (Weaver, 1956), the method is hardly justified.

Aggregation and testing will be very much speeded when computer programmes are developed for rapidly checking all possible number of ways in which contiguous units can be combined and re-combined. In view of the enormously large number of possible combinations it is uncertain that the combinations used so far are the optimum ones in terms of uniformity, of size, shape, and number.

Grid-Type Solutions

The difficulties of dealing with aggregate areas is that they are themselves highly irregular in shape if not in size. Attempts have therefore been made to collect information not in areal units but in regular frames or grids.

An outstanding example of this type of work is the *Atlas of the British Flora* (Perring and Walters, 1962) where field data on the occurrence of British vascular plants was collected for the 100 square kilometre grid-squares of the British National Grid System. This grid system was also used by Johnson (1958) in a study of the location of factory population in the West Midlands. Grid systems have the great advantage of ease of mapping and the flora maps were all directly plotted by computer. This has very considerable merit in an era where more maps are being produced directly from punched-tape data (Tobler, 1959) and allows very ready comparison between the original data and controlling factors. It also allows micro-analysis by breaking down the original squares into smaller ones or macro-analysis by combining such squares into larger units, and on the lines suggested by filter sampling above.

In both examples cited, data was either collected on a regular grid pattern or was precisely located and could therefore be assigned to a grid. Where data only exists for irregular administrative areas the transfer to a grid is more complex. Robinson, Lindberg and Brinkman (1961, p. 214) used a regular hexagonal grid in a study of population trends in the Great Plains of the United States. County data was transferred to the grid by measuring how much of each hexagon was made up by any one county and multiplying this value by the population density of the county. The sum of all the county parts gave the average value for the hexagonal unit. This principle was first used by Thiessen in 1911 for calculating the average rainfall over watersheds and its accuracy clearly hinges on two principles. First, the degree to which population density (or any similar measure) can logically be regarded as uniform over the areal sub-division and, second, the number of such sub-divisions which make up the regular unit. Where each hexagon contains a number of undivided counties the assumptions under the first principle become less limiting as the 'split' counties contribute less to the total value. Again the problem is one of optimizing through linear programming both the reliability of each grid-unit (by increasing its size) and the number of such grid-units (by decreasing their size).

Grid-free Solutions

Clearly the two foregoing methods, aggregation and grid-type solutions must involve some loss in detail in that the revised units are fewer than the original. Attention has recently been directed to the

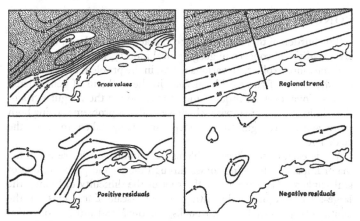

FIG. 9.7. *Regional trend surface with positive and negative residuals: south-east Brazil. (source: Haggett, 1964).*

problem of how generalized maps can be made which retain all the original data. This problem first came to light through geophysical problems, e.g. through meteorology where general weather patterns have to be mapped from irregular and often highly localized weather recording stations and through petroleum prospecting where basin and facies characteristics may have to be mapped from irregular well and bore records. Krumbein (1959) has illustrated how computers may be used to derive an algebraic formula which gives the average surface which 'fits' the irregular control points best. This 'best fit' polynomial surface uses all the available records and builds them into a generalized picture. It is of particular importance where gaps in the areal spread of records are found and has been used by Whitten (1959) to fill in the 'ghost stratigraphy' of crystalline areas.

Use of these methods in landform analysis has been undertaken by Chorley (1964) in a study of the topography of the Lower Greensand, while simpler first-order trend surfaces, i.e. simple planes (Figure 9.7) have been used by Robinson and Caroe (Garrison, 1964) in a study of population in Nebraska, and by Haggett (1964) in a study of forest distribution in south-east Brazil. There is every indication that they will be more widely used in the future (Chorley and Haggett, In Press).

CONCLUSION

The burden of this chapter has been that scale problems, while they are a traditional concern of geographical inquiry, have in recent years become more explicit. In various forms, in the problem of overcoming the sheer size of the earth's surface, in the problem of generalizing over various levels of inquiry and in overcoming the irregularities of the administrative areas with which we deal, the problem is being both more acutely recognized and steps are being taken to meet the challenge.

While each of the three problems has significance for research the major impact on teaching comes through the second problem, that of scale-linkage. Here we are at the root of the difficulty of linking the large-scale sample study with the small-scale lineaments of the continental pattern. Some textbooks, notably Mead and Brown (1962) in their study of North America, go some way towards a solution by reproducing air photographs and portions of 1/62,500 maps alongside regional transects. Ideas that factors may change with scale may be introduced for familiar phenomena. The location of a light manufacturing plant in north-west London might be explained in terms of regional factors (location within industrial north-west Europe), local factors (access to the London and Midlands market), and site factors (land characteristics such as bearing capacity of soils for foundations and heavy machinery). Each explanation nests within the other and operates within the general restraints set by the next highest factor. This has the advantage of resolving some of the apparently conflicting hypotheses by restricting each to a particular level of generalization. It also accords with the known practice of locational decision-making (McLaughlin and Robock, 1949).

It is clear that open recognition of the problem of working within a scale continuum clarifies some problems and raises others. While it is undoubtedly more comfortable to work on in happy or perverse oblivion of the problems its recognition brings, there are encouraging signs that with recognition comes part-solutions; from such part-solutions we hope some final answers may emerge.

References

BERRY, B. J. L., 1962, 'Sampling, Coding and Storing Flood Plain Data', *U.S. Dept. Agric. Farm Econ. Div., Agric. Hdbk.*, **237**, 1–27.

BIRD, J., 1956, 'Scale in Regional Study: Illustrated by Brief Comparisons between the Western Peninsulas of England and France', *Geog.*, **41**, 25–38.

BOX, G. E. P., 'The Exploration and Exploitation of Response Surfaces', *Biometrics*, **10**, 16–30.

BRONOWSKI, J., 1960, *The Common Sense of Science* (London).

CHISHOLM, M. D. I., 1960, 'The Geography of Commuting', *Ann. Assn. Amer. Geog.*, **50**, 187–8 and 491–2.

CHORLEY, R. J., 1964, 'An Analysis of the Areal Distribution of Soil Size Facies on the Lower Greensand Rocks of East-central England by the use of Trend Surface Analysis', *Geol. Mag.*, **101**, 314–21.

CHORLEY, R. J. and HAGGETT, P. In Press. 'Trend-surface Mapping in Geographical Research', *Trans. Inst. Brit. Geog.*

COCHRAN, W. G., 1953, *Sample Survey Techniques* (New York).

COLWELL, R. L. (Ed.), 1960, *Manual of Photographic Interpretation* (New York).

COPPOCK, J. T., 1960, 'The Parish as a Geographical-Statistical Unit', *Tijds. v. econ. soc. Geogr.*, **51**, 317–26.

DICKINSON, G. C., 1963, *Statistical Mapping and the Presentation of Statistics* (London).

DICKINSON, R. E., 1957, 'The Geography of Commuting: the Netherlands and Belgium', *Geog. Rev.*, **47**, 521–38.

DUNCAN, O., CUZZORT, R. P. and DUNCAN, B., 1961, *Statistical Geography: Problems in Analysing Areal Data* (Glencoe, Ill.).

FENNEMAN, N. M., 1916, 'Physiographic Divisions of the United States', *Ann. Assn. Amer. Geog.*, **6**, 19–98.

FLORENCE, P. S., 1944, 'The Selection of Industries Suitable for Dispersal into Rural Areas', *Jour. Roy. Stat. Soc.*, **107**, 93–116.

GARRISON, W. L. (ed.), 1964, *Quantitative Geography* (New York).

GINSBURG, N., 1961, *Atlas of Economic Development* (Chicago).

GREIG-SMITH, P., 1964, *Quantitative Plant Ecology* (London).

HAGGETT, P., 1963, 'Regional and Local Components in Land-use Sampling: a Case Study from the Brazilian Triangulo', *Erdkunde*, **17**, 108–14.

— 1964. 'Regional and Local Components in the Distribution of forested areas in South-East Brazil: a Multivariate approach', *Geog. Jour.*, 130, 365-80.

HAGGETT, P., CHORLEY, R. J. and STODDART, D. R., 1965, 'Scale Standards in Geographical Research: A New Measure of Areal Magnitude'; *Nature*, **205**, 844–7.

HALL, P., 1962, *The Industries of London since 1861* (London).

HARTSHORNE, R., 1959, *Perspective on the Nature of Geography* (London).

HIGHSMITH, R. M., HEINTZELMAN, O. H., JENSEN, J. G., RUDD, R. D. and TSCHIRLEY, P. R., 1961, *Case Studies in World Geography* (New York).

JAMES, P. E., JONES, C. F. and WRIGHT, J. K. (eds.), 1954, *American Geography: Inventory and Prospect* (Syracuse).

JOHNSON, B. L. C., 1958, 'The Distribution of Factory Population in the West Midlands Conurbations', *Trans. Inst. Brit. Geog.* Pub. **25**, 209–23.

KRUMBEIN, W. C., 1959, 'Trend Surface Analysis of Contour-type Maps with Irregular Control-point Spacing', *Jour. Geophys. Res.*, **64**, 823–34.

— 1960, 'The "Geological Population" as a Framework for Analysing Numerical Data in Geology', *Liv. and Man. Geol. Jour.*, **2**, 341–68.

KRUMBEIN, W. C. and SLACK, H. A., 1956, 'Statistical Analysis of Low Level Radioactivity of Pennsylvanian Black Fissile Shale in Illinois', *Bull. Geol. Soc. Amer.*, **67**, 739–62.

KÖSTLER, J., 1956, *Silviculture* (Edinburgh).

LANGBEIN, W. B. and HOYT, W. G., 1959, *Water Facts for the Nation's Future: Uses and Benefits of Hydrological Data Programmes* (New York).

LATHAM, J. P., 1963, 'Methodology for an Instrumented Geographic Analysis', *Ann. Assn. Amer. Geog.*, **53**, 194–209.

LINTON, D. L., 1949, 'The Delimitation of Morphological Regions', *Trans. Inst. Brit. Geog. Pub.* **14**, 86–87.

LOPIK, J. R. VAN, 1962, 'Optimum Utilization of Airborne Sensors in Military Geography', *Photogramm. Eng.*, **28**, 773–8.

MᶜCARTY, H. H., HOOK, J. C. and KNOS, D. S., 1956, 'The Measurement of Association in Industrial Geography', *Univ. Iowa. Dept. Geogr. Rept.*, **1**, 1–143.

MᶜLAUGHLIN, G. E. and ROBOCK, S., 1949, *Why Industry moves South* (Kingsport).

MEAD, W. R. and BROWN, E. H., 1962, *The United States and Canada* (London).

YEMNEN, E., 1960, *Orbis geographicus 1960* (Wiesbaden).

MURRAY, M., 1962, 'The Geography of Death in England and Wales', *Ann. Assn. Amer. Geog.*, **52**, 130–49.

NETTLETON, L. L., 1954, 'Regions, Residuals and Structures', *Geophysics*, **19**, 1–22.

OLDHAM, C. H. G. and SUTHERLAND, D. B., 1955, 'Orthogonal Polynomials: Their Use in Estimating the Regional Effect', *Geophysics*, **20**, 295–306.

OLSON, J. S. and POTTER, P. E., 1954, 'Variance Components of Cross-bedding Direction in Some Basal Pennsylvanian Sandstones of the Eastern Interior Basin: Statistical Methods', *Jour. Geol.*, **62**, 26–49.

PERRING, F. H. and WALTERS, S. M., 1962, *Atlas of the British Flora* (London).

PHILBRICK, A. K., 1957, 'Principles of Areal Functional Organization in Regional Human Geography', *Econ. Geog.*, **33**, 299–336.

PLATT, R. S., 1942, *Latin America: Countrysides and United Regions* (New York).

— 1959, 'Field Study in American Geography: the Development of Theory and Method exemplified by Selections', *Univ. Chicago. Dept. Geogr. Res. Pap.*, **61**, 1–405.

ROBINSON, A. H., 1956, 'The Necessity of Weighting Values in Correlation of Areal Data', *Ann. Assn. Amer. Geog.*, **46**, 233–6.

ROBINSON, A. H., LINDBERG, J. B. and BRINKMAN, L. W., 1961, 'A Correlation and Regression Analysis applied to Rural Farm Population Densities in the Great Plains', *Ann. Assn. Amer. Geog.*, **51**, 211–21.

SAUER, C. O., 1952, *Agricultural Origins and Dispersals* (New York).

TOBLER, W., 1959, 'Automation and Cartography', *Geog. Rev.*, **44**, 534–44.

UNSTEAD, J. F., 1933, 'A System of Regional Geography', *Geog.*, **18**, 175–87.

WEAVER, J. C., 1956, 'The County as a Spatial Average in Agricultural Geography', *Geog. Rev.*, **46**, 536–65.

WHITTEN, E. H. T., 1959, 'Composition Trends in a Granite: Modal Variation and Ghost Stratigraphy in Part of the Donegal Granite', *Jour. Geophys. Res.*, **64**, 835–48.

WHITTLESEY, D., 1956, 'Southern Rhodesia: an African Compage', *Ann. Assn. Amer. Geog.*, **46**, 1–97.

WOOD, W. F., 1955, 'Use of Stratified Random Samples in a Land-use Survey', *Ann. Assn. Amer. Geog.*, **45**, 350–67.

YATES, F., 1960, *Sampling Methods for Census and Surveys* (London).

CHAPTER TEN

Field Work in Geography, with Particular Emphasis on the Role of Land-Use Survey

C. BOARD

Lecturer in Geography, London School of Economics

In the space of a year geography has lost two of its greatest advocates of field work. Mainly as a result of the activities of Professor Wooldridge and Geoffrey Hutchings field work has progressed far in the last half century, so that it is a universally respected approach to the study of geography. The generation of geographers trained by Wooldridge is today playing a major part in training yet more geographers in the same well tried methods. It is the wish of the pioneers in this approach that 'other teachers will carry on the method' (Wooldridge and Hutchings, 1957, p. xi). This suggests that the time is not inopportune for an appraisal of the position of field work. In looking back to the development of the methods characteristic of this approach and by comparing its influence in British geography with its role in North American geography, it should be possible to arrive at an evaluation of the distinctive part that this approach has played in the progress of geography.

WOOLDRIDGE AND 'REAL FIELD WORK'

The kind of field work as practised by Wooldridge is essentially British, and represents the trade-mark of much British geographical writing. It has its roots in the observation and records of naturalists such as Gilbert White of Selborne (1720–93), an uncompromising protagonist of first-hand observation. The careful collection of specimens and painstaking recording of occurrences from Nature are characteristic approaches of the field sciences, botany, geology, and zoology. Geographical field methods owe much to those of geology;

in fact Wooldridge's training as a field man was as a geologist. It is in physical geography that the influence of field geologists like Archibald Geikie is felt most of all. In an essay on 'Science in Education' (1905), Geikie pointed out that deficiencies of 'literary methods' could be overcome by cultivating the faculty of observation. He realized that everyone was not equally endowed with this faculty and that training was required so that the student could see much more in the world around him 'than is visible to the uninstructed man' (Geikie, 1905, p. 296). There is little doubt that views such as these inspired Wooldridge, whose pronouncements on field work are in sympathy with those of Geikie and the great field geologists. Indeed, it may be said that because much of the early field work in geography was done by geologists, who had become interested in landforms, the currently accepted form of geographical field work concentrates on the natural landscape. Anxious that geography was becoming town centred and too concerned with man rather than land, Wooldridge (1949 A, p. 14) enters an impassioned plea for the teaching of 'natural history geography'. For Wooldridge (1949 B, p. 3) the Weald was one of the true field laboratories, used by field scientists of all breeds. Although most commentary on field work in geography has been made by Wooldridge and Hutchings, it has also become fashionable to include references to the necessity for field work in the inaugural addresses by new occupants of chairs of geography (Linton, 1946; Balchin, 1955; Edwards, 1950; Monkhouse, 1955; Pye, 1955). Such comments stressing the value of field work in geographical education point out other praiseworthy advantages ranging from an open-air life to informality between instructor and instructed. There is clearly little need now to defend the position of field work in geography.

Lest it be supposed that all geographical field work is of this didactic kind, it must be asserted here and now that American geography usually reflects a different brand of field study. This will be distinguished by the name field research. At the same time, it must be said that field research has also been characteristic of British geography. Nevertheless, it will be contended here that field teaching, as distinct from field research, has had a peculiarly powerful influence upon British geographical work. Wooldridge and East (1958, p. 163) in fact recognize this distinction, insisting on the need for training before undertaking research in the field. They give as an example Platt's specimen studies of landscapes in Latin America (1942), whose 'objective is to provide an *interpretative description* of the countrysides and regions of a vast, diversified and on the whole little known continent'. *Real field work* as Wooldridge and East (1958, p. 161) regard it is 'the close examination and analysis in the field of an accessible piece of country, showing one or more aspects of areal differentiation'. Their exposition of its methods demonstrates that by this is meant training in field work. The danger is that these should be equated with field work as a whole.

It would be equally misleading to suggest that all field work in British geography was influenced by the principles of field teaching. Morphological mapping as developed by Linton and Waters (1958) has all the hallmarks of field research, in that 'it is capable of universal application' (Waters, 1958, p. 15). By careful observation and plotting, the indivisible morphological units of the earth's surface can be mapped empirically. No prior knowledge of geology or geomorphology is presupposed and little room is left for subjective interpretation (Waters, 1958, p. 16). Such detailed morphological mapping owes much to similar but scattered work done by other geomorphologists, whose intention was not primarily didactic (Sparks, 1949, pp. 167–8; Hare, 1948, pp. 301–7).

In order to see the extent to which field teaching has affected field research, or is likely to affect it in the future, it is necessary to pay more detailed attention to the aims and methods of field teaching in Britain. In the nineteenth century, it must have seemed to those who looked at the writings of geographers that they shared a strange long-sightedness with Mrs Jellyby whose eyes appeared to 'see nothing nearer than Africa' (Dickens, 1890 edn., p. 30). This cult of 'other-whereitis' as it was termed by Wooldridge (1950, p. 9), was gradually broken down by geographers and others interested in education. Their pleas for making geography more realistic (Branford, 1915–16, p. 97; Fairgrieve, 1937, p. 16; Hunt, 1953, p. 277; Hutchings, 1962; Wooldridge, 1955, pp. 78–9) are amply recorded in the pages of the *Geographical Teacher* or *Geography*. 'The only true geographical laboratory is the world outside the classroom' (Incorporated Association of Assistant Masters, 1954, p. 180). This statement, and others like it, owe much to the notion that by making use of one's immediate surroundings, the raw material for teaching geography could be made convincing. Fairgrieve's advocacy of the study of the 'home region' as 'the only criterion by which the rest of the world may be judged' (1937 B, p. 251) stressed the second-hand nature of much geographical knowledge. He went on, 'the standards which are known to the children themselves are the only ones which may be used to measure other places'. This theme is taken up by Wooldridge (1955, p. 80) when he makes one of his frequent references to G. K. Chesterton's belief that 'to make a thing real you *must* make it local'. Apart from the possibility offered by local field work, the only way of making far-off lands more real was by using material published in the form of detailed first-hand accounts of 'sample areas' (Roberson and Long, 1956). These, however, are still no substitute for the pupils' own first-hand study of the ground.

 After several decades of field teaching, Wooldridge and Hutchings in their guide to field excursions around London (1957, pp. xi and xii) state their view that the aim of geographical fieldwork 'is regional synthesis and though this depends largely upon labours in the library and map room, it cannot dispense with sensitive field observation'. This places field work in true context as but one way of collecting information.

THE METHOD OF FIELD TEACHING

Both Wooldridge and Hutchings have recently provided a fairly comprehensive account of the aims and methods of field teaching. Wooldridge (Wooldridge and East, 1958, p. 161) dismisses as not being real field work, surveying, visits to farms and factories, and censuses. For him the starting point is the comparison of the map with the ground (Wooldridge, 1948 B, p. 2). This is an essential process in map reading and is a way by which the student may gain an appreciation of the scale of phenomena (Hutchings, 1962, p. 6). Since 'the ground, not the map, is the primary document' (Wooldridge and East, 1958, p. 162), the student should work from the ground to the map. Wooldridge (1948 B, p. 4) also points out '. . . the fact that, over a great range of studies, reality is in the field'. Since the map is frequently deficient, because of the exigencies of convention and scale, a second principle of geographical field work is to make 'significant additions' to it, from observations in the field (Wooldridge and East, 1958, p. 165). Hutchings (1962) points out the need for making sketch maps and annotated field sketches of landscapes. Both of these are considered as adjuncts to the description of landscape which forms a major part of work in the field. In reviewing current practice in one university geography department, Wise (1957, p. 20) maintains that students are introduced to areas of progressive complexity in their three-year training to develop an eye for country. Some areas are in fact so 'subtle' or difficult to interpret, that Wooldridge and Hutchings (1957, pp. xiii and xiv) consider them 'not really suited to those beginning to learn the craft of field observation'. Much of the drift covered country of low relief in Eastern England falls into this category. In still another way, it can be seen that the difficulties of field work are not minimized when Wooldridge (1948 B, p. 3) pointed out to laboratory-bound scientists:

'I wish sometimes they could come with us and practise thinking on their feet in all weathers when rain or perspiration drips from one's person and the bar-parlour with its insidious temptation to spirituous theorizing insistently beckons.'

When one turns to the interpretation of field evidence, relatively little guidance is offered. 'The high Art is to teach the learner to use his eyes and draw his own conclusions' (Wooldridge, 1960, p. 3). Hutchings (1960, p. 6), in insisting that a landscape sketch is in in itself an interpretation, writes, 'The geographical draughtsman weaves into his drawing some of his *knowledge* of the things he is depicting.' There is no doubt that one of the main advantages deriving from field study lies in the appreciation of interrelationships of things in space (Wooldridge, 1960, p. 3). This, however, has a corollary in which lie pitfalls for unwary, lesser minds. Field teaching quite properly concentrates on 'observable field data' (Wooldridge, 1955, p. 79), but the explanations for visible patterns are not always, nor ever entirely, to be found by visual inquiry. It is difficult to imagine undertaking a survey of a market town in order to explain its present character, without resorting to some form of interviewing, or the close inspection of the interior of buildings. It is Hutchings' view that, although 'the story of a village, and especially of its social and economic working, is by no means legible in visible signs' (1962, p. 10), not much of it can be considered geography, and therefore in field work it is better not to interrogate the local inhabitants. This would confine geography to a study of the visible landscape. Wooldridge and Goldring (1953, p. 4) are perfectly well aware that the facts of modern economic and social geography are not amenable to the methods of field teaching. They justify the exclusion of the patterns of modern times from their study of the Weald because 'their study involves distinct methods and ways of thought which are not those of the "field man".'

THE NATURE OF FIELD RESEARCH

Before turning to examine the effect field teaching has had on the progress of geographical research, it is instructive to consider the characteristic methods of field research. Many non-descriptive works base their initial hypotheses upon chance remarks and marginal observations of other workers, not necessarily geographers. In other instances the field itself is held to be 'the primary source of inspiration and ideas' (Wooldridge, 1948 B, p. 2). In many cases both field evidence and previous literature have combined to suggest new topics of research. Wooldridge's own work on the Pliocene history of the London Basin (1927) clearly indicates his debt to Barrow (1919, and Barrow and Green, 1921) for its point of departure. Another work finding inspiration from the landscape but the point of departure in previous literature is McCann's study of raised beaches in Western Scotland (1963). Stevens (1959) attempts to illustrate Wooldridge and Linton's contention that geomorphology is of geographical significance by showing that morphological differences which have their roots in geological history have a profound effect on the land-use pattern in north-east Hampshire. Linton (1955), Palmer (1956) and Palmer and Neilson (1962) provide studies of tors, which are good examples of features of the physical landscape that have given rise to speculation and thorough investigation. In the field of historical geography Mead's account (1954) of ridge and furrow in Buckinghamshire shows his indebtedness to a vision of the countryside as well as to the stimulus of a remark by an economic historian. An analysis of the distribution of strip-lynchets by Whittington (1962) is largely inspired by the evidence of so-called cultivation terraces in various parts of Britain. There is no doubt that 'a large part of the evidence in our various subjects and specialisms (in field sciences) is evidence obtained in the field' though supplemented by work in the laboratory (Wooldridge, 1948 B, p. 2). It is less certain that the field 'inspires a great part of both the matter and the *method* [my italics] of our subjects' (Wooldridge, 1948 B, p. 2). Although in economic geography the field evidence may provide the initial problem, as in the case of Hodder's account of rural markets in Nigeria, the methods of field investigation are not those usually associated with the physical field sciences. They demand 'techniques more commonly associated with the social anthropologist' (Hodder, 1961, p. 158). Undoubtedly, in any profound studies of contemporary human geography visual observation has to be supplemented by inquiry. The frequent use of the specimen farm in the reports of the Land Utilisation Survey of Britain is a case in point (Stamp, 1948, pp. 335–50). In some cases, geographical hypotheses may be suggested by the contemplation of patterns on maps, such patterns not being visible in the field. Much of the interpretative work of the Land Utilisation Survey follows this pattern. Still other land-use studies have stemmed from the visual recognition of areal correspondence such as that between pasture land and flood plain (Berry, 1962).

THE ROLE OF THE VISIBLE LANDSCAPE

Visible elements in the landscape form a large part of the subject matter for both geomorphologists and historical geographers who make use of the genetic method. We are reminded that 'geomorphology is the historical geography of the physical landscape' (Wooldridge, 1948 A, p. 28). Some of the dangers that befall historical geographers, who rely too heavily on legacies in the present landscape as a guide to the interpretation of its evolution are expressed by Gulley (1961, p. 308). Can one be sure that conditions, both physical and social, in the past would have the same value as today? (Kirk, 1952, pp. 156–7, 159–60). Would the identification of a series of relicts from former landscapes be an adequate basis for an explanation of its evolution? Some features survive better than others. Strip-lynchets for instance are more likely to be preserved on more resistant rock formations. Ridge and furrow patterns clearly survive best when undisturbed by the plough (Mead, 1954, p. 36), whose long use may have destroyed them in the chalk belt of Buckinghamshire. Such considerations threaten the value of the palimpsest concept. In 1893, Geikie compared the surface of the country to a palimpsest bearing the writing of earlier centuries (Geikie, 1905, p. 56) and was followed by Maitland (1897, p. 15) who called the Ordnance Survey 'One Inch' map 'that marvellous palimpsest' from which it was possible to distinguish at least two distinctive forms of village settlement pattern. Wooldridge and Hutchings (1957, p. 44) urge 'we must need attempt the same *genetic* method if we are to understand our country in any real sense' and warn of missing the essential and interesting content of geography if one allows the dominance of London in south-east England to obscure legacies from the past or the fact 'that much of the pattern seen on the ground or the map is of ancient establishment'. Such an attitude plainly will not serve contemporary economic and social geography. Even if the historical geographer does attempt to interpret the legacies of former landscapes in the present landscape, the historian rarely substitutes for his body of material the fragmentary evidence provided by the faded writing beneath the latest script on the palimpsest. The archaeologist in the field has, perforce, to employ the technique of piecing together evidence of a previous culture from its material remains. By extending those methods to the study of the present landscape, with only the evidence of that landscape on which to base one's conclusions, some geographers have encouraged the neglect of evidence not visible in the field. Wooldridge and East (1958, pp. 30–31) have commented on attempts to treat of cultural elements of landscape which are 'unbearably trite' chiefly because they ignore the invisible aspects of the American farm or the city. Unfortunately the morphological approach to economic and social geography has been encouraged by the powerful influence of Sauer (1925, p. 32) who saw 'the cultural landscape as the culminating expression of the organic area' and 'the geographic area in the final meaning' (p. 46). Sauer interpreted Vidal de la Blache and Brunhes to American geographers, representing that they were important for their studies of the application of morphology to the works of man in the landscape. The morphological approach adopted by Brunhes is seen stripped to its bare essentials in his *Human Geography* (1952 edn., p. 36) – 'ridding our mind of knowledge of man, let us try to see and note the essential facts of human geography with the same eyes and in the same way as we discover and disentangle the morphological, topographical and hydrographical features of the earth's surface'. The restrictive view of the landscape morphologist has frequently been adopted by the geographical field teacher, with the result that an imperfect picture and explanation of the region under study is inevitable.

Field work in American geography seems to have been identified much more with research than with teaching (Davis, 1954). Jones and Sauer (1915, p. 522), insist that 'field work raises many questions which must be solved, if at all, after leaving the field'. They see field work as a method of testing preconceived theories. For Sauer (1924, p. 20) field work was to be equated with the systematic survey method. Platt, another great exponent of field methods, said of geography: 'In the discipline a major approach has been through field study, in which geographers go directly to the source of all geographical knowledge and confront the raw and undisturbed phenomena with which they have to deal' (1959, p. 1). Not all

American geography was so systematic, as Strahler points out. W. M. Davis' qualitative approach to landscape 'appealed then, as it does now, to persons who have had little training in basic physical sciences, but who like scenery and outdoor life' (Strahler, 1950, p. 209). Sauer (1956, p. 295), reviewing his own professional life, bewailed the abandonment of geomorphological work by geographers, as this provided a strong incentive to field observation and training the eye. On the need for such a training he is in no doubt, saying, 'there are those who never see anything until it is pointed out to them' (Sauer, 1956, p. 290).

REGIONAL SURVEYING AND FIELD WORK

Geographical field work in Britain came in the 1920's and 1930's to be dominated by the regional survey movement. Although its practitioners claimed to be inspired by Frédéric le Play, they felt his influence largely through the indefatigable and versatile Patrick Geddes who interpreted Le Play, adding some of his own characteristic ideas and techniques. In spite of the fact that much of Le Play's original work (1855, p. 22) relied heavily upon interviewing, the methods of the Le Play Society, judging from their published reports of field work (e.g. Fleure and Evans (eds.) 1939) only rarely included systematic interviewing. Le Play's succinct recipe for survey method, *Lieu, Travail, Famille*, was transformed by Patrick Geddes into *Place, Work, People*, or *Folk*. Although Geddes provided a framework for social investigation, in the hands of geographers in foreign lands the intimate contact of the investigator with the family on the land was forgotten.

An earlier manifestation of the regional survey movement sprang from summer meetings organized by Geddes in Edinburgh from 1887. Allied to these discussions which drew such scholars as Herbertson, Fleure, Fawcett, George Chisholm and Elisée Reclus (Fagg, 1928–9), was the 'Outlook Tower' which provided through its exhibits, stained-glass windows and *camera obscura* a visual synthesis of the surrounding region (Boardman, 1944, pp. 177–92). The regional survey method emanating from this source bore some similarities to the field techniques of the naturalists. They both relied essentially on visual observation, training novices to perceive features and relationships not seen by ordinary men. The regional survey of the 1920's which was inspired by Geddes (Fagg and

Hutchings, 1930, p. 35) was held by Fagg (1915, p. 24) to be 'the organized study of a region and its inhabitants . . . from every conceivable aspect, and the correlation of all aspects so as to give a complete picture of the region'. A key step in collecting information about a region is the undertaking of a surface utilization survey (Fagg and Hutchings, 1930, Ch. V). Even the collection of more data than we actually require was justified on the grounds that specialized workers may have missed their significance at first (Fagg and Hutchings, 1930, p. 64).

After recording the facts of the use to which the land is put the next step is interpretation. This is difficult without instruction in the methods of field geography, such as geographical landscape drawing, where the very act of drawing a landscape is an exercise in interpretation (Hutchings, 1960, p. 6). Nevertheless, there still remains the danger of ignoring factors and interpretations invisible even to the trained eye. Buchanan (1952, p. 4) warns against the risk of thinking along with the landscape purist 'that the landscape carries with itself the answers to the questions it asks'. It would be a mistake to explain a land-use pattern simply in terms of altitude, relief, soil texture and drainage, important though they be. Just as important are the size and scale of enterprise of the farms whose land-use pattern is being examined. Furthermore, the characteristics of the farmer himself may well affect his decisions to plough, to sow a straw crop or a cash root such as potatoes (Butler, 1960, p. 36). Comparing the map with the ground will hardly lead us to the right answers in this case. The map is a generalization of the landscape, with features such as names and boundaries added. Even the landscape is only a selection of reality, the universe with which we must deal. Field teaching which trains us to squeeze more out of the landscape and to use the map to help synthesize features of a wider area than we can see at any one time, is an essential preliminary to field research. It should not be mistaken for it. In field research the map is merely a tool, one of several useful for solving problems.

FIELD WORK AND LAND-USE SURVEYS

The effects of the influence of field teaching on geographical research can be seen in much British work on land use. In contrast, much American land-use work had a definite purpose from the outset, whereas the origins of British land-use studies may be traced to the regional survey, which had a purely general academic interest until the town planners discovered its value (Pepler, 1925, p. 80). Although some of the earliest land-use surveys in the United States were academic, in origin they sprang from purposeful attempts to test hypotheses on the nature of the relationship between the environment and the cultural landscape (Whittlesey, 1925, p. 187). In an early paper on land-use survey, Sauer (1919, p. 51) points out that such work can provide the geographer with a definite purpose and that 'it eliminates facile generalizations and the plausible and insecure reconnaissances that constitute too large a part of geographical literature'. In Britain on the other hand, the surface utilization survey provided a systematic method of accumulating a mountain of information about various aspects of a region, and could be undertaken by schoolchildren and students. Thus it fitted very neatly into field teaching as an exercise that could be done with a minimum of supervision. It concentrated for obvious reasons on the visible aspects of the landscape. It followed that with the influence of field teaching so strong and with the prevalent enthusiasm for man/land relationships, much but not all of the interpretation was in terms of the distribution of elements of the natural environment, and occasionally socioeconomic or what were vaguely termed historical factors. The chief method of interpretation seems to have been the comparison of the land-use map with maps of other distributions (Stamp, 1960, p. 56).

It is instructive to trace the development of land-use studies in Britain from the days of pilot surveys often done by individual members of the Geographical Association to the flowering and final fruition of the Land Utilisation Survey of Britain. When Stamp became the Chairman of the Geographical Association's Regional Survey Committee he was able to channel a good deal of enthusiasm for regional survey into the systematic survey of land use of the whole of Britain (Stamp, 1948, pp. 3–4). What is often overlooked is the growing interest in the countryside during the inter-war era. Whether the townsman sought out bucolic pleasures through hiking, cycling, the Youth Hostel movement, or through the increasingly popular small family car, the effect was the same. Some like Sir George Stapledon were urging that this invasion of the countryside should furnish an opportunity for instruction. Conditions called for a national policy for land use (Stapledon, 1935, p. 7). The exigencies of planning for post-war reconstruction did more than anything else to confer upon land-use survey the status of a primary and fundamental source of information for physical planning of future development (Stamp, 1951, p. 5). The motives for the second national survey of land use are stated to be mainly intellectual curiosity, a desire to know and understand the face of our country and especially its regional contrasts (Coleman, 1960, p. 347). Furthermore, the first survey is now out of date and needs revision if it is to be of value for planning.

However, it would be quite misleading to suggest that American work in land-use survey did not have an academic purpose or did not spring from a curiosity in the countryside. An active band of young geographers was led by Sauer, who pioneered a land-use survey in Michigan before 1920 (Sauer, 1917). He saw in intensive field work, or geographic survey, a way of making 'the sprightly sketchiness of observation of the geographer-traveler' more profound and objective (Sauer, 1924, p. 25). The major task of this method was to represent the way in which land was utilized. The keenness to study the 'super-ficially familiar scene' (Sauer, 1924, p. 32) was characteristic of other Mid-Western geographers and is reflected in the pages of the *Annals of the Association of American Geographers* in the 1920's. Indeed, Jones and Finch (1925) record the activities of a group of like-minded geographers who met to study these questions in the field, in 1924 and 1925. They agreed that the synthesis of economic life and natural environment which was achieved in land-use survey and recorded on a map of land use and physical features 'compels the observer to group together in the field phenomena which occur together and thus is much superior to synthesis in the office of related facts' (Jones and Finch, 1925, p. 151). A number of other experimental studies were undertaken, for example by Whittlesey (1925). But, in the United States, it did not become possible for even these enthusi-astic geographers to command sufficiently large resources for field mapping to complete a countrywide land-use survey. In any case, coverage of topographic maps of suitable scale was not available. Nor is there any evidence that such a complete survey was ever contem-plated. In fact there are very early signs that it was realized that inten-sive field study was only possible in restricted areas 'of strong geographic individuality and high interest' (Sauer, 1924, p. 31). This necessitated field work at two levels, the less intensive for large areas.

In Britain, on the other hand, Stamp saw that the completion of a standardized survey of land use was perfectly within the bounds of possibility. The good topographic map coverage and goodwill of local education authorities, followed by the generosity of local patrons

o

were not insignificant in allowing the survey and its publication to go forward. Thus, the very different conditions under which land-use survey was being pioneered in America and Britain, were almost bound to affect its later development as a geographical field technique. This situation had two main effects. American work was the first to develop methods to save time, effort and money by the increasing use of sampling techniques and similar statistical devices. But, in Britain, there was no such incentive to use these economical techniques, so that the major effort went into completing the mapping. In this lay the great danger that the use of new methods to interpret the patterns thus produced were completely overshadowed by the sheer exhaustion of completing and checking the map. Interpretation was thus relegated to second place. Although American work on land use was not to find newer methods of explaining patterns until the 1950's, sampling techniques were no new thing. Sauer (1956, p. 298), when reflecting on the training of geographers, is impelled to point out the disadvantages of wasting valuable field time. 'Mapping soon runs into diminishing returns. . . . Routine may bring the euphoria of daily accomplishment as filling in blank areas. . . . Time consuming precision of location, limit, and area is rarely needed.' He placed the 'unit area' scheme of mapping 'below almost any other expenditure of effort'. These trends of thought exemplify the willingness with which American workers have sought new attitudes to and methods for land-use survey. As a result their field and mapping techniques have been equally applicable to research and teaching. This is far from true of the conventional attitudes to land-use survey among British geographers.

THE EFFECT OF REGIONAL SURVEY ON LAND-USE STUDIES

From our mid-century vantage point we can now see that British work on land-use survey came out of the regional survey stable and was primarily concerned with the production of an inventory of the facts of land occupance. In this way the British effort was spread out over a relatively wide front, the effort being spent much as the power of storm waves is dissipated on a wide, shallow sandy beach. Such an attack on the problems of agricultural geography failed to bring to light any methods suited to the interpretation of some of the fundamental aspects of agricultural patterns. It actually held back research in this part of geography. That land use and agriculture were regarded as interchangeable terms is clear from an examination of the reports of the Land Utilisation Survey and other work, for instance in New Zealand (Fox, 1956, p. 9). In particular, land-use regions were being differentiated on a basis of agricultural criteria of wider application than the form that use of land took. A notable exception to this tendency is found in Wooldridge's report (1945) on land use in the North Riding of Yorkshire, where the land-use regions are clearly delineated on a basis of land-use patterns.

The author's own work in South Africa (Board, 1962, p. 170) although paying lip service to the need for such a procedure goes on to point out that some boundaries of land-use regions were drawn so as to 'coincide with the transition from one type of farming to another'. The names of several of the land-use regions (e.g. Chalumna Native Trust Agricultural Area) are patently more closely related to the interpretation of the land-use pattern than to the pattern itself. This illustrates the essential difficulty – the impossibility of interpreting land-use patterns in any terms other than farming type areas, or agricultural regions. Only where the type of farming is homogeneous over the whole area of study (a very rare occurrence) is there a chance of the interpretation being made principally in terms of, say, the physical environment. In effect in the Border Region (Board, 1962) the distribution of different types of farming was the subject of interpretation. The land-use pattern was merely one manifestation of that distribution; perhaps the best single guide, but certainly not the only one. At the time of the completion of that survey the author, like many others, was unaware of the importance of work in America which was to make possible the more satisfactory, rigorous interpretation of land-use patterns.

The two reasons why land-use survey held back progress in research into the geography of agriculture were, first, this difficulty of finding objective methods by which to interpret the pattern of land use, and secondly, the excessive concentration on morphological aspects of land use, to the exclusion of the processes which moulded the agricultural landscape. This latter was particularly the inheritance of the regional survey movement.

These trends are combined in the difficulty which faces all workers intending to explain the pattern of land use in a given area. Standard practice was to analyse elements of the pattern, one by one, mapping each separately. Major factors thought to be responsible for the distribution were then isolated and compared with the land-use pattern. Stamp (1960, p. 56) points out that this was achieved by mapping these factors and comparing the succession of maps. 'It is often difficult to do this with a series side by side, and hence a very common device is to use a series of transparencies, where certain facts can be printed on transparent paper and two or more maps placed one above the other.' This procedure was a very useful by-product of the preparation of maps to illustrate the county reports (Wilson, 1964), the separation drawings of individual elements being drawn at the scale of 1:63,360 on tracing paper. The method of interpretation, however, remained subjective.

A further problem was introduced by the fact that interpretation had to be made of the static distribution of a constantly changing pattern. Changing patterns could be studied only by re-mapping (which could scarcely be contemplated), or by case-studies using maps of land use at different times. This could easily lead to short-term responses to changing economic conditions being ignored. Such changes were studied chiefly through the medium of agricultural statistics, as published for whole counties, or more rarely for individual parishes. It was naturally quite difficult to integrate such figures into a pattern of land use at a particular time. Coppock's studies of agricultural changes in the Chilterns (1957, 1960) include parish maps showing changes over comparatively short periods. But these can be related to land-use changes only in the text and by implication. Maps of changes in land use appear separately (Coppock, 1954, p. 130). The period of change is, however, a relatively long period of time and is entirely conditioned by dates when surveys happened to be done. Such information is of course very useful, but it tends to favour the identification of long-term trends. Classic examples of the use of historical material on land use are seen in Willatts' report (1937) on Middlesex and the London Region. Nevertheless, the difficulty remains that the explanation of changes, or indeed of the pattern at any one time, depends upon a subjective process of visual comparison, even though it is simplified by the close juxtaposition of maps of geology, relief and land-use elements. Thus, it is difficult to show how the land-use pattern is a reflection of the interaction of the farmer's resources, choice, ability to adapt to changing technology and price levels. This problem has usually been overcome by the parallel study of the specimen farm (Stamp, 1948, p. 348).

THE IMPORTANCE OF VISIBLE FEATURES IN LAND-USE STUDIES

Allied to these disadvantages of traditional methods is over-concentration on visible relationships between land use at the time of survey and elements of the physical environment. It is contended that this derives in part from the methods adopted by the regional survey school of geographers. It is also related to this difficulty of relating land use to the less obvious factors affecting it. If complete mapping is the objective of a land-use survey, there is rarely time for a thoroughly adequate inquiry into the practices and motives of in-individual farmers. By concentrating on the factors which are relatively easy to map and see in the field, the land-use surveyor runs the risk of underestimating the importance of social and economic factors, which generally operate in a more subtle way. This was well appreciated by the forester Bourne (1931, p. 52), whose classic work on regional survey recognizes the need for 'farm survey' as well as 'regional survey' because 'from the point of view of the farm manager this simple Regional Survey of a developed area does not bring to light all the combinations of conditions which constitute existing farms'. It cannot even be accepted that every farmer is motivated by a desire to maximize his profit (Butler, 1960, p. 45). Even if he is, his knowledge of improved techniques and his ability to introduce them into his existing farming system, or into a modified farming system, will vary from farm to farm independently of the natural environment. Similarly, changes in relative price levels, as for instance between wheat and barley, may lead to the cultivation of barley on soils normally regarded as too heavy and more suited to wheat. Buchanan (1959, p. 10) has reminded us that: 'The crop that best fits the physical conditions will be the preferred crop if, and only if, its advantage in physical yield over its competitors is reflected in an advantage in financial yield.' The ploughing up of much heavy land west of Cambridge which carried the ridge and furrow indicative of an earlier phase of cultivation, was made possible on the one hand by the system of deficiency payments for cereals and, secondly, by the employment of heavier, crawler tractors financed by forward-looking farmers. The wheat and barley acreages are adjusted from season to season, according to prevailing prices. A final illustration will suffice to emphasize that the study of land use cannot be made without taking into account the traditional restraints exercised in some areas by a land tenure system. It has been argued that the legislation to protect the crofting system, conferring a large measure of security of tenure, has been responsible for the failure to make use of the holdings of absentee tenants. In some townships land is left uncultivated for this reason when in general the size of the holdings of crofters is too small for them to be viable (Scotland, Cmd. 9091, 1954, pp. 40, 41). Such an explanation for uncultivated arable land would be obtainable only by inquiry, and generalizations only by survey or commission procedure. It is heartening to see that some modern surveys, under the auspices of the Geographical Field Group (Moisley, 1961, p. 30), are carrying out surveys of land use supplemented by surveys of farmers (crofters in this case) and their families. What is perhaps also interesting about these surveys is that the Group responsible for their organization is a lineal descendant of the Le Play House student group which had the blessing and encouragement of Geddes (Geographical Field Group, 1962).

One further problem facing those who seek to interpret land-use patterns is not peculiar to British studies. In the attempt to analyse that aspect of pattern which concerns the shape of individual elements of the pattern, there has been little advance beyond the subjective approach indicated by Stamp (1948, pp. 88–89), where complementary patterns of arable land and permanent pasture are examined side by side. The use of words for description puts a strain on the interpreter because of their limited range and the subjective way in which they can be applied. The search for a more consistent method of describing shape in land-use patterns has already begun in America (Latham, 1958 and 1963).

So far British work on land-use survey and mapping has reached a cul-de-sac. Although complete surveys are feasible, we are not much nearer to being able to interpret the patterns thus obtained, so that the relative importance of different causal factors can be weighed.

THE SEARCH FOR SPEED AND EFFICIENCY IN AMERICAN LAND-USE STUDIES

In contrast with British work, it is contended that conditions in America were conducive to the early development of new methods, devised to overcome some of these obstacles and to take land-use studies out of the purely descriptive, or speculative, phase into a phase of greater certainty in the field of explanation. Because it was not tempted by the possibility of producing complete surveys,

American work turned to sampling as a way of accumulating more information about wider areas. Even Finch in his classic work on the Montfort region (1933, p. 9) pointed out that the value of statistics of land use determined from a field-by-field study 'seemed hardly to warrant the expenditure of the time necessary for their preparation. A more ambitious project might recommend the use of Hollworth [*sic*; for Hollerith?] machines or other mechanical devices in a more facile accomplishment of this tedious statistical analysis.' The geologist Trefethen (1936) quickly took this matter up and, using the method developed by Rosiwal (1898) by line traverses in two directions at right angles to each other, was able to produce land-use statistics for 50 to 100 square miles for a week's work. This he contrasted with Finch's 47 square miles in 120 man-days. Thus a technique first used for petrographic analysis was put to use in geography. It was not long before Proudfoot (1942) and Osborne (1942) continued this work. Proudfoot compared the accuracy of traverse line sampling as against area measurement of land-use type by planimeter. Osborne's attention was drawn to the relative efficiency of types of line traverse sample – the stratified random and the systematic, concluding that the latter were more so. Some years later, Steiner (1958) also compared the efficiency of systematically and randomly located area samples, by comparing land-use statistics thus obtained with quantities known from other sources. Wood (1955) has made use of a stratified random sample of areas so that aspects of land use may be mapped in the field over a limited area representative of Eastern Wisconsin. The data for these areas indicated that another complete source of land-use statistics could be used to make a map of the whole area with an ascertainable reliability. Dot maps of different land-use types were thus produced by a combination of sample field mapping and correlation with available statistical data. Anderson (1961) has pointed out that the United States National Inventory of Soil and Water Conservation Needs (1962) has made use of a 2 per cent area sampling rate to speed the field collection of data on soil and land use. In this way, the whole country may be covered by a uniform survey. Arising from the programme of investigation into flood-plain problems being carried out by Chicago University, Berry (1962) has compared the relative efficiency of different sampling systems for accumulating land-use statistics. He concludes that the stratified, systematic, unaligned sample of points at which a record of land use is obtained is the most efficient. Berry, however, makes the interesting comment (1962, p. 14) that improved, speedier methods of field work are not provided by sampling. Nevertheless, this method does provide a quick and unambiguous way of estimating the area of a particular category of land use, it side-steps the problem of the detailed delimitation of the actual area of separate categories of land use, and most important of all, it is versatile in that it facilitates the interpretation of the land-use pattern through a study of spatial associations.

BRITISH STUDIES TAKE UP THE SEARCH

Work in Britain has been principally concerned with the assessment of woodland area by sampling. The Forestry Commission have been carrying out some experiments with different types of random samples in order to establish to within ±2 per cent the area of woodland in different regions of Britain (Locke, 1963). Haggett (1963) sampled woodland cover shown on 1:63,360 Ordnance Survey maps to test the relative efficiency, both from the point of view of accuracy and time taken for obtaining area measurements. Furthermore, the tests were designed to indicate the most appropriate sampling methods for both regional and local levels of sampling. Haggett found that at the regional level, the area of woodland cover could be estimated to within ±1 per cent by using sixteen cells of 5 square kilometres per one 100 km by 100 km square, selected by stratified random sampling. At the local level, 40-line traverse samples were preferred, since they gave a mean error of < ±1 per cent forest cover.

THE INTERPRETATION OF LAND-USE PATTERNS

It has been emphasized that the use of sampling in many cases enables more rapid accumulation of land-use statistics. Over wider areas, where land-use patterns are of coarser grain this may be more important still. Furthermore, as Berry (1962) noticed, sampling designs help in the progression to the next stage, that of explaining the patterns of land use. It is always possible to sample the maps which are the product of a complete survey, in order to test the relationship between aspects of that pattern and factors thought to have influenced it. But unless there is an actual need for the map of land use, for whatever purpose, there seems little point in laboriously collecting and plotting the distribution of land use, if sufficient data can be collected in another way. As yet, there have been comparatively few attempts to explain land-use patterns in a consistent and non-subjective fashion. Wood's study of land use in Wisconsin refers to the inverse relationship between crop land and farm land in trees, which enabled him to predict the area of the first from the second. He comments significantly: 'neither of these two ingredients of the landscape may be expected to cause the other; nevertheless they are closely related' (1955, p. 364). Zobler appears to be one of the first to have suggested a method of relating natural factors to land use, 'the key to the understanding of the geographic adjustment to agricultural resources' (1957, p. 89). Zobler obtained land-use acreages from air photographs in part of New Jersey. The area was divided into physiographic regions and a number of other factors likely to have influenced land use were also noted. Included in the series of χ^2 tests, is one to see whether the acreages of different types of land use differ significantly from purely chance variations as between all possible pairs of physiographic regions. It was possible to see that in some cases an apparent difference in physiography was not reflected in differences in land use. Another series of tests were made to see whether land-use variations occurred with reference to type of soil. Berry (1962) goes further than Zobler by expressing the degree of the association between one land-use category and an aspect of the natural environment. He makes use of frequencies (presence or absence of permanent pasture) obtained by a point sampling design. Relating these frequencies to the presence or absence of flood plain, and employing the coefficient ϕ to express the degree of association, he applies the χ^2 test to check the significance of the relationship. In this case the value of ϕ ranges between 0 and 1. If the coefficient Q (Hagood and Price, 1960, p. 358–70) were to be used, by extending the range of values of the cofficient to -1 through to $+1$, one is able to allow for unexpected relationships (negative values) as well as expected relationships (positive values). Although Berry does not map the occurrences of pasture which do not fall on flood-plain, or vice versa, it would be quite feasible and indeed illuminating to map such so-called anomalies. The field investigator then may well feel that concentrations of exceptions such as these warrant further inspection. In Britain, only Haggett (1964) and Reid (1963) have so far used statistical techniques to interpret land-use patterns. Both used multiple regression in an attempt to assess the relative influence that different factors had on proportions of different kinds of land use. Reid obtained his data by sampling land-use maps. Haggett was examining the factors responsible for the extent of deforestation in part of Brazil. With the results of a second land-use survey now being published in Britain, it is important that the methods used by Haggett and particularly Reid should be widely appreciated. Since high-speed electronic computers are becoming part of the equipment of modern geographical investigations (Coppock, 1962 and 1964) even in Britain, it is not too much to hope that complicated exercises such as these will become commonplace methods of dealing with the multiple explanations of land-use patterns.

FARM SURVEYS AS AN ALTERNATIVE TO LAND-USE SURVEYS

Sampling and statistical analysis have long been employed in another field of agricultural investigation. The economic surveys of farming, normally based upon a sample of farms, is well known both in America and in Britain. It is only relatively recently that such techniques have been applied to geographical investigations of patterns of farm-type. Birch's study of the Isle of Man (1954) is almost the only British example. There are a number of similar studies which have been carried out by agricultural economists in recent years. These include Bennett Jones' study of the pattern of farming in the East Midlands (1954), Mitchell's work on the dairy farms of the Somerset Levels (1962) and Jackson, Barnard and Sturrock's study of the distribution of types of farm in Eastern England (1963). These studies are not fundamentally interested in land use, but in the pattern of the functional units of the agricultural landscape (Birch, 1954, p. 144). Sample surveys of the type developed by agricultural economists may well be used to supplement land-use surveys. They certainly provide a depth of understanding rarely achieved in the more subjective interpretation of land-use patterns and in the use of specimen farms. At least it is possible to say that the data from most sample surveys is representative of a certain area, or class of farms; the selection of case studies, whether specimen parishes, or farms, is frequently conditioned by the availability of data, or the willingness of the farmer to provide information. At the same time, it is difficult to see how the two approaches can be fused into one method of describing and interpreting agricultural patterns. The land-use survey, with its emphasis on area *qua* area is difficult to marry with the farm survey approach, whose emphasis lies on the operating unit. The one is more characteristic of regional geography, the other, of economic geography. Both seem equally valid in their particular fields.

The problem may be illustrated by examining what would happen if a stratified, systematic, unaligned point sample of land use were made of a certain area. Each occurrence of a category of land use could be associated with a continuous distribution of, say, a natural phenomenon, such as relief. It would not be possible to associate these measures of land use with size of farm, sampled on the same basis, because of the strong likelihood of double counting. Some way would have to be found to generalize the information on farm size, and perhaps land use too, so that the two patterns may be associated by virtue of their distributions. Reid (1963) suggests that mapping of sample data of land use by point sampling is less successful than line sampling, and this is also Dahlberg's impression (1963). Line sampling of both land uses would yield a map of proportions of different uses, but line sampling of farms to relate to that distribution would produce some very curious results because of the fragmented character of many farms, quite apart from their irregular shape. What would be considered the area of a farm could best be obtained by inquiry from the operator himself. It would appear that line sampling would not be an efficient way of collecting data on farm size. The most obvious method would be to generalize from a systematic point sample of farms, with the object of producing a map of farm size of a certain reliability; but it would have the disadvantage of over-representing the large farms. In view of these difficulties, one is forced back to the traditional method of sampling farms, from lists of farms or farm operators. Farm size, if mapped on a basis of this sample, would be valid only for the area from which the sample was drawn, but it would be representative of the whole farm population.

There is ample scope here for pilot investigations into the problems of mapping and interpreting sample data over area. Whatever the future has in store in the way of new techniques of mapping and interpretation, there will still be a place for the traditional, complete mapping of land use. The difficulties encountered by Latham (1958, 1963) in an attempt to work out satisfactory measures of shape and orientation for elements in land-use patterns reinforce this view. Although Latham demonstrates that it is possible to express such aspects of land-use patterns, by means of 'uninterrupted distance' measurements along rotated sample traverse lines, such measures are useful only for regional summaries (Latham, 1958, p. 254). There is, as yet, at the detailed local scale, no substitute for the field by field land-use map. But the analysis of such maps has been considerably aided by the development of modern methods.

CONCLUSION

Although it may be fairly claimed by Hunt that land-use survey is 'a means of carrying economic geography into the field' (1953, p. 284), it is more extravagant to insist that, apart from the land-use map, 'there is no other way of showing the distribution of *all* significant activities in their true spatial context' (Hunt, 1953, p. 285). Herein exists the confusion that has arisen between field teaching and field research. Hunt is in fact arguing that land-use survey is of great value in field teaching, but this is not to say that land-use survey has no place in field research. Similarly, it is contended that field teaching is an essential preliminary to field research. No one can expect to embark on field work for a research project unless he has first taken the trouble to understand the significance of the features which are visible in the landscape. As Marsh put it 'Sight is a faculty; seeing, an art' (1864, p. 10). One cannot map river terraces or temporary pasture until one can recognize them. If field research is to be regarded as the preserve of past masters in the art, instruction through field teaching is the means by which one may aspire to membership of the brotherhood. It is only by the exercise to the fullest extent of our powers of intuition, observation and discrimination that we shall achieve a deeper understanding of patterns on the surface of our earth.

References

ANDERSON, J. R., 1961, 'Toward More Effective Methods of obtaining Land Use Data in Geographic Research', *Prof. Geog.*, **13** (6), 15–18.
BALCHIN, W. G. V., 1955, 'Research in Geography' (Swansea), 23 pp.
BARROW, G., 1919, 'Some Future Work for the Geologists' Association', *Proc. Geol. Assn., Lond.*, **30**, 1–48.
BARROW, G. and GREEN, J. F. N., 1921, 'Excursion to Wendover and Buckland Common Near Cholesbury', *Proc. Geol. Assn., Lond.*, **32**, 32–46.
BEAVER, S. H., 1962, 'The Le Play Society and Field Work', *Geog.*, **47**, 226–39.

BERRY, B. J. L., 1962, *Sampling, Coding and Storing Flood Plain Data*, U.S. Department of Agriculture Handbook No. 237 (Washington, D.C.), 27 pp.

BIRCH, J. W., 1954, 'Observations on the Delimitation of Farming Type Regions with Special Reference to the Isle of Man', *Trans. Inst. Brit. Geog.*, Pub. No. **20**, 141–58.

— 1960, 'A Note on the Sample-Farm Survey and Its Use as a Basis for Generalized Mapping', *Econ. Geog.*, **36**, 254–9.

BOARD, C., 1962, *The Border Region: Natural Environment and Land Use in the Eastern Cape* (Cape Town), 238 pp.

BOARDMAN, P., 1944, *Patrick Geddes, Maker of the Future* (North Carolina), 504 pp.

BOURNE, R., 1931, 'Regional Survey and Its Relation to the Stocktaking of the Agricultural and Forest Resources of the British Empire', *Oxf. For. Mem.*, **13**, 169 pp.

BRANFORD, V. V., 1915–16, 'The Regional Survey as a Method of Social Study', *Geog. Teach.*, **8**, 97–102.

BRUNHES, J., 1952, *Human Geography* (translated from the French abridged edition of 1947) (London), 256 pp.

BUCHANAN, R. O., 1952, 'Approach to Economic Geography', *Ind. Geogr. Jour.* (Madras), Silver Jubilee Souvenir Volume, 1–8.

— 1959, 'Some Reflections on Agricultural Geography', *Geog.*, **44**, 1–13.

BUTLER, J. B., 1960, *Profit and Purpose in Farming. A Study of Farms and Smallholdings in Part of the North Riding*, Leeds, Economic Section, Department of Agriculture, Leeds University.

COLEMAN, A. M., 1960, 'A New Land-use Survey of Britain', *Geog. Mag., Lond.*, **33**, 347–54.

COPPOCK, J. T., 1954, 'Land Use Changes in the Chilterns, 1931–51', *Trans. Inst. Brit. Geogr.*, Pub. No. **20**, 113–40.

— 1957, 'The Changing Arable in the Chilterns, 1875–1951', *Geog.*, **42**, 217–29.

— 1960, 'Crop and Livestock Changes in the Chilterns, 1931–51', *Trans. Inst. Brit. Geog.*, Pub. No. **28**, 179–96.

— 1962, 'Electronic Data Processing in Geographical Research', *Prof. Geog.*, **14**(4), 1–14.

— 1964, 'Crop, Livestock, and Enterprise Combinations in England and Wales', *Econ. Geog.*, **40**, 64–81.

DAHLBERG, R. E., 1963, Personal Communication.

DAVIS, C. M., 1954, 'Field Techniques', Ch. 24 in James, P. E., Jones, C. F. and Wright, J. K. (eds), *American Geography: Inventory and Prospect* (Syracuse).

DICKENS, CHARLES, 1890 edn, *Bleak House* (London).

EDWARDS, K. C., 1950, *Land, Area and Region* (Nottingham), 20 pp.

FAGG, C. C., 1915, 'The Regional Survey and the Local History Societies', *South East Nat.*, **20**, 21–30.
— 1928–9, 'The History of the Regional Survey Movement', *South East Nat.*, **33**, 71–94.
FAGG, C. C. and HUTCHINGS, G. E., 1930, *An Introduction to Regional Surveying* (Cambridge), 150 pp.
FAIRGRIEVE, J., 1937 A, 'Can We Teach Geography Better?', *Geog.*, **21**, 1–17.
— 1937 B, *Geography in School* (London), 417 pp.
FINCH, V. C., 1933, 'Geographic Surveying', in Colby, C. C. (ed.), Geographic Surveys, *Bull. Geog. Soc.* Chicago, **9**, xiii, 75 pp.
FLEURE, H. J. and EVANS, E. E. (eds), 1939, *South Carpathian Studies, Roumania*, II, Le Play Society, 60 pp.
FOX, J. W., 1956, *Land Use Survey. General Principles and a New Zealand Example* (Auckland), 46 pp.
GEIKIE, A., 1905, *Landscape in History and Other Essays* (London), 352 pp.
GEOGRAPHICAL FIELD GROUP, 1962, *A Short History of the Geographical Field Group* (Nottingham), 2 pp. (Roneoed).
GULLEY, J. L. M., 1961, 'The Retrospective Approach in Historical Geography', *Erdkunde*, **15**, 306–9.
HAGGETT, P., 1963, 'Regional and Local Components in Land-use Sampling: A Case-study from the Brazilian Triangulo', *Erdkunde*, **17**, 108–14.
— 1964, 'Regional and Local Components in the Distribution of Forested Areas in South-East Brazil: A Multivariate Approach', *Geog. Jour.*, **130**, 365–380.
HAGOOD, M. J. and PRICE, D. O., 1960, *Statistics for Sociologists* (New York), 575 pp.
HARE, F. K., 1947, 'The Geomorphology of a Part of the Middle Thames', *Proc. Geol. Assn., Lond.*, **58**, 294–339.
HODDER, B. W., 1961, 'Rural Periodic Day Markets in Part of Yorubaland, Western Nigeria', *Trans. Inst. Brit. Geog.*, Pub. No. **29**, 149–59.
HUNT, A. J., 1953, 'Land-use Survey as a Training Project', *Geog.*, **38**, 277–86.
HUTCHINGS, G. E., 1960, *Landscape Drawing* (London), 134 pp.
— 1962, 'Geographical Field Teaching', *Geog.*, **47**, 1–14.
INCORPORATED ASSOCIATION OF ASSISTANT MASTERS IN SECONDARY SCHOOLS, 1954, *The Teaching of Geography in Secondary Schools* (London), 512 pp.
JACKSON, B. G., BARNARD, C. S. and STURROCK, F. G., 1963, *The Pattern of Farming in the Eastern Counties. A report on Classification of Farms in eastern England.* Occasional Papers, Farm Economics Branch, School of Agriculture, Cambridge University, No. 8, 60 pp.

JONES, R. BENNETT, 1954, *The Pattern of Farming in the East Midlands* (Sutton Bonington), University of Nottingham School of Agriculture, Department of Agricultural Economics, 176 pp.

JONES, W. D. and FINCH, V. C., 1925, 'Detailed Field Mapping in the Study of the Economic Geography of an Agricultural Area', *Ann. Assn. Amer. Geog.*, **15**, 148–57.

JONES, W. D. and SAUER, C. O., 1915, 'Outline for Field Work in Geography', *Bull. Amer. Geog. Soc.*, **47**, 520–5.

KIRK, W., 1952, 'Historical Geography and the Concept of the Behavioural Environment', *Ind. Geogr. Jour.* (Madras), Silver Jubilee Souvenir Volume, 152–60.

LATHAM, J. P., 1958, *The Distance Relations and Some other Characteristics of Cropland Areas in Pennyslvania: An Experiment in Methodology for Empirically Analyzing, Regionalizing and Describing Complexly-Distributed Areal Phenomena*, Philadelphia University of Pennsylvania, Wharton School of Finance and Commerce. Technical Report No. 4, NR No. 389–055, Office of Naval Research.

— 1963, 'Methodology for an Instrumented Geographic Analysis', *Ann. Assn. Amer. Geog.*, **53**, 194–211.

LE PLAY, P. G. F., 1855, *Les Ourvriers Européens* (Paris), 301 pp.

LINTON, D. L., 1946, *Discovery, Education and Research* (Sheffield), 17 pp.

— 1955, 'The Problem of Tors', *Geog. Jour.*, **121**, 470–87.

LOCKE, G. M. L., 1963, Personal Communication.

MᶜCANN, S. B., 1963, 'The Late Glacial Raised Beaches and Re-Advance Moraines of the Loch Carron Area, Ross-shire', *Scot. Geog. Mag.*, **79**, 164–9.

MAITLAND, F. W., 1897, *Domesday Book and Beyond* (Cambridge), 527 pp.

MARSH, G. P., 1864, *Man and Nature; or Physical Geography as modified by Human Action* (London), 560 pp.

MEAD, W. M., 1954, 'Ridge and Furrow in Buckinghamshire', *Geog. Jour.*, **120**, 34–42.

MITCHELL, G. F. C., 1962, 'The Central Somerset Lowlands: the Importance and Availability of Alternative Enterprises in a Predominantly Dairying District', *Selected Papers in Agricultural Economics* (Bristol), **7**, 295–453.

MOISLEY, H. A., 1961, *Uig, a Hebridean Parish*; Parts I and II (Nottingham), Geographical Field Group, iv, 55 pp.

MONKHOUSE, F. J., 1955, *The Concept and Content of Modern Geography* (Southampton), 31 pp.

OSBORNE, J. G., 1942, 'Sampling Errors of Systematic and Random Surveys of Cover-type Areas', *Jour. Amer. Statist. Assn.*, **37**, 256–64.

PALMER, J., 1956, 'Tor Formation at the Bridestones in North-east Yorkshire, and Its Significance in Relation to Problems of Valley-side Development and Regional Glaciation', *Trans. Inst. Brit. Geogr.*, Pub. No. **22**, 55–71.

PALMER, J. and NEILSON, R. A., 1962, 'The Origin of Granite Tors on Dartmoor, Devonshire', *Proc. Yorks. Geol. (Polyt.) Soc.*, **33**, 315–40.

PEPLER, G., 1925, 'Regional Survey as a Preliminary to Town Planning', *South East Nat.*, **30**, 81–89.

PLATT, R. S., 1942, *Latin America: Countrysides and United Regions* (New York), 564 pp.

— 1959, *Field Study in American Geography, The Development of Theory and Method exemplified by Selections*, Chicago, University of Chicago, Department of Geography, Research Papers No. 61, 405 pp.

PROUDFOOT, M. J., 1942, 'Sampling with Transverse Traverse Lines', *Jour. Amer. Statist. Assn.*, **26**, 265–70.

PYE, N., 1955, *Object and Method in Geographical Studies* (Leicester), 19 pp.

REID, I. D., 1963, *The Application of Statistical Sampling in Geographical Studies* (Liverpool University, M.A. thesis), 112 pp.

ROBERSON, B. S. and LONG, M., 1956, 'Sample Studies: The Development of a Method', *Geog.*, **41**, 248–59.

ROSIWAL, A., 1898, 'Ueber geometrische Gesteinen analysen. Ein einfacher Weg zur ziffermassigen Feststellung des Quantitatsverhaltnisses der Mineralbestandtheile gemegter Gesteine', *Verh. geol. Reichs Anst. Wien*, 1898 (5 & 6), 143–175.

RUSSELL, E. J., 1925–6 (1926), 'Regional Surveys and Scientific Societies', *Geographical Teacher*, **13**, 439–47.

SAUER, C. O., 1917, 'Proposal of an Agricultural Survey on a Geographical Basis', *19th Annual Report of the Michigan Academy of Science*, pp. 79–86.

— 1919, 'Mapping the Utilization of the Land', *Geog. Rev.*, **8**, 47–54.

— 1924, 'The Survey Method in Geography and Its Objectives', *Ann. Assn. Amer. Geog.*, **14**, 17–33.

— 1925, 'The Morphology of Landscape', *Univ. Calif. Pub. Geog.*, **2** (2), 19–53.

— 1941, 'Foreword to Historical Geography', *Ann. Assn. Amer. Geog.*, **31**, 1–24.

— 1956, 'The Education of a Geographer', *Ann. Assn. Amer. Geog.*, **46**, 287–99.

SCOTLAND, 1954, *Report of a Commission of Enquiry into Crofting Conditions* (Cmd. 9091), 100 pp.

SPARKS, B. W., 1949, 'The Denudation Chronology of the Dip-slope of the South Downs', *Proc. Geol. Assn., Lond.*, **60**, 165–215.

STAMP, L. D., 1948, *The Land of Britain: Its Use and Misuse* (London), 507 pp.

— 1960, *Applied Geography* (Harmondsworth, Penguin Books, Ltd.), 208 pp.

STAPLEDON, R. G., 1935, *The Land Now and Tomorrow* (London), 323 pp.

STEINER, R., 1958, 'Some Sampling of Rural Land Uses' (abstract), *Ann. Assn. Geog.*, **48**, 290.

STEVENS, A. J., 1959, 'Surfaces, Soils and Land Use in North-east Hampshire', *Trans. Inst. Brit. Geogr.*, Pub. No. **26**, 51–66.

STRAHLER, A. N., 1950, 'Davis' Concepts of Slope Development Viewed in the Light of Recent Quantitative Investigations', *Ann. Assn. Amer. Geog.*, **40**, 209–13.

TREFETHEN, J. M., 1936, 'A Method for Geographic Surveying', *Amer. Jour. Sci.*, **32**, 454–64.

UNITED STATES OF AMERICA, DEPARTMENT OF AGRICULTURE, 1962, *Agricultural Land Resources: Capabilities; Uses; Conservation Needs.* Agricultural Information Bulletin, No. 263 (Washington, D.C.), 30 pp.

WATERS, R. S., 1958, 'Morphological Mapping', *Geog.*, **43**, 10–17.

WHITTINGTON, G., 1962, 'The Distribution of Strip Lynchets', *Trans. Inst. Brit. Geog.*, Pub. No. **31**, 115–30.

WHITTLESEY, D. S., 1925, 'Field Maps for the Geography of an Agricultural Area', *Ann. Assn. Amer. Geog.*, **15**, 187–91.

WILLATTS, E. C., 1937, *Middlesex and the London Region*, The Report of the Land Utilisation Survey of Britain, Part 79, 117–304.

— 1951, 'Some Principles of Land Use Planning', in *London Essays in Geography* (Rodwell Jones Memorial Volume), eds: Stamp, L. D. and Wooldridge, S. W., 289–302.

WILSON, E., 1964, Personal Communication.

WISE, M. J., 1957, 'The Role of Field Work in the University Teaching of Geography', *Journal for Geography* (Stellenbosch), **1**(1), 17–23.

WOOD, W. F., 1955, 'Use of Stratified Random Samples in a Land-Use Survey', *Ann. Assn. Amer. Geog.*, **45**, 350–67.

WOOLDRIDGE, S. W., 1927, 'The Pliocene History of the London Basin', *Proc. Geol. Assn., Lond.*, **38**, 49–132.

— 1945, *The North Riding of Yorkshire*; The Report of the Land Utilisation Survey of Britain, Part 51, 351–417.

— 1948 A, 'The Role and Relations of Geomorphology' (Inaugural lecture at King's College, London), reprinted in Stamp, L. D. and Wooldridge, S.W. (eds), 1951, *London Essays in Geography*, 19–31.

— 1948 B, *The Spirit and Significance of Field Work*; Address at the Annual Meeting of the Council for the Promotion of Field Studies, 8 pp.

— 1949 A, 'On Taking the Ge- out of Geography', *Geog.*, **34**, 9–18.

— 1949 B, 'The Weald and the Field Sciences', *Adv. Sci.*, **6**(21), 3–11.

WOOLDRIDGE, S. W., 1950, 'Reflections on Regional Geography in Teaching and Research', *Trans. Inst. Brit. Geog.*, Pub. No. **16**, 1–11.

— 1955, 'The Status of Geography and the Role of Field Work', *Geog.*, **40**, 73–83.

— 1960, *Field Studies Council: Retrospect and Prospect*; Address at the Annual meeting of the Field Studies Council, 4 pp.

WOOLDRIDGE, S. W. and EAST, W. G., 1958, *The Spirit and Purpose of Geography*, 2nd Edn (London), 186 pp.

WOOLDRIDGE, S. W. and GOLDRING, F., 1953, *The Weald* (London), 276 pp.

WOOLDRIDGE, S. W. and HUTCHINGS, G. E., 1957, *London's Countryside Geographical Field Work for Students and Teachers of Geography* (London), 223 pp.

ZOBLER, L., 1957, 'Statistical Testing of Regional Boundaries', *Ann. Assn. Amer. Geog.*, **47**, 83–95.

— 1958, 'The Distinction between Relative and Absolute Frequencies in using Chi Square for Regional Analysis', *Ann. Assn. Amer. Geog.*, **48**, 456–7.

CHAPTER ELEVEN

Field Work in Urban Areas

M. P. COLLINS

Lecturer in Town Planning, University College London

The existence of urban settlements can be traced back to the fourth millennium B.C. when they played an important part in facilitating advances in learning, the arts, technology and social organization by providing a relatively secure and economically stable environment. The emergent cultures became localized in the towns, and in time nations became known by the splendour and achievements of their cities. By virtue of their function cities have seldom remained static in either size or form for very long and now they have grown to such an extent that the contiguous built-up area constitutes a distinct region in its own right. The advent of the city region has been accompanied, however, by a host of problems which have still to be resolved. London, for example, continues to expand and attract both industry and population from other parts of the country despite the timely warning contained in the Barlow Report, and in this context it is very disturbing to learn that the metropolis is still committed to developments which will provide employment for an additional 400,000 office workers (Hookway *et al.*, 1963). Whilst the regional and national consequences of this trend have still to be evaluated, it is probable that over two-thirds of the country's population will be living in the South of England by the year A.D. 2000 (Childs, 1962).

TOWN CLASSIFICATION

Despite the urgency and complexity of these problems very little more is known about the forces which control the present-day composition, morphology and functions of our cities than is known about such ancient cities as Mohenjo-Daro and Babylon. In the field of urban studies little progress has been made in establishing definitive techniques, let alone laws, which stand the acid test of universal application. Most of the present day studies of towns and cities tend to be descriptive rather than interpretative and fail to explain the dynamics of urban evolution. Even the more sophisticated approach of the *Centre for Urban Studies* fails to come to grips with the subject when examining the character and composition of 157 urban areas (Moser and Scott, 1961). This study examines the urban areas with a population of over 50,000 inhabitants in 1951 and employs fifty-seven separate indices to establish the basic character of each urban area; such as the size and changing structure of the population, the number of households, the type and condition of housing, the social class, health, education and voting habits. This material is compared, analysed, and classified mathematically, using correlation coefficients to measure the closeness of the relationship between any given pair of variables (there being 1,596 possible pairings). Many of the relationships seem to be straightforward and predictable, whilst others are apparently irrational. Having completed this basic correlation, the results are summarized by means of four new indices which describe the characteristics of each urban area almost as well as the original fifty-seven factors. This condensation involves the use of component analysis to calculate the four components which account for at least 60 per cent of the basic differences between the 157 urban areas. Component one reflects differences in social class; component two the differences in the amount of intercensal population change; component three the differences in post-1951 developments; and component four the housing conditions. The end-product is a classification which distinguishes three main categories of urban areas: viz.

Group A: Comprised of resorts, administrative and commercial towns;
Group B: Comprised mainly of industrial towns; and
Group C: Comprised of suburbs and suburban type towns.

The use of these statistical techniques to measure the degree of association between observable phenomena removes the subjective element from the task of analysis and synthesis, but it is important to remember that they only establish the intensity of the association without attempting to define or explain it (Chadwick, 1961). In other words, a causal relationship can only be inferred with varying degrees of probability depending on the calculated value of the correlation coefficient. Viewed in this light it is difficult to subscribe to the present enthusiasm for techniques which regard 'association' as synonymous with 'causality'.

LAND USE IN URBAN AREAS

Although geographers have made laudable attempts to examine the use of land within the United Kingdom (Stamp, 1950; Best and Coppock, 1962) there has been no similar effort to examine the detailed pattern of land use within the built-up areas of towns and cities. Most of the present detailed land-use studies of urban areas derive from the Town and Country Planning Act of 1947 (Section 5, sub-section 1), which required local planning authorities to undertake a detailed survey of their respective administrative areas and keep it up to date. The Ministry of Town and Country Planning advised the local planning authorities as to the best method of conducting the survey, classifying the land uses and of presenting the results in map form (Ministry of Town and Country Planning, 1949 A). The land-use surveys which have finally emerged, however, vary considerably and do not submit readily to comparison due to differences in presentation, content and even sometimes in the basic classification. As the land-use survey constitutes a vital part in any urban study it is essential that the classification should be sufficiently broad to give a valid picture of the distribution of functional elements within the built-up area. This information should be plotted on a series of large scale maps, preferably at a scale of 1/1250, with an accompanying schedule of explanatory data, noting the date at which the survey was conducted. The following suggested classification indicates the main categories of land use which should be distinguished, together with points of detail, whilst conducting the actual survey.

1. Residential Use: This includes hotels, boarding houses, residential clubs, hostels, welfare homes and nursing homes. It is important to note the type of dwelling (i.e. house, maisonette or flat), the number of storeys, whether the building is occupied or vacant, and to make a subjective assessment of the physical condition of the building. When this material has been analysed it can be amplified further by examining the information contained in the electoral rolls, which will confirm whether the dwellings are in multi-occupation and also provide the basis for establishing the probable population of any particular area or district. These population estimates are arrived at by multiplying the figures obtained from the electoral rolls by a conversion factor which represents the ratio between the total population and the total registered electors for the same administrative area. By checking the yearly incidence of surnames it is possible to assess the

'neighbourhood potential' of the area in so far as a stable population infers the existence of a defined sense of place and strong local allegiances.

2. Open Space: This use can be classified further as follows:

(*a*) Public open space, i.e. parks, recreation grounds, commons and heathlands which are freely available for the enjoyment of the public at large. The number of games pitches, the number and capacity of the play apparatus, together with an estimate of the number of persons actually using these facilities should also be noted. The provision of public open space can be measured in terms of acres per 1,000 persons as is the case in London, for example, where the London County Council has prepared an open space deficiency map which indicates both the lack and mal-distribution of public open spaces (London County Council, 1951 and 1961).

(*b*) Private open space, i.e. allotments, cemeteries, woodlands, golf-courses, sports-grounds, school playing fields and the grounds of private institutions.

3. Public Buildings: These can be classified further as follows:

(*a*) Places of assembly, i.e. museums, art galleries, churches, exhibition halls, chapels, missions and other halls, sunday schools, cinemas, theatres, concert halls, music halls and dance halls, non-residential clubs, gymnasia, skating rinks, small exhibition and amusement arcades, and fun fairs.

(*b*) Other public buildings, i.e. libraries, baths, clinics, dispensaries, police stations, courts of law, and fire brigade stations.

(*c*) Institutions, i.e. hospitals, health and social centres, and schools.

4. Industry: This primary use can be sub-divided into the twenty-four standard categories which form the basis of the Industrial Census Tables (Central Statistical Office, 1958). Later these industries can be grouped into three major categories which reflect their probable impact upon the environment of the surrounding residential areas, e.g.

(*a*) Light Industry, viz., buildings in which the machinery and processes carried on are such as could continue in any residential area without detriment to amenity, having regard to noise, vibration, fumes, smoke, ash, dust or grit.

(*b*) General Industry, viz. buildings in which the machinery and processes carried on are considered to be detrimental to residential amenity having regard to noise, vibration, fumes, etc.

(*c*) Special Industry, viz. buildings in which the machinery and processes carried on are considered to be obnoxious because of poisonous fumes, unpleasant odours, the need to treat products in an offensive condition with consequent health hazards to neighbouring dwellings (Ministry of Housing and Local Government, 1963 A).

Wherever possible an attempt should be made to interview the factory manager in order to ascertain the following additional information:

(*a*) When was the factory founded?

(*b*) If the factory was not founded on this site why did it move here?

(*c*) What are the assets and drawbacks of the present site?

(*d*) Where do the raw materials come from and how are they transported?

(*e*) What is the nature of the finished product, where is it marketed and how is it transported there?

(*f*) How many people does the factory employ, where do they live and how do they travel to work?

(*g*) Is the factory tied to the local area or region?

5. Commerce: This use includes buildings used as warehouses, wharves, depositaries, stores, garages, builders' yards, timber stores, contractors' yards (including local authority and Ministry of Works Depots), post office sorting and parcels offices, grain silos, etc. As in the case of the industrial uses, additional information should be solicited by interviews, in an attempt to establish the relationship between site and function with its consequential impact upon the surrounding region.

6. Offices: This use includes buildings used primarily for administrative purposes including banks, post offices, central and local government offices, buildings occupied by registrars and other public and quasi-public servants. Local offices providing professional services but still employing general clerical workers, e.g. solicitors, accountants, estate agents, travel agents, etc., should all be included in this category. Information should be solicited by interviews to establish the number of persons employed in the building, their place of residence and method of travelling to work.

7. Shops: In order that a proper study can be made of the pattern of retail trading it will be necessary to study this particular use in detail, and the following points should be noted:

(*a*) The total number of shops.

(*b*) The type of shop, e.g. foodstuffs, household goods, luxury goods, clothing, and establishments providing personal services, noting any significant variations in the provision of these facilities in centres of varying sizes.

(*c*) The number of department stores, multiple and chain stores, etc.

8. Statutory Undertakers: Including the basic utilities, viz. Gas, Electricity and Water, the railways, and the land under the jurisdiction of various government departments.

9. Vacant and derelict buildings, cleared land, etc.

When this survey material has been collected and presented in a series of analytical maps which highlight the more salient patterns of incidence and distribution, it is worth remembering that there is an accepted colour notation for these main land-use categories, e.g. residential use is indicated by a light red-brown, public open space by green, private open space by yellow-green, public buildings by red, industry by purple, commerce by grey, offices by light blue, shops by dark blue, and vacant land by yellow. There is no standard notation for statutory undertakers, although Gas and Electricity Undertakings are usually coloured purple, railways grey, and the service departments by a red verge around the boundary of the land in question. This basic land-use survey can be carried out in greater detail depending upon the size of the built-up area and the time and number of persons available. In the Central Business Area of larger towns it will be found that the buildings commonly encompass a wide variety of these land-use categories, necessitating the use of a letter subnotation to augment the basic colour notation. Various combinations have been tried by local authorities with only a limited degree of success and this problem presents a challenge for all who are interested in cartography.

This land-use survey will establish the geographical complexities of the urban way of life, by indicating the daily journeys which have to be made before the inhabitants can find employment, shop, and utilize the educational, social, recreational, cultural and commercial facilities. Further investigation will be needed, however, before it is possible to attempt to delimit the neighbourhood and communal structure of the built-up area. Although the geographer has demonstrated his ability to analyse the constituent elements which comprise a particular area, he has not evidenced a similar proficiency in the more complex art of synthesis with a view to establishing the criteria for delimiting urban, let alone geographical, regions. Whether this failure is due to the untenability of the concept of the region or the paucity of techniques is open to debate, but it cannot be denied that even in the more restricted field of urban studies only limited progress has been made in defining the catchment areas of neighbouring service centres.

COMMUNITIES AND URBAN FIELDS

Reference has been made to the existence of a neighbourhood and community structure within the build-up area of urban settlements, and in this context the term 'community' refers to a primary 'face to face collectivity' which is comprised of an admixture of formal and informal social groups. Occasionally it may possess a civic identity which is distinct from that of its component social groupings, but in essence it describes a mass of individuals who live in a conscious though undefined physical association (Anderson, 1960). The precise nature of these human relationships is difficult if not impossible to define. It varies from town to town, but there is little doubt that its intensity is inversely proportional to the actual size of the town in question. The inhabitants of a large town are interdependent in all sorts of specialized and external ways, yet often they lack the personal intimacy and emotional security which springs from a central unifying urban value or sense of place. The focal point of the urban community is usually the town centre with its concentration of shops, entertainment facilities, restaurants, cultural facilities, and professional and commercial services. Nearly all towns possess only one such centre and its sphere of influence encompasses the whole of the built-up area, which constitutes the urban community in place. The urban field concept rightly stresses the importance of the town as the seat for centralized services, and attempts to delimit its sphere of influence by superimposing a series of generic regions which reflect particular functions of the town (Smailes, 1944). Needless to say the results are cartographically disastrous and provide only an indication of the 'zones of divided loyalty' where the inhabitants can choose between equally accessible and hence rival town centres. This procedure cannot readily be applied in the larger cities such as London where there are numerous urban communities with their respective regional service centres. In these larger cities the distribution of specialist institutions such as secondary and grammar schools, hospitals, technical college libraries, etc., is sporadic in appearance, for it is dictated by the differing tributary populations that are needed to sustain each type of institution. Inevitably these institutions serve metropolitan rather than regional needs and their catchment areas tend to cut right across the urban community structure as expressed in terms of tribute to the regional service centre. In other words the generic or functional regions cannot be superimposed as a means of delimiting the metropolitan community structure, for they are not related to a common point of urban nodality. But perhaps the greatest defect of the urban field concept is its failure to take account of the quantitative use of central area facilities, despite the fact that this is, perhaps, a more sensitive index for assessing the regional importance of rival town centres (Herbert, 1961).

THE NEIGHBOURHOOD STRUCTURE

As soon as an attempt is made to define the urban community in spatial terms it usually becomes necessary to do so on a neighbourhood basis, for the 'neighbourhood' usually has more precise physical associations (Forshaw and Abercrombie, 1943). By definition the neighbourhood concept implies an intensity of social communion which results from the close juxtaposition of a group of families or individuals within a prescribed physical setting. To some extent it reflects the historical evolution of the urban area, for the resultant morphology tended to create relatively self-contained residential enclaves. Attempts to define the neighbourhood in terms of residential proximity are useful for ensuring that residential services and amenities are properly distributed, as is usually the case in the 'New Towns'. However, it does not always fit the geographical realities of urban friendships now that people tend to be drawn together by common interests rather than by the fact that they live in the same street or district.

In most towns the neighbourhood structure can be delimited in terms of the existing land-use pattern, for it highlights the potential physical barriers within the built-up area. Railway viaducts and cuttings, canals with their associated belts of industry and commerce, main roads and the grounds of large institutions all tend to inhibit movement and create inward-looking and relatively self-contained residential enclaves. Usually these 'enclaves' possess a local shopping complex comprised of half a dozen food shops, a chemist, a draper, an ironmonger, a hairdresser, a shoe repairer, a branch post office, and sometimes a bank and a branch library. These local sub-centres are an essential complement to the regional service centre and the acquisition and further intensification of commercial services in no way enables them to compete with the main centre during the week-end period.

A cursory examination of the residential areas in any large town often fails to reveal the identity of any particular locality, and this is most certainly the case in London where the residential areas have tended to assume the identity of the nearest service centre. In those areas where this process of 'urban adoption' has proved impracticable, the task was accomplished by the Postmaster General who used district numbers. Needless to say, these areas possess a diversity of character and often a very real sense of place for the inhabitants who were born there, but the uniformity of housing types and layouts (often the result of bye-law control) present a formidable challenge to newcomers who attempt to establish a local 'sense of place'. This term has been used frequently and it implies something which is experienced, almost subconsciously, a feeling of security which is engendered by the familiarity of everyday surroundings. It usually implies a degree of kinship which springs from having shared the same experiences – perhaps from childhood to old age. These experiences are set against the back-cloth of buildings, streets, trees, parks and, most important of all, of friends and relations. Each building and element in the urban scene acquires a subjective character which cannot be related to its aesthetic qualities, until gradually the whole area is seen in this light by its inhabitants. This process marks the emergence of the neighbourhood in place, and its formal name invokes memories, nostalgia and a sense of kinship and belonging. Sometimes, however, an area can assume a character or atmosphere which is not obvious to its actual inhabitants, as in Soho where the residents are more conscious of the squalor and poor housing conditions than its 'cosmopolitan or continental' character (*The Times*, 1963).

SERVICE CENTRES

As stated previously, the residential areas have often tended to establish their identity in terms of the nearest service centre, which is the most significant place of assembly. The service centre is very susceptible to changes in the economic potential of the surrounding area, and redevelopment at higher net residential densities has led to a renewal of prosperity in centres which were previously deemed to be decaying. This is especially true in London despite the fact that there is no causal relationship between many of the suburbs and the service centres which supply their daily needs. Indeed, most of the suburban areas are dependent upon the concentration of economic opportunity within the central area, and have only utilized the historic pattern of settlement as a convenient framework for siting additional commercial services and urban amenities. Consequently the historic townships in Greater London have acquired large populations, within a relatively short period of time, without developing the economic facilities to support them. For five days in each week these historic townships play a minor role in the economic life of the community, and it is only on Saturdays that they assume their rightful role as regional centres.

The regional function of the service centres presents an absorbing field for investigation and their geographical distribution, size, composition, functions and probable catchment areas are already being studied in some detail (Carruthers, 1957 and 1962; Collins, 1960 A and B; Lomas, 1964; Parker, 1962; Smailes and Hartley, 1961; Manchester University, 1964). It has been established that a hierarchy of service centres exists within most regions, but the criteria which have been used to rank these centres, viz. the provision of specified facilities within the central area together with an assessment of their utilization, as reflected by rateable values and the nodality of bus services, is inadequate and the actual rankings which result are questionable (Carruthers, 1962). The subtleties of this particular classification were not supported by a series of survey visits which indicated that there is a marked degree of similarity in the composition of service centres in South London, and that the differences were quantitative rather than qualitative (Collins, 1960 A and B). The distribution of such stores as Woolworth's, Marks and Spencer's, Littlewood's, British Home Stores, Times Furnishings, Dolcis, Saxone's, Sainsbury's, Boots the Chemists, together with that of cinemas, shows a remarkable consistency of incidence within the service centres in South London.

The land-use survey will provide much useful information about the composition of such service centres and the variations in the basic morphology will provide some indication of the probable regional status of any particular centre. This material can be considered under the following headings:

1. The number of shops which comprise the service centre.

2. The number of shops which specialize in the five major categories, viz. foodstuffs, clothing, household goods, luxury goods and commercial services such as hairdressing, cleaning, etc.

3. The number of department, variety, chain and multiple stores.

4. The number of establishments providing professional services such as solicitors, accountants, estate agents, etc.

5. The number of banks.

6. The number of local and national government offices.

7. The number of commercial offices.

8. The number of places of entertainment.

More detailed information can be obtained by measuring the actual allocation of floor space between competing land uses. These figures can be ascertained by taking measurements from the 1/1250 ordnance survey sheets which are quite adequate for this purpose, providing that a fairly detailed field survey has been carried out and, they will serve as a guide to the probable employment potential of the service centre. The rating lists also provide some indication of the regional importance of the centre in as much as the current rateable value reflects the amount of capital which has been invested in the centre, and by inference its importance as seen through the eyes of entrepreneurs.

I have attempted to carry out this type of study for South London, and it revealed the existence of a hierarchy of service centres which is essentially complementary despite the fact that the larger centres have attracted most of the major redevelopment schemes. So far four categories of service centres have been distinguished for the convenience of presentation, and this somewhat arbitrary classification is based solely on the provision of central area facilities with no assessment of their use. The first category is comprised of 9 centres with over 300 shops (Figure 11.1A), ranging from 301 shops in Bromley to 425 shops in Croydon. Even in this category over 70 per cent of the total shops are concerned with satisfying the basic needs of the resident population. The accessibility of these 9 centres has been assessed in terms of bus services and it would appear that most residents of South London can reach one of these centres after completing a twenty-minute journey; although this criteria is of less significance within the County of London due to the closer spacing of the centres. The second category is comprised of 12 centres with between 200 to 300 shops (Fig. 11.1B), ranging from 215 shops in Bexleyheath to 270 shops in Deptford. The centres in the outer

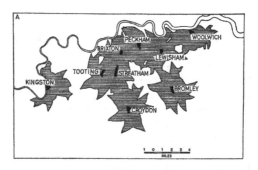

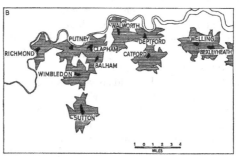

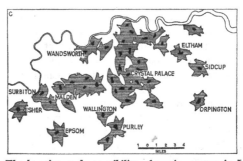

FIG. 11.1. *The location and accessibility of service centres in London south of the Thames.* A: *Centres with 300 or more shops. Shaded areas show zones within the twenty-minute isochrone (as determined by bus, walking, or a combination of both).* B: *Centres with 200 to 300 shops. Shaded areas show zones within the fifteen-minute isochrone.* C: *Centres with 100 to 200 shops. Shaded areas show zones within the ten-minute isochrone.*

suburbs exert a considerable regional influence (e.g. Dartford, Richmond, Sutton and Wimbledon), whereas those nearer to Central London tend to serve local rather than regional needs (e.g. Balham, Catford and Deptford). The range of shops, services and entertainment facilities mirrors that found in the larger centres, and most of South London can reach one of these centres after completing a fifteen-minute journey. The third category is comprised of 39 centres with between 100 to 200 shops (Figure 11.1C), ranging from 102 shops in Malden to 195 shops in Crystal Palace. It was noted that these centres tended to attract fewer weekend shoppers, despite the fact that their basic composition mirrors that of the centres within the second category. The number of shops engaged in the sale of food-stuffs, however, tends to dominate the character of these centres and often amounts to 50 per cent of their total composition. Most of South London can reach one of these centres after completing a ten-minute journey. The fourth category is comprised of 23 centres with less than 100 shops, ranging from 45 shops in Carshalton to 98 shops in Petts Wood. Most of these centres are located in the outer suburbs, and with one exception they serve purely local needs. The exception is the Elephant and Castle which is in course of being redeveloped as a regional centre with 200,000 square feet of new shops and nearly 700,000 square feet of new office accommodation; and it may curb the regional aspirations of neighbouring centres.

As is to be expected, the multiple and variety chain stores have paid considerable attention to the probable catchment population of neighbouring centres in South London before deciding to locate a new branch store. The following table indicates the estimated population requirements of several well-known firms:

	catchment population	
Boots the Chemists	10,000	
Mac Fisheries	25,000	
Barratts Ltd. (Shoes)	20–30,000	
Sainsbury's (Groceries)	60,000	(For a medium-sized self-service store)
Marks and Spencer's	50–100,000	
John Lewis	50,000	(For a super-market)
John Lewis	100,000	(For a department store)

These figures do not tell the whole story, however, and other important factors such as the presence of a department store, the social and economic character of the surrounding area, the levels of rentals, the availability of central sites that are ripe for redevelopment, and the problems of integrating the new branch into the firm's existing administrative and supply systems, are all taken into account before the final decision is taken. The exigencies of retail competition are so severe, on the other hand, that rival combines cannot afford not to be represented in the larger service centres and this accounts, in part, for the similarities in basic composition which have been noted already.

When examining the composition of neighbouring service centres it is important to take account of the current planning policies which govern their expansion and redevelopment, for often these explain variations in the regional distribution of shops, offices and other commercial services. An attempt should also be made to investigate the dynamics of redevelopment and establish the regional potential of rival towns as revealed, for example, by the number of applications seeking planning permission to redevelop land in the central area. Each local planning authority is required to keep a register of these applications, and these statutory registers are always available for public inspection.

The failure to define quantitatively the use of central area facilities is an outstanding weakness in the present studies of service centres, for potentially this could be a significant yardstick for measuring the regional importance of any town. The Census of Distribution gives some indication of the regional importance of the shopping centre of the town, in as much as it provides the basis for estimating the number of persons who visit the town in order to patronize the shopping facilities. These estimates are based upon the average sales per head of the resident population, which establishes the percentage of the total sales that is derived from the resident population, whilst the balance comprises the percentage which derives from the surrounding catchment area (Waide, 1963).

The catchment areas of the service centres within the metropolitan built-up area are difficult to define in precise terms, moreover, due to the increasing mobility of the resident population. Admittedly the morphology of the metropolis tends to delimit areas which look inwards for services, but the main roads now constitute the arteries of the circulation pattern. Public transportation still plays an important part in determining the probable catchment area of any one centre, and isochrones can be drawn to establish the zones of divided loyalty where personal preference rather than accessibility dictates the choice of centre. These zones require further detailed study and the residents should be questioned as to their shopping habits. This affords an opportunity to establish whether these zones possess a geographical 'sense of place'. The residents should also be questioned as to the name of the district in which they live, noting whether the discrepancies are due to the fact that the persons concerned tend to patronize different service centres. When the results of the questionnaire are plotted in map form it should be possible to define the effective catchment areas of the respective service centres. It is worth noting in this context that the local newspapers tend to confuse the issue due to the overlapping of circulation areas. Local news items and private advertisements do, however, provide some indication of the community structure, but the commercial advertisements of retail stores tend to reflect their regional *aspirations* rather than the geographical *realities* of their catchment areas. This also holds true for the retail delivery areas, due to the exigencies of modern competition.

In essence these catchment areas are zones of convenience within which the residents will probably establish their geographical sense of place in terms of the nearest service centre. However, this does not necessarily engender the urban community, as is the case in most towns, due to the close proximity of rival centres. The distribution of specialist urban institutions such as libraries, grammar and secondary schools, technical colleges and welfare centres, which all play their part in furthering the creation of communal loyalties, does not complement the distribution of service centres. This lack of unity in the functional framework of the metropolis reflects its monolithic character and has undermined the historic community structure. Exceptions do occur and usually they are to be found in areas where the 'village character' of the original settlement has been preserved. Some examples of this phenomenon can be seen in London (e.g. Dulwich, Highgate, and Hampstead), but they are noteworthy only in indicating the barren nature of the surrounding built-up area.

JOURNEY TO WORK

Having examined the basic morphology of the built-up area and the journeys which have to be made in order that the residents can shop, and utilize the social, cultural and entertainment facilities located therein, the next step is to examine the daily journey to work which dominates the life of every urban community. The census on *Usual Residence and Workplace* provides quantitative information as

Q

to where people travel in order to seek employment, but it does not indicate the means of effecting the journeys or the type of employment that is being sought. The *Industrial and Occupational Tables of the Census* will provide rudimentary data regarding the availability and demand for specified categories of employment, but it is difficult to establish a precise correlation which accounts in full for the large number of journeys that take place daily. Most of the statistical information relates to local government areas which no longer reflect the geographical realities of the urban community structure, with the result that little is certain about the journey-to-work pattern, although much can be inferred. In the case of South London, for example, certain movements are of particular interest and indicate the 'uneven' distribution of industry, commerce, offices, etc. In 1951 over 327,000 travelled daily into the three central boroughs of the City, Holborn and Westminster in order to seek employment. This movement accounted for over one-third of the total journeys which originated within South London. In six of the suburban boroughs (e.g. Banstead, Beckenham and Bromley) over 40 per cent of the persons engaged in the daily journey to work travel into these three central boroughs. Only in Crayford, Dartford and Erith does the percentage of persons thus engaged fall below 20 per cent. This is due here presumably to the predominance of heavy and more specialized industries such as paper-making, electrical engineering, chemicals and munitions which have attracted a resident labour force as a result of the manifest difficulties of travelling across South London. The outer suburban belt is especially dependent upon the employment facilities located within the County of London. In the case of eight boroughs, over 70 per cent of the persons engaged in the daily journey to work in 1951 travelled into the County of London in order to seek employment.

This type of analysis is limited to those journeys which involve crossing administrative boundaries and are hence depicted in the Census. It takes no account of the considerable number of 'hidden journeys' which take place within the confines of the respective administrative districts. In 1951 in South London some 596,150 persons lived and worked within their respective administrative districts (i.e. 39 per cent of the total number of persons seeking employment), indicating the scale and importance of these hidden journeys. Whilst the land-use survey will establish the 'desire lines' which are fundamental to the journey to work, this information will need the support of local origin and destination studies to determine how these journeys are made. In essence these studies are based upon the selection of a 'cordon line', preferably related to physical barriers which limit the number of interview or observation stations (Wood, 1963). For normal purposes simple observational studies are adequate, requiring that the registration number of each vehicle, the time, and the number of occupants should be noted at the various observation stations. When this material has been correlated it is possible to establish the number of journeys which originate and terminate within the cordon area. In order to determine the actual routes it is necessary to establish a further series of observation stations along the main thoroughfares, particularly at the major intersections. This type of study requires a considerable amount of labour and can be carried out satisfactorily by local schools, which can also conduct private questionnaire studies amongst friends and neighbours. This type of local field work can provide a valuable insight into the daily journey to work, the mobility of labour, and the employment needs of local areas.

URBAN RENEWAL

An analysis of the present morphology of the built-up area leads to the need to ascertain the morphological changes which are currently taking place. Urban renewal is a constant process which can change the existing social and economic character of an area, or simply confirm the historic pattern of land use. The business advertisements in the property journals and daily newspapers constitute a valuable source of information in this respect, for they indicate where new office, industrial and commercial developments are taking place, together with the rentals which each area can command on the open market. These developments are an important feature of the changing morphology of the built-up area and quite often they indicate the changing regional status of the local service centres. In greater London many large-scale office developments are taking place in the outer suburbs in such centres as Croydon, Harrow and Tolworth (Figure 11.2).

The question of future changes is to some extent already determined by the statutory development plans of the respective Local Planning Authorities. These plans dictate the present, as well as attempting to forecast the future, use of land. When the Town Map proposals are examined in detail it will be seen that provision is made for the future employment, educational, shopping, residential and recreational needs of the community at large. A comparison of the land-use map and the Town Map will indicate whether the Town Map has confirmed the historic pattern, whilst providing for the eradication of the more blatent absurdities, or has attempted to reshape the basic functional structure of the built-up area. This comparison may also reveal the existence of a number of factories, warehouses, builders' yards, etc., which are deemed to be badly sited from the viewpoint of residential amenity and will need to be re-located when redevelopment takes place. In most cases the economic and social consequences of these proposals have not been fully evaluated. Little is known about the regional and local significance of these dispersed centres of employment, and geographical research in this field can contribute towards a fuller understanding of the dynamics of industrial expansion and location generally. Once again the local questionnaire, which forms an integral part of the land-use survey, will be of considerable assistance in assessing the economic vitality of the local area.

THE HISTORICAL EVOLUTION OF TOWNS

Having analysed the existing land-use pattern, considered the geographical implications of the resultant morphology, delimited the zones of 'divided loyalty' and regional tribute, and studied the daily journey to work, the time has come to consider the historical evolution of these phenomena. This is in contrast to the customary geographical study of a town which strives to date the first signs of human occupa-tion of the site in the Stone Age and follows a tortuous path through the Bronze and Iron Ages until it eventually reaches the comparative safety of the Domesday Survey. The scholarship behind such en-deavour is beyond reproach but one may be pardoned for questioning whether this exercise should not be left to the archaeologist who will establish the time-scale with some degree of precision, leaving the geographer free to explain why these events took place in this particu-lar area. All too often these geographical studies evidence an alarming preoccupation with historic succession and attempt to analyse the 'Town Plan' in terms of the changing form of the buildings and layout of the streets, without proper reference to the functions of the town (Conzen, 1960). Surely the geographical significance of these changes lies in the fact that they reflect the changing functions of the town, which in turn reflect the changing needs and character of the surrounding catchment area. This historical study should attempt to highlight the factors which sustain and change the pattern of urban land use, and not just describe the phases of its evolution. Whilst hesitating to deny the fundamental significance of the questions 'Where?' and 'When'?, I would submit that foremost in the mind of every geographer should be the question 'Why?'. This is particularly relevant when examining the evolution of towns, with special reference to their transition from an agricultural to an industrial economy (Dyos, 1961; Randle, 1962). The dynamics of urban growth

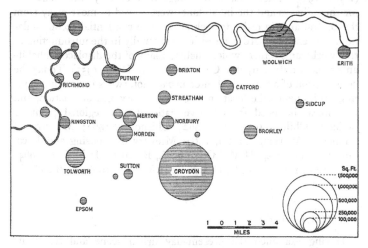

FIG. 11.2. *The location and size of office developments in London, mainly south of the Thames, 1964 (including proposals which have been approved in principle).*

are still something of a mystery, and recourse to Local Acts of Parliament, Vestry Records, Rating Returns, etc., will illuminate the processes and phases of growth without necessarily explaining them. There are manifold dangers in this type of analysis, for it is always tempting to seek geographical reasons for discernible patterns of growth and incidence. Often this amounts to hindsight, and many of the relationships which appear to be established so convincingly in terms of geographical determinism are in reality spurious. This is especially true when examining the mid-nineteenth-century growth of South London (Dyos, 1961). For often it was the *availability* of a strategically-sited country estate, due to the death or bankruptcy of its owner, which precipitated speculative housing developments.

The advent of the railway era also left its mark during this period, by facilitating the increasing separation of workplace and residence (Carter, 1959; Dyos, 1961; Sekon, 1938; Simmons, 1961; Thomas, 1928). The actual construction of the railways in South London made this separation inevitable, for they were deliberately routed through the densely-peopled poorer districts, where the land values were low, in order to reduce the cost of establishing a right-of-way (Dyos, 1955). Thousands of dwellings were destroyed to make way for the railways and the inhabitants were forced to move away from the central area in order to seek alternative residential accommodation. The ensuing hardship resulted eventually in the introduction of the 'workman's return' which enabled many of the displaced inhabitants to commute daily into Central London (Dyos, 1953). Today the construction of a network of inter-urban motorways is already beginning to influence the location of both industry and population, and reinforces the need for effective national and regional planning policies (Childs, 1961 and 1962; Chisholm, 1962; Ministry of Transport Working Group, 1963; Ministry of Housing and Local Government, 1964; H.M.S.O., 1963 A, B and C; H.M.S.O., 1964; H.M.S.O., 1965; A, B, C, D and E; H.M.S.O., 1966, A, B and C.)

CONCLUSION

Having examined the present-day urban scene and traced its historical evolution, all that remains is to identify any major defects in the present local and regional distribution of land use with a view to suggesting improvements in the basic pattern. This is, perhaps, the point where urban geography merges into town and country planning, for the planner is primarily concerned with re-shaping the existing morphology of urban settlements into a more efficient and socially acceptable form, whilst providing for their future as well as their present needs. The planner has tended to be more than somewhat conservative in his approach to this task and has attached undue significance to the historic pattern of land use. The 'Buchanan Report' (Ministry of Transport Working Group, 1963) has sounded a warning note by highlighting the hazards of attempting to determine present, let alone future, patterns of redevelopment in terms of outmoded concepts and an outdated urban inheritance. The call for a more dynamic approach cannot be allowed to remain unanswered now that the Standing Conference on London Regional Planning has forecasted the scale of future events, with an estimated increase of 550,000 jobs and 800,000 inhabitants by 1971 (London County Council, 1963). The recently published *South-East Study: 1961–1981* (Ministry of Housing and Local Government, 1964) has revealed the dearth of both national and regional planning thought in this country, and evidences the need for further research into the very nature of urban growth.

The geographer, by virtue of his training and basic philosophy, is well equipped to undertake the regional studies which are absent in current government publications dealing with the redevelopment of town centres (Ministry of Housing and Local Government, 1962, 1963 B, and 1963 C). Unfortunately, as discussed earlier, the geographer has failed to establish definitive techniques which stand the test of universal application, and further research is of paramount importance for events in the south-east of England indicate that time is short. Local field work studies will do much to reveal the composition, functions and character of the respective urban areas, and provide a valuable insight into the spatial relationships of the component land uses. Too much reliance has been placed on the decennial census reports which are years out of date by the time they are published. The government has at last recognized this defect, however, and announced its intention to hold a census every five years starting from 1966. A more serious drawback is the fact that the census statistics tend to mask the local changes which take place within the confines of the local government areas. The investigations which have been described in this chapter can all be undertaken by urban schools, and the pupils can conduct the land-use surveys, traffic counts and random sample surveys with only a limited amount of direction and supervision. Field work is an essential part of any geographical study and the pupils will learn much by having to devise survey techniques, methods of presentation, questionnaires, and analyse the resultant findings. These local field work studies are essential to a fuller understanding of the dynamics of urban regionalism, for only when the geographer has come to grips with the urban microcosm can he hope to arrive at any significant conclusions regarding the nature of towns and the areas that they serve.

References and Further Reading

ANDERSON, N., 1960, *The Urban Community* (London).

BEST, R. H. and COPPOCK, J. T., 1962, *The Changing Use of Land in Britain* (London).

CARRUTHERS, W. I., 1957, 'A Classification of Service Centres in England and Wales', *Geog. Jour.*, **123**, 371–85.

— 1962, 'Service Centres in Greater London', *Town Planning Review*, **33**, 5–27.

CARTER, E. F., 1959, *An Historical Geography of the Railways of the British Isles* (London) (see especially pp. 352–3).

CENTRAL STATISTICAL OFFICE, 1958, *Standard Industrial Classification* (revised edn) (H.M.S.O.).

CHADWICK, J. G., 1961, 'Correlation between Geographical Distributions', *Geog.*, **46**, 25–30.

CHILDS, D. R., 1961, 'Urban Balance', *Jour. Town Planning Inst.*, **47** (6), 170–1.

— 1962, 'Counterdrift', *Jour. Town Planning Inst.*, **48** (7), 215–25.

CHISHOLM, M., 1962, 'The Common Market and British Industry and Transport', *Jour. Town Planning Inst.*, **48** (1), 10–13.

COLE, H. R., 1966, 'Shopping Assessments at Haydock and Elsewhere', *Urban Studies*, **3**, No. 2, 147–57.

COLLINS, M. P., 1960 A, 'A Study of Urban Complexes within the Southern Half of the Greater London Region', in *Report of the Royal Commission on Local Government in Greater London, 1957–60*, V. 5, 731–7 (Cmnd. Paper No. 1164) (H.M.S.O.).

— 1960 B, 'South London: Its Metropolitan Evolution and Town Planning Requirements, (Unpublished Thesis, Dept. Town Planning, Univ. Coll., London).

CONZEN, G., 1960, 'Alnwick, Northumberland', *Trans. Inst. Brit. Geog.*, Pub. No. 27 (London).

DIAMOND, D. R. and GIBBS, E. B., 1962, 'Development of New Shopping Centres: Area Estimation', *Scottish Journal of Political Economy No.* 9.

DYOS, H. J., 1953, 'Workmen's Fares in South London 1860–1914', *Jour. Transport Hist.*, **1**, 3–19.

— 1954, 'The Growth of a pre-Victorian Suburb; South London 1580–1836', *Town Planning Review*, **25**, 59–78.

— 1955, 'Railways and Housing in Victorian London', *Jour. Transport Hist.*, **2**, 11–21 and 90–100.

— 1961, *The Victorian Suburb: a Study of the Growth of Camberwell* (London).

FORSHAW, H. J. and ABERCROMBIE, P., 1943, *County of London Plan* (London) (see paras. 96–111, pp. 26–29).

HERBERT, D. T., 1961, 'An Approach to the Study of the Town as a Central Place', *Sociological Rev.*, **9** (3) (NS), 273–92.

HERBERT, D. T., 1963, 'Some Aspects of Central Area Redevelopments', *Jour. Town Planning Inst.*, 49 (4), 92–99.

H.M.S.O., 1963 A, *The North-East: a Programme for Regional Development and Growth* (Cmnd. Paper No. 2206).

— 1963 B, *Central Scotland: a Programme for Development and Growth* (Cmnd. Paper No. 2188).

— 1963 C, *Growth of the U.K. Economy 1961–1966*, National Economic Development Council.

— 1964, *The Growth of the Economy*, National Economic Development Council.

— 1965 A, *The National Plan* (Cmnd. Paper No. 2764).

— 1965 B, *The West Midlands—A Regional Study*, D.E.A.

— 1965 C, *The Problems of Merseyside*, D.E.A.

— 1965 D, *Wales 1965* (Cmnd Paper No. 2918).

— 1965 E, *The North-West—A Regional Study*, D.E.A.

— 1966 A, *Investment Incentives* (Cmnd. Paper No. 2874).

— 1966 B, *The Scottish Economy* (Cmnd. Paper No. 2864).

— 1966 C, *A New City*, Ministry of Housing and Local Government.

HOOKWAY, R. J. S. *et al.*, 1963, 'Current Practice Notes', *Jour. Town Planning Inst.*, 49 (7), 234–6.

HUDSON, A. C., 1963, 'Brixton Central Area Redevelopment (Map)', *Jour. Town Planning Inst.*, 49 (8), 277.

LOMAS, G. M., 1964, 'Retail Trading Centres in the Midlands', *Jour. Town Planning Inst.*, 50 (3), 104–19.

LONDON COUNTY COUNCIL, 1951, *Administrative County of London Development Plan* (London) (see 'Analysis', Chap. 7).

— 1957, *A Plan to Combat Congestion in Central London* (London).

— 1961, *Administrative County of London Development Plan: First Review*, County Planning Dept., Vol. 1, Chap. 11.

— 1963, *Reports on Population, Employment and Transport in the London Region*, Standing Conference on London Regional Planning, dated 28 March, 31 May, 3 October and 4 December.

MADIN, JOHN H. D. AND PTNRS., 1964, *Worcester Expansion Study*, published by Ministry of Housing and Local Government.

MANCHESTER UNIVERSITY, 1964, *Regional Shopping Centres—A Planning Report on N.W. England*, Department of Town Planning.

MINISTRY OF HOUSING AND LOCAL GOVERNMENT, 1962, *Town Centres – Approach to Renewal*, Planning Bull. No. 1 (H.M.S.O.).

— 1963 A, *The Town and Country Planning (Classes) Order*, Statutory Instrument No. 708 (H.M.S.O.).

— 1963 B, *Town Centres – Cost and Control of Redevelopment*, Planning Bull. No. 3 (H.M.S.O.).

— 1963 C, *Town Centres – Current Practice*, Planning Bull. No. 4 (H.M.S.O.).

— 1964, *The South-East Study: 1961–1981* (H.M.S.O.).

MINISTRY OF TOWN AND COUNTRY PLANNING, 1949 A, *Report of the Survey*, Circular No. 63 (H.M.S.O.).
— 1949 B, *The Redevelopment of Central Areas* (H.M.S.O.).
— 1951, *Reproduction of Survey and Development Plans*, Circular No. 92 (H.M.S.O.).
MINISTRY OF TRANSPORT WORKING GROUP, 1963, *Traffic in Towns* ('Buchanan Report') (H.M.S.O.).
MOSER, C. A. and SCOTT, W., 1961, *British Towns: a Study of Their Social and Economic Differences* (Edinburgh).
NATIONAL CASH REGISTER COY. LTD. (undated), *Thoughts on Future Shopping Requirements*. Issued by Modern Merchandising Methods Dept., N.C.R. Co. Ltd., 206–216 Marylebone Road, London, N.W.1.
PARKER, H. R. 'Suburban Shopping Facilities in Liverpool', *Town Planning Review*, April 1962, **xxxiii**, 197–223.
RANDLE, P. H., 1962, 'The Uses of Historical Data', *Jour. Town Planning Inst.*, *48* (8), 247–50.
REYNOLDS, D. J., 1961, 'Planning, Transport and Economic Forces', *Jour. Town Planning Inst.*, 47 (9), 282–6.
ROYAL COMMISSION, 1940, *Report on the Distribution of the Industrial Population* ('Barlow Report') (Cmnd. Paper No. 6153) (H.M.S.O.).
SEKON, G. A. (Pseud. for NOKES, G. A.), 1938, *Locomotion in Victorian London* (London).
SIMMONS, J., 1961, *The Railways of Britain: an Historical Introduction* (London).
SMAILES, A. E., 1944, 'The Urban Hierarchy in England and Wales', *Geog.*, **29**, 41–51.
— 1953, *The Geography of Towns* (London) (see Chap. 8).
SMAILES, A. E. and HARTLEY, G., 1961, 'Shopping Centres in the Greater London Area', *Proceedings of Institute of British Geographers* No. 29, 201–213.
STAMP, L. D., 1950, *The Land of Britain, Its Use and Misuse*, 2nd Edn (London).
THOMAS, J. P., 1928, *Handling London's Underground Traffic* (London Underground Railways Publication).
TIMES, THE, 1963, 'Few Mourners as Soho Buries the Past', 18 November, p. 7.
— 1965, 'The Future Economic Shape of Britain, 15th September, p. 16.
WAIDE, W. L., 1963, 'The Changing Shopping Habits and Their Impact on Town Planning', *Jour. Town Planning Inst.*, 49 (8), 254–64.
WELLS, HENRY W., *Peterborough—An Expansion Study*, 116 Kensington High Street, London, W.8.
WOOD, P., 1963, 'Studying Traffic in Towns', *Jour. Town Planning Inst.*, *49* (8), 265–71.

CHAPTER TWELVE

Quantitative Techniques in Urban Social Geography

D. TIMMS

Lecturer in Sociology, University of Queensland

Geography has often been accused of barrenness in terms of the meaningful generalizations and models it has produced. Much of this infertility may be attributed to the crude and subjective measuring instruments which have been generally employed by geographers. In the absence of precise and objective measurement and statement, accurate comparison and abstraction become impossible, and without abstraction the construction of explanatory models can be little more than guesswork.

The last two decades have witnessed a considerable increase in studies concerned with areal variation. The traditional geographical interest in areal studies has been joined by that of several other disciplines. The incursion of human ecologists, regional scientists, and students from many of the systematic sciences, has resulted in the development of a wide range of measures for the accurate description, analysis, and generalization of the facts of areal variation. Human ecology has been the most fertile source of new techniques concerned with the measurement of the areal pattern of urban social phenomena, but significant contributions have also come from regional science and from psychology. All the techniques to be described have as their aim the accurate and objective description and analysis of the areal variation of sociological phenomena within and between urban areas.

THE DESCRIPTION OF GEOGRAPHICAL DISTRIBUTIONS

In *The Nature of Geography* Hartshorne (1949, p. 376) states that: 'The scientific ideal of certainty commands that the terms and concepts of description and relationships be made as specific as possible – we cannot develop a sound structure on a marsh foundation. . . .'

Any research concerned with the spatial structure of urban areas is confronted by the necessity of providing a systematic description of various areal distributions and of their interrelationships. Three main groups of techniques have been developed in response to this need: the various cartographic techniques, measures of spatial association, and the so-called centrographic measures.

CARTOGRAPHIC TECHNIQUES

Maps are often referred to as *the* geographical tool, although their use is by no means confined to geographers. It is assumed that readers are familiar with the elements of cartographic representation and with the considerable degree of elaboration possible therein. Much research identified as social geography has consisted of the compilation of massive inventories of the observable characteristics of community life and of the plotting of these facts on various types of dot or choropleth maps. With this operation 'analysis' has often ceased, the final report of such research generally consisting of no more than a set of maps with an accompanying descriptive text. But the plotting of distributions is only the first step in the analysis of areal variation. Analysis which is based on rule-of-thumb assessments of the similarities revealed on successive maps or on the degrees of overlap present in two distributions is open to grave errors of approximation. Subjective judgement may produce comparisons which are not real and may ignore 'awkward' relationships. As illustrative devices maps have few peers, but as analytical instruments they suffer from many fundamental weaknesses. Mapped distributions can only provide the raw material for analysis and the success of this operation depends on the use of more concise and specific measures of distribution, which are capable of quantitative statement and allow precise comparison.

MEASURES OF SPATIAL ASSOCIATION

The most widely-used and in many ways the most useful instruments yet devised for the quantitative description of geographical patterns are the various measures derived from the *Index of Dissimilarity* (I_D), a statement of the evenness of distribution of two statistical populations.[1] The index is calculated from data giving for both populations the percentage of the total living in each areal sub-unit. The index of dissimilarity is then one-half the sum of the absolute differences between the two populations, taken area by area. In the hypothetical example (Table 9) the index of dissimilarity between the 'A' and 'B' populations is 25 per cent.

Table 9. Hypothetical Data for Computation of Index of Dissimilarity

Area	'A' Population	'B' Population	Difference
1	15%	10%	5%
2	20	10	10
3	20	20	0
4	30	20	10
5	15	40	25
Totals:	100	100	50

The index of dissimilarity may be interpreted as a measure of net displacement, showing the percentage of the one population who would have to move into other areas in order to reproduce the percentage distribution of the other population. Thus, in the hypothetical example, 25 per cent of the 'A' population would have to move in order to reproduce the percentage distribution of the 'B' population.

The basic formula for the index of dissimilarity is given in formula 1, with x_i representing the percentage of the 'x' population in the i'th areal sub-unit, y_i representing the percentage of the 'y' population in the i'th sub-unit, and the summation being over all the k sub-units making up the given universe of territory, such as a city.

$$\text{Formula 1:} \quad I_D = \frac{1}{2} \sum_{i=1}^{k} \left| x_i - y_i \right|$$

The basic form of the index of dissimilarity may be applied to a wide variety of phenomena, depending on the definition given to the two populations. One of its most well-known guises is the so-called *Index of Concentration*, a measure of the degree of correspondence between population units and area. Applied to urban population, x_i becomes the percentage of city population living in the i'th areal sub-unit and y_i becomes the percentage of city area contained in that unit. If the population is distributed evenly through the city

[1] Throughout this chapter the term population is to be interpreted in its statistical sense as a number of distributed objects.

area, each territorial division will contain a proportion of population equal to its proportion of total area. In this case the index of concentration will be o. Conversely, if all the population is concentrated into one small area, the remaining parts of the city being uninhabited, the index of concentration will equal almost 100 (specifically 100 minus the percentage of city area contained in the populated sub-unit). Another application of the index of dissimilarity yields the *Index of Redistribution*, the net percentage of population which would have to change its area of residence in any given year in order to reproduce the distributional pattern of an earlier year. In this case x_i is the percentage of city population contained in the i'th sub-unit in the earlier year and y_i is the percentage in the given year. In conjunction with data on the demographic characteristics of the city population, the index of redistribution is of considerable value in studies of intra-urban migration. Florence's (1948) Coefficient of Localization, a measure of the relative regional concentration of a given industry compared to all industry, is functionally identical to the Index of Dissimilarity, but uses a different divisor.

If the index of dissimilarity is computed between a sub-group of the population and the remainder of that population (i.e., total population minus those in the specified group) the resulting measure is referred to as an *Index of Segregation* (I_S) and shows the extent to which the specified sub-group is residentially separated from the rest of the population. The most convenient means of computing the index is given in formula 2, where I_D represents the index of dissimilarity between the sub-group and the total population (including the sub-group), $\sum x_{ai}$ represents the total number of the sub-group in the city, and $\sum x_{ni}$ represents the total population of the city.

$$Formula\ 2: I_S = \frac{I_D}{1 - \dfrac{\sum x_{ai}}{\sum x_{ni}}}$$

A large number of indices have been developed to deal with various aspects of segregation, but few possess the general utility and clear meaning of the present measure. For an exhaustive discussion of the various segregation indices the reader is referred to a recent methodological article by O. D. and B. Duncan (1955 A).

The index of dissimilarity provides a description of the association between two distributions over the several sub-areas comprising a city. The relative concentration of a population within any one of these sub-areas is most readily measured by the *Location Quotient* (L_Q), an index with which most geographers are doubtless familiar. The location quotient is simply defined as the ratio between the percentage of one population occurring in a given area and the percentage of another population in that area. In formula 3, x_i is the percentage of the total 'x' population occurring in the i'th area and y_i is the percentage of the 'y' population in the i'th area.

$$Formula \ 3: \ L_Q = \frac{x_i}{y_i}$$

A location quotient of one indicates that the two populations are proportionately equally represented in the area. An index of less than one indicates an under-representation of the 'x' population, while an index of more than one indicates an over-representation of the 'x' population.

All measures depending on a comparison of proportionate distributions over a series of areal sub-units are very dependent on the nature of those sub-units. In their book *Statistical Geography*, Duncan, Cuzzort and Duncan (1961, pp. 80–94) devote considerable space to a consideration of the problems raised by this dependence. There is no way of 'adjusting' index values to give a value which is independent of the territorial subdivision used. In general, the smaller the average size of areal unit, the larger is the index value. More precisely, if one system of sub-units is derived by the subdivision of the units of another system, then the index computed for the former can be no smaller than the index for the latter and will generally be larger. In Brisbane, the index of net redistribution of population between 1954 and 1961 is 24.7 when calculated on the basis of census collector's districts, which have an average population of 760, and 18·0 when calculated on the basis of statistical divisions, having an average population of nearly 10,000. No unique value can ever be attached to any measure which is derived from the comparison of percentage distributions and a statement as to the nature of the sub-units on which they are based should always accompany any description of index values.

CENTROGRAPHIC AND POTENTIAL MEASURES OF DISTRIBUTION

Centrographic and potential measures of distribution are formally more satisfying than the index of dissimilarity and its variants, but, as yet, have proved operationally more lean. The great advantage which they possess is that they are largely independent of the territorial subdivisions on which they are computed. Their values, in general, will be more precise if based on many small units rather than a few large ones, but otherwise the significance of the values is unaltered by changes in territorial base.

The centrographic technique consists of the application of conventional statistical measures to the data of areal distributions. A distribution may differ from other distributions in two basic ways, in addition to variations in the number of units involved in its population. First, it may differ in terms of the position or positions around which it tends to cluster (i.e., its centre), and second, it may differ in the way its component units are scattered around this centre. Both features are readily amenable to measurement (Warntz and Neft, 1960).

The average position or *Mean Centre* of a distribution is the exact equivalent of the arithmetic mean of conventional statistics. It is the 'balancing point' or 'centre of gravity' of the distribution. Several alternative methods are available for the computation of the mean centre of a distribution, but the most convenient system depends on the use of grid co-ordinates. A square grid is superimposed on a map showing the base areal units for which data is available. On the assumption of even distribution within each base area, the geographical centre of each unit is determined and expressed in terms of the two co-ordinates of the grid, x and y. Each x and y measurement is weighted by the population of the base unit concerned. The sum of these terms for the whole city, divided by the total city population, gives weighted mean positions along each of the co-ordinates. The mean centre of the distribution may be found at the intersection of the mean of the x's and the mean of the y's (formula 4).

Formula 4:

$$\bar{x} = \frac{\sum\limits_i (x_i P_i)}{\sum\limits_i P_i},$$

$$\bar{\Delta} =$$

$$\bar{y} = \frac{\sum\limits_i (y_i P_i)}{\sum\limits_i P_i}$$

where x_i = vertical co-ordinate of the i'th area

P_i = population of the i'th area

y_i = horizontal co-ordinate of the i'th area

Two measures are available to describe the scatter of a distribution around its centre. Both may be computed with reference to any given point, e.g. the centre of the Central Business District or the point of highest land value. The simple measure, the *Mean Distance Deviation* (MD_Δ), is an expression of the average distance separating the units of the distribution from the given centre. The formula for computing the mean distance deviation is given in formula 5, where d_i represents the linear distance between the geographical centre of the i'th base unit and any given point, O, and P_i is the population of the i'th unit.

$$\text{Formula 5:} \quad MD_\Delta = \frac{\sum_i (P_i d_i)}{\sum_i P_i}$$

The second measure of dispersion, the *Standard Distance Deviation* (s_Δ), is strictly analogous to the standard deviation of conventional statistics. If a system of grid co-ordinates has been used to compute the mean centre of a population it is a simple matter to extend the tabulation in order to derive the standard distance by either version of formula 6:

$$(a) \; s_\Delta = \sqrt{\frac{\sum(P_i[x_i - \bar{x}]^2)}{\sum P_i} + \frac{\sum(P_i[y_i - \bar{y}]^2)}{\sum P_i}}$$

Formula 6:

$$(b) \; s_\Delta = \sqrt{\frac{\sum P_i x_i^2}{\sum P_i} - \bar{x}^2 + \frac{\sum P_i y_i^2}{\sum P_i} - \bar{y}^2}$$

Formula 6 gives both the standard distance, s_Δ, and the two component 'latitude' and 'longitude' deviations, s_x and s_y, which comprise it. The standard distance itself is invariant under the rotation of co-ordinate axes, but this is not true of its x and y components. For this reason no truly satisfactory means has yet been devised for plotting the standard distance on a map, although a possible solution to this problem, involving respectively the maximization and minimization of the x and y deviations, has recently been proposed (Bachi, 1963).

An interesting theorem connects the standard deviation to the average root-mean-square deviation of a total frequency distribution. The proof of the theorem can be readily generalized to include areal distributions and it may then be shown that the *General Standard*

R

Distance Deviation (S_Δ) is equal to the standard distance deviation times the square root of two (formula 7). Thus, the amount of scatter revealed in the distances separating all the units of a distribution can be shown to be related to the amount of scatter of those units around the mean centre of the distribution.

$$Formula\ 7: S_\Delta = s_\Delta \sqrt{2}$$

Closely related to centrographic measures of distribution are the concepts variously known as population potential, market potential, and workplace potential. For a full discussion of the population potential concept the reader is referred to an article by Stewart and Warntz (1958 B). The *Population Potential* (V_o) at a point may be considered as a measure of the influence of total population on that point. The initial formulation of the concept was closely modelled on analogies with mechanics, but its most general use has been as a measure of general accessibility. According to its strict definition, the potential at a point, *O*, is obtained by measuring the distance from that point of each individual in the population inhabiting the universe of territory under study, computing the reciprocals of each distance, and summing the reciprocals. In practice a simpler procedure is followed. The universe of territory is subdivided into a manageable number of sub-units, the population of each unit is ascertained and, on the assumption that this population is concentrated at a single central point within each unit, the distance is measured between this centre and point *O*. The population potential at point *O* is then given by formula 8.

$$Formula\ 8: V_o = \sum_{I}^{k} \frac{P_i}{d_i} \qquad \text{Where } d_i \text{ is the distance of the } i\text{'th areal unit from point O}$$

Conversion of population potential into market potential is simply attained by the use of an income or expenditure multiplier. Workplace potential (Duncan and Duncan, 1960) is defined, analogously to population potential, as the sum over all workplaces in the city of the reciprocals of the distances separating each workplace from any given point. The measure is interpreted as an index of the accessibility of a point to workplaces, on the assumption that accessibility declines as distance increases. Exploratory work using the workplace potential in the analysis of intra-urban residential structure is part of the Chicago Urban Analysis Project. The population potential measure has primarily been used in inter-urban studies of economic activity, but as a measure of accessibility and of 'sociological intensity' offers several possibilities for intra-urban research.

THE EMPIRICAL VALUE OF MEASURES OF DISTRIBUTION

The various percentage, centrographic and potential measures used in the description of areal distributions, have as their aim the identification of general regularities underlying the details of local spatial arrangement. The utility of all the measures mentioned has been illustrated in a number of empirical studies. The use of a few of the measures can be illustrated by a brief survey of the residential distribution and social status of various birthplace and occupation groups in the area of Greater Brisbane. The study is modelled, in part, on a paper by Duncan and Duncan (1955 B), in which they examine the residential pattern of occupation groups in Chicago.

Four main aspects of distribution are dealt with. The first aspect is the degree of residential dissimilarity between each of the birthplace groups and each of the occupation categories. In this way an indication of the residential social status of each birthplace group is obtained. The second aspect to be considered is the degree of residential segregation of each birthplace and occupation group, the extent to which each given group is residentially distinct from the rest of the Brisbane population. The third aspect is the degree of residential dissimilarity between pairs of birthplace and occupation groups, the extent to which they isolate themselves from one another. The fourth aspect of distribution to be considered is the centralization of each group, the 'mean' distance separating the group from the centre of Brisbane, taken to be a point in the city's C.B.D. In each case attention is focused on the relationship between residential distribution and social status.

The data of the study are from a 6 per cent household sample carried out for the Brisbane City Council in January 1960. The territorial subdivisions used as the basis of the calculations are a series of quarter-mile zones concentric to the city centre.

Although the local newspapers devote much space to the topic, the social ranking of the various birthplace groups in Brisbane is far from obvious. An indication of the hierarchy, as it operates residentially, is provided by the data of Table 10, which show the indices of residential dissimilarity between each of the birthplace groups and those occupational categories generally accepted as providing a measure of social status. The index values represent the net percentage of the one group who would have to move from their present zones of residence in order to reproduce the percentage distribution of the other group.

Table 10. Indices of Dissimilarity between Birthplace Groups and Occupations

	Prof.	Manag.	Sales/ Cler.	Skilled	Manual Workers Semi- skld.	Un- skilled	Pers. Serv.
Rest Queensland	12	12	7	7	10	20	28
Brisbane-born	13	15	11	9	12	19	30
Rest Australia	14	10	10	11	13	19	28
U.K.-born	22	20	15	12	11	18	23
Europe-born	46	44	40	34	32	26	20

The smaller the index of dissimilarity, the more similarly are the two populations distributed over Brisbane. A net total of 12 per cent of the 'rest of Queensland-born' would have to move from their present zones of residence into other zones in order to give the group the same proportionate distribution as that of the professional workers. To achieve the same balance, some 46 per cent of the European-born group would have to move. Within the limits of the sample, no significant difference can be observed in the indices of dissimilarity of the three Australian-born groups, but the United Kingdom-born and the Europeans are each clearly distinguishable. The suggested social ranking has the Australian groups at its head and the European group at its base.

The social hierarchy evident in the residential distribution of the various birthplace groups is similar, but not identical, to that revealed in their occupational structure (Table 11). The main difference revealed in the two tables is the place of the United Kingdom birthplace group. On the basis of its occupational structure the group has the same social status as the three Australian-born groups. On the basis of its residential pattern, however, the group has to be assigned to a lower status level. Although the United Kingdom-born residents have a high status occupational structure, they tend to live in areas of low status. The difference is of considerable import in the social

Table 11. *Occupation of Employed Males as a Percentage of Those in Birthplace Group*

	Prof.	Manag.	Sales/ Cler.	Skilled	Manual Workers Semi- skld.	Un- skilled	Pers. Serv.
Rest Queensland	9	11	23	25	20	7	5
Brisbane	8	11	21	30	20	8	2
Rest Australia	6	17	18	25	17	10	4
U.K.-born	9	13	14	35	16	10	4
Europe-born	6	9	4	39	16	21	4

ecology of Brisbane and also, presumably, in the life of the persons concerned.

In their work on Chicago, Duncan and Duncan show that a clear relationship exists between social status and residential segregation. Evidence from Brisbane largely substantiates the Duncans' conclusion. Table 12 shows the indices of segregation for the various birthplace and occupation groups in Brisbane. The higher the index the more segregated is the group concerned.

Notable features of Table 12 are the U-shaped pattern of the occupation indices, repeating a Chicago finding, and the high degrees of segregation experienced by the European-born and by those employed in the personal service occupations.

On the assumption that spatial distance parallels social distance, the greater the social disparity between any two populations the more dissimilar should be their residential distribution. The hypothesis is

Table 12. *Indices of Residential Segregation for Birthplace and Occupation Groups*

	Rest Queensland-born	Brisbane-born	Rest Australia	U.K.-born	Europe-born
$I_s =$	5	12	10	14	40

	Prof.	Manag.	Sales/ Cler.	Skilled	Semi- skld.	Un- skilled	Pers. Serv.
$I_s =$	16	14	7	5	9	17	26

readily tested by computing the index of dissimilarity between pairs of the birthplace and occupation categories (Tables 13 and 14).

The evidence clearly supports the hypothesis that spatial distance and social distance are closely related. A notable feature of the tables is the very distinct pattern revealed for the European-born and for the personal service category.

Table 13. Indices of Residential Dissimilarity for Birthplace Groups in Brisbane

	Rest Queensland	Brisbane	Rest Australia	U.K.	
Brisbane	9	—			Index show %
Rest Australia	9	17	—		needing to move
U.K.	14	17	15	—	into other zones to
Europe	40	42	38	32	balance distribu-
					tions

Table 14. Indices of Residential Dissimilarity for Occupation Groups in Brisbane

	Prof.	Manag.	Sales/ Cler.	Skilled M.	Semi- Skil. M.	Unskilled Manual.
Managerial	11	—				
Sales/Clerical	12	11	—			
Skilled/Manual	17	16	11	—		
Semi-skilled	19	18	13	8	—	
Unskilled	27	25	20	16	14	—
Pers. Service	32	28	25	24	23	20

The final aspect of distribution to be considered is the relative centralization of the various groups. According to the well-known Burgess zonal hypothesis of city growth there is an upward gradient in socio-economic status with increasing distance from the city centre. The application of the model to Brisbane may be tested by computing the mean distance deviation of the birthplace and occupation groups from the city centre (Table 15).

The data of Table 15 are clearly at variance with the Burgess hypothesis. The arrangement of the occupational groups even suggests an inverted form of the model. More detailed knowledge of the status structure of the city in relationship to the 'ideal' models developed by human ecology must await the conclusion of further studies.

No amount of cartographic skill could elicit the information about the association between residential distribution and social status revealed by the indices of dissimilarity and segregation and by the mean distance deviation. The first two measures are implicitly dependent on the territorial subdivisions on which they are computed, but this has very little effect on their empirical value. It is unlikely that a sufficient range of centrographic measures will be developed to render the indices of association empirically redundant. In the meantime, the index of dissimilarity and its derivatives have a very considerable potential as descriptive-analytical tools for research in urban social geography.

Table 15. Mean Distance Deviations for Birthplace and Occupation Groups in Brisbane

U.K.-born	Brisbane	Rest Queensland	Rest Australia	Europe-born
4·29 miles	4·22 miles	4·08 miles	3·96 miles	3·68 miles

Semi-Skilled Man.	4·15	*Managerial*	3·87	
Skilled	4·07	*Sales and Clerical*	3·83	
Unskilled	4·02	*Professional*	3·76	

GEOGRAPHICAL ANALYSIS AND THE PROBLEM OF REGIONALIZATION

The recognition of valid regions is the basis of geographical research. Urban social geography is dependent on the delimitation of meaningful social areas, regions characterized by a stated combination of social conditions. An aggregation of social areas forms the total social structure and provides the framework for further analyses which attempt to relate the distribution of individual phenomena to the characteristics of the local community.

The choice of a regional system appropriate to the project in hand is the greatest single problem which faces the investigator of urban social distributions. The concept of region as an abstract or as a partial model of reality has been widely explored by geographical theorists, but much less attention has been given to the definition of methods by which the requirements of a formally-valid regional system may be satisfied. The criteria by which the meaningfulness of a region may be evaluated are in dispute and largely depend on the methodological position adopted by the investigator concerned. At the same time, the requirements of a regional system delimited for illustrative purposes or for the description of 'regional character' are very different from those of a system conceived primarily as an analytical device. The present discussion is largely concerned with regions in the latter category, in particular as they are relevant to the analysis of urban social structure and as they provide a frame for the study of urban areal variation.

In order to satisfy the requirements of analytical manipulation, a system of regions should fulfil three main conditions: first, the individual regions should be strictly comparable one with another; secondly, each region should be so formulated as to provide a maximum of external variation and a minimum of internal variation; thirdly, the item or items on which the regions are delimited should be directly related to the distributional problem being studied.

The traditional geographical approach to regional delimitation has been based on the comparison of mapped distributions. The relevance of the criteria to the problem being studied or the statistical validity of the eventual system have generally been left to the subjective judgement of the individual investigator. This judgement has doubtless often been correct, but the dangers of the approach are manifold. It is in this respect that the recent interest of students from other disciplines in the methodology of regionalism has been of greatest benefit to the progress of geographical research. The problems of 'regionalization' have been brought into new prominence and the concern of human ecologists and regional scientists to provide a meaningful framework for the analysis of spatial structure has not only revived many old arguments about the reality of regions, but has also greatly increased the amount of attention accorded to the formulation of regional systems which are formally satisfactory.

The most promising new techniques of regionalization in an urban context are the various multi-variable systems associated with the names of E. Shevky, R. C. Tryon, and C. F. Schmid. All three workers have attempted to identify social area 'types'[1] which are of primary significance in the social structure of the urban population.

---

[1] In all cases concern is with the stratification of the variables into meaningful classes. Although the regional types have an areal base and can be mapped, the factor of areal contiguity is absent from their formulation.

The eventual regional typologies produced by the various methods are very similar, but there are important differences in their procedures.

SOCIAL AREA ANALYSIS

The theoretical basis of the social area typology has been ably outlined in a monograph by E. Shevky and W. Bell (1955). Urban aggregations are viewed not as unique phenomena, with their own organizing principles, but as component parts of the wider modern society. From a review of the major structural trends taking place within that society, the authors of the social area typology identify a number of constructs which serve as discriminatory factors in the analysis of urban differentiation and stratification. The scheme lends especial emphasis to the postulates of C. Clark, on changes in the structure of productive activity and of employment, of L. Wirth, on the increasing scale of society and of social interaction, and of W. F. Ogburn *et al.*, on changes in the importance and role of the family. The subsequent argument can best be followed with the aid of Table 16 (from Shevky and Bell, 1955). From the initial postulates, a series of interrelated trends is recognized (column 2), which are believed to be major underlying factors of modern society. At particular points in time, any given social system may be described in terms of its differential relationships to these major trends (column 3). The temporal argument can be applied with equal validity, at any one time, to sub-groups within society. Three constructs can therefore be identified (column 4) which allow a classification and stratification of population sub-groups.

Thus, from certain broad postulates concerning modern society and from the analysis of temporal trends, we have selected three structural reflections of change which can be used as factors for the study of social differentiation and stratification at a particular time in modern society (Shevky and Bell, in *Theodorson*, 1961, p. 227).

The three factors identified in the social area typology were termed by Shevky; social rank, urbanization, and segregation. Bell prefers the more explicit terms social status, family status, and ethnic status, respectively. From a list of sample statistics believed to be related to the three constructs (column 5) a brief selection is derived which most adequately define the indices (column 6). Census tract populations

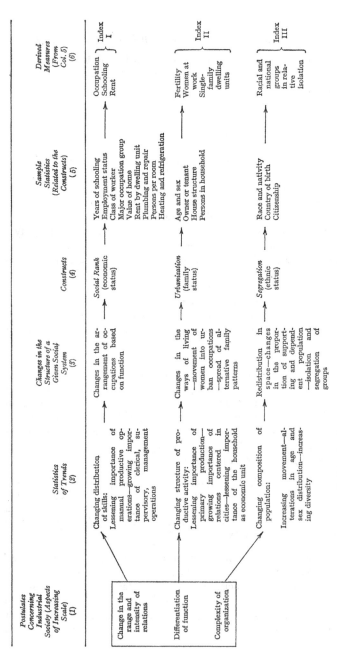

Postulates Concerning Industrial Society (Aspects of Increasing Scale) (1)	Statistics of Trends (2)	Changes in the Structure of a Given Social System (3)	Constructs (4)	Sample Statistics (Related to the Constructs) (5)	Derived Measures (From Col. 5) (6)
Change in the range and intensity of relations	Changing distribution of skills: Lessening importance of manual productive operations—growing importance of clerical, supervisory, management operations	Changes in the arrangement of occupations based on function	Social Rank (economic status)	Years of schooling / Employment status / Class of worker / Major occupation group / Value of home / Rent by dwelling unit / Plumbing and repair / Persons per room / Heating and refrigeration	Occupation / Schooling / Rent } Index I
Differentiation of function	Changing structure of productive activity: Lessening importance of primary production—growing importance of relations centered in cities—lessening importance of the household as economic unit	Changes in the ways of living—movement of women into urban occupations—spread of alternative family patterns	Urbanization (family status)	Age and sex / Owner or tenant / House structure / Persons in household	Fertility / Women at work / Single-family dwelling units } Index II
Complexity of organization	Changing composition of population: Increasing movement—alterations in age and sex distribution—increasing diversity	Redistribution in space—changes in the proportion of supporting and dependent population—isolation and segregation of groups	Segregation (ethnic status)	Race and nativity / Country of birth / Citizenship	Racial and national groups in relative isolation } Index III

Table 16. Steps in Construct Formation and Index Construction

can be described in terms of the indices and grouped into types on the basis of similar configurations of scores in the three-dimensional attribute space formed by the three status factors.

> Thus, the typological analysis . . . is a logically demonstrable reflection of those major changes which have produced modern, urban society (Shevky and Bell, in *Theodorson*, 1961, p. 230).

The social area typology is based on *a priori* reasoning from certain broad postulates concerning the nature of modern urban society. The formal validity of the system has been demonstrated by a series of American studies which suggest that not only is the typology generally applicable to urban society (Van Arsdol Jr. *et al.*, 1958), but that its key indices are both sufficient and necessary to account for the observed variation occurring in social phenomena (Bell, 1952; Tryon, 1955).

Less agreement has been forthcoming on the theoretical under-pinnings of the typology or on the integration of its key indices with that theory (e.g. Hawley and Duncan, 1957). None the less, the technique has attracted wide attention as a stratification instrument and as a frame for the design and interpretation of urban sub-area field studies (Bell, 1958; Greer, 1956; Schmid, 1960 B; Shevky and Bell, 1955, pp. 20–22).

The general significance and utility of the social area typology can only be established by an extension of comparative studies, but it is readily apparent that the technique represents one of the most promising attempts yet available to provide a coherent and logically-demonstrable frame for the analysis of urban social structure.

FACTORIAL DESIGNS

In contrast to the social areas derived by the Shevky technique, regional systems developed on the basis of factor analysis are largely free of *a priori* considerations. Factor analysis allows the identification of the underlying order in a set of data and can reduce a large number of interrelated variables to a few basic independent factors which are sufficient in themselves to account for practically all the observed variation in the phenomena concerned. The technique of factor analysis is rigorous and demands a considerable degree of mathematical expertise. Readers interested in applying the technique should consult any of the standard texts listed in the bibliography.

Factor analysis was originally developed in psychology, for the purpose of identifying the principal dimensions or categories of mentality, but it has since proved of general application to a wide range of scientific problems. The application of factorial design to the delineation of geographical regions was pioneered by Hagood (1941 and 1943) who utilized data on population and agriculture. Price (1942) used the technique to evolve a classification of metropolitan centres. Application of factor analysis to the problems of intra-urban regionalization is primarily associated with the work of C. F. Schmid. The development of Schmid's technique can be followed in a series of articles dealing with various aspects of urban structure (Schmid, 1950, 1958, 1960). The most advanced form of his technique appears in a paper on the crime areas of Seattle (1960). An original correlation matrix containing twenty variables relating to crime and eighteen indices of demographic, economic, and social conditions, is reduced to eight principal factors which account for almost all the variation observed in the original data. Standard scores for each of the factors, computed tract by tract, form the basis of a classification system used to derive a sample of 'typical' tracts for further intensive study. At the same time, Schmid demonstrates that certain combinations of the principal factors 'represent the urban crime dimension par excellence'.

Closely related to factorial design is the technique of cluster analysis developed by R. C. Tryon (1939, 1955) who states that its purpose, when applied to urban social structure, is to identify the number, type and location of the sub-cultural groups which form the basis of the structure. The analysis reduces a large number of observations to three social dimensions which are both necessary and sufficient to describe all that generally differentiates the observed census tracts. In an application of the technique to San Francisco, Tryon shows that the 243 census tracts of the city fall into about eight general social areas.

> Within each social area the neighbourhoods are relatively homogeneous in terms of their values in these social dimensions, and each social area is described by its set of values in the social dimensions. The total configuration of all the social areas provides a final metric, objective description of the social structure of the population [Tryon, 1955, p. 3].

The stratification is primarily based on demographic and economic characteristics, but Tryon believes that the social areas are far more than merely regions of homogeneous demographic phenomena:

There are . . . theoretical grounds and empirical evidence to support the belief that areas of people substantially different in demographic patterns . . . are also critically different in patterns of social ways [Loc. cit.].

The advantages of factorial and multi-variable designs are many and important. All the techniques are firmly based on empirical observations and are readily verifiable. Regional systems which are based on the factorial analysis of a set of areal data are, by formulation, basic to the understanding of the variation occurring within the data. A multitude of observations can be reduced to a manageable number of factors. In the process, previously unsuspected relationships may be revealed, while others, previously assumed, may be shown to be spurious. Particularly in the case of the social area typology, the analytical technique is intimately related to a body of more general theory and allows the formulation of readily-tested structural hypotheses. Finally, the use of certain types of factorial designs allows the objective measurement and statement of the amount of detailed variation in the data which has to be sacrificed at successive levels of regional generalization (Berry, 1961).

The disadvantages of multi-factor analysis are largely a result of its conceptual and operational complexity. The techniques demand a considerable mathematical prowess. Particularly in urban areas, they also demand a sophistication of data which is rarely attained outside the United States and Scandinavia. Perhaps the most difficult problem of all to overcome is that factorial designs depend on the use of costly computer time and presuppose, as Berry (1961) points out, considerable financial assistance. This is rarely available to the individual student of areal variation.

The adoption of factorial methods of regionalization is obviously desirable if urban social geography is ever to attain the 'scientific ideal of certainty'. In view of the demands of the techniques, however, their widespread adoption must inevitably be long delayed.

SCALOGRAM ANALYSIS

In those parts of the world which are less well endowed with statistics than the United States there is considerable scope for a method of regionalization which, while being as objective and meaningful as possible, is less complicated than factorial and social area designs and can utilize less sophisticated data. A technique which may satisfy some of these requirements is based on a modification of the Guttman scalogram method, which has been widely used in attitude surveys (see References). The technique was devised in the course of a study dealing with the geographical pattern of criminality and mental illness in two British cities (Timms, 1962).

In essence the scalogram technique is a means of ordering ranked data in such a way that a single unidimensional scale is produced along which effective measurement is possible. As in multi-factor analysis a number of observations may be reduced to a single dimension, which may then be used as a stratification and correlative instrument. The criterion of unidimensionality lies in the pattern of scale responses, the rank position of the respondents in each of the items included in the scale. A perfect scale takes the form of a parallelogram (Table 17).

Table 17. Dichotomized three item scalogram

	High Values (Score 1)			Low Values (Score 0)			Scale Type.	Scale Score
Items:	A	B	C	A	B	C		
Area 1	x	x	x				I	3
Area 2		x	x	x			II	2
Area 3			x	x	x		III	1
Area 4				x	x	x	IV	0

x indicates area response to item.

Knowledge of the scale type or score of an area allows a complete description of all the attributes which it possesses. Two areas with the same score have identical characteristics in terms of the scale items and categories. Random departures from the perfect form are allowable in so far as they comprise less than 10 per cent. of all responses. Certain other criteria have also to be satisfied before a scale can be accepted as valid (Ford, 1950).

The scalogram technique constitutes a valuable analytical instrument. One of its most important properties is that a correlation between scale scores and any external variable is equivalent to a multiple correlation between all the items constituting the scale dimension and whichever external variable is concerned. The scale can be constructed from data which can be ranked but not precisely measured. At the same time, a sample of items from a valid scale dimension reproduces the scale pattern of the whole.

The use of the technique in the delimitation of regions may be demonstrated by an example from Luton.

A trichotomized ranking of a number of socio-economic and social defect data, available street by street, was used to evolve two series of scale types. The initial selection of items for inclusion in the dimensions was based on a random area sample of 150 base units. From the sample two valid scalograms were constructed containing (1) the variables rateable value per adult, net population density, and percentage of jurors,[1] and (2) the variables adult and juvenile crime rate and mental illness rate. In both cases all the criteria of scalability were met.

Once the scale content was determined by the sample data, the process of regionalization was straightforward. Information was tabulated for each street in terms of the scale items and categories previously defined. An example of tabulation and of the subsequent identification of regional types is given in figure 12.1 and its associated tables.

Each of the scales yielded seven regional types and these were utilized as the cells for a conventional correlation analysis with sub-area populations entered as the frequencies. A generalized map of the resulting two-factor regional types is given in figure 12.2.

The procedure of regionalization was relatively objective. Given the prior decisions to define regional types in terms of certain dimensions relevant to the project and to use a particular division of the data, the only subjective element in the scheme concerns the allocation of sub-areas exhibiting non-conforming scale patterns. In these circumstances the allocation is rarely left ambiguous after a consideration of the scale pattern of contiguous areas and of the internal ordering of the scale items.

The scale analysis technique provides a convenient and precise means for studying the internal structure of urban areas. The technique is by no means perfect, but it is a considerable improvement over the more conventional tools of regionalization.

---

[1] The statistic percentage of jurors has been widely used as an idex of socio-economic status, viz. Gray, Corlett and Jones, *The Proportion of Jurors as an Index of the Economic Status of a District*, Govt. Social Survey, London, 1951.

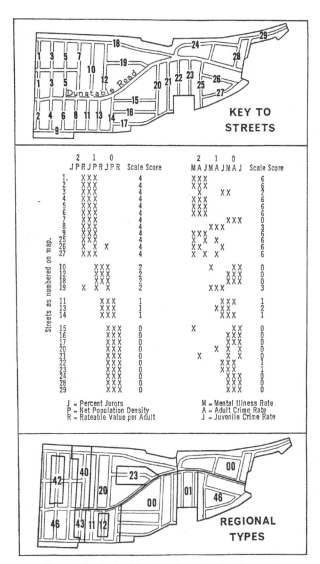

FIG. 12.1. *Luton, England: Identification of regional types based on the trichotomized ranking of a number of socio-economic and social defect data.*

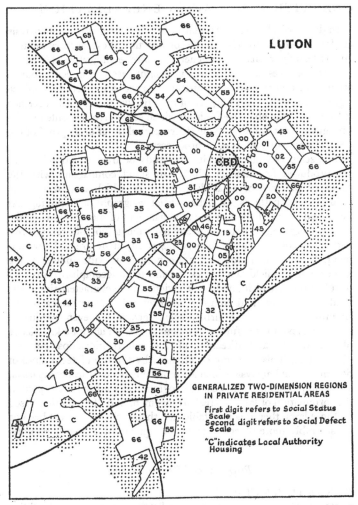

FIG. 12.2. *Luton, England: Generalized map of two-factor residential status/defect regions.*

QUANTIFICATION AND GENERALIZATION

In the classical scheme of scientific inquiry, description and analysis are followed by synthesis and the construction of explanatory models. In the leap from analysis to generalization the subjective element of insight is a necessary condition, but insight based on insecure foundations and divorced from empirical evidence is unlikely to produce significant results. The purpose of quantitative techniques in science is to provide the objective basis on which subjective elements may be brought to bear and to provide the empirical proof or disproof of the generalizations which insight produces.

The sciences concerned with the study of areal variation have as yet produced few models which can stand comparison with the observed patterns of areal distribution or which can be used to predict those patterns. In urban social geography a beginning has been made in prediction by the establishment of models showing the areal arrangement of population and of certain socio-cultural features, and in the construction of ideal forms of city structure and development. Prediction, however, is not synonymous with understanding and even in this field much work remains to be accomplished in order to relate the role of individual behaviour to the characteristics of the local community, and to fully comprehend the dynamics of the situation. Elsewhere in urban social geography even prediction remains a task for the future. Prediction rests on accurate knowledge of the degree and direction of the interrelationships between phenomena. This can only be attained by the use of techniques of description and analysis which are amenable to statistical comparison and manipulation. If the goal of geographical studies be accepted as the formulation of laws of areal arrangement and of prediction based on those laws, then it is inevitable that their techniques must become considerably more objective and more quantitative than heretofore.

References and Further Reading

GENERAL TEXTS

DUNCAN, O. D., CUZZORT, R. P. and DUNCAN, B., 1961, *Statistical Geography: Problems in Analysing Areal Data* (Glencoe, Ill.).

ISARD, W., 1960, *Methods of Regional Analysis* (Cambridge, Mass.) (see Chapters 7 and 11).

REYNOLDS, R. B., 1956, 'Statistical Methods in Geographical Research', *Geog. Rev.*, **46**, 129–31.

INDICES OF GEOGRAPHICAL ASSOCIATION

DUNCAN, O. D., 1957, 'The Measurement of Population Distribution', *Population Studies*, **11**, 27–45.

DUNCAN, O. D. and DUNCAN, B., 1955 A, 'A Methodological Analysis of Segregation Indices', *American Sociological Review*, **20**, 210–17.

— 1955 B, 'Residential Distribution and Occupational Stratification', *American Journal of Sociology*, **60**, 493–503.

DUNCAN, O. D. and LIEBERSON, S., 1959, 'Ethnic Segregation and Assimilation', *American Journal of Sociology*, **64**, 364–74.

FLORENCE, P. S., 1948, *Investment, Location and Size of Plant* (Cambridge).

PETERS, W. S., 1946, 'A Method of deriving Geographic Patterns of Associated Demographic Characteristics within Urban Areas', *Social Forces*, **35**, 62–8.

ROBINSON, A. H., 1957, 'A Method for describing Quantitatively the Correspondence of Geographical Distributions', *Ann. Assn. Amer. Geog.*, **47**, 379–91.

CENTROGRAPHIC TECHNIQUES

BACHI, R., 1963, 'Standard Distance Measures and Related Methods for Spatial Analysis', *Papers and Proceedings of the Regional Science Association*, **10**, 83–132.

HART, J. F., 1954, 'Central Tendency in Geographical Distributions', *Econ. Geog.*, **30**, 48–59.

SVIATLOVSKY, E. E. and EELLS, W. C., 1937, 'The Centrographical Method and Regional Analysis', *Geog. Rev.*, **27**, 240–54.

WARNTZ, W. and NEFT, D., 1960, 'Contributions to a Statistical Methodology for Areal Distributions', *Journal of Regional Science*, **2**, 47–66.

POPULATION POTENTIAL

ANDERSON, T., 1956, 'Potential Models and Spatial Distribution of Population', *Papers and Proceedings of the Regional Science Association*, **2**, 175–82.

CARROLL, J. D., 1955, 'Spatial Interaction and the Urban-Metropolitan Description', *Papers and Proceedings of the Regional Science Association*, **1**, 0.1–0.14.

CARROTHERS, G., 1956, 'An Historical Review of the Gravity and Potential Concepts of Human Interactions', *Journal American Institute of Planners*, **22**, 94–102.

DUNCAN, B. and DUNCAN, O. D., 1960, 'The Measurement of Intra-city Locational and Residential Patterns', *Journal of Regional Science*, **2**, 37–54.

STEWART, J. Q., 1947, 'Empirical Mathematical Rules concerning the Distribution and Equilibrium of Population', *Geog. Rev.*, **37**, 461–85.

— 1948, 'Demographic Gravitation: Evidence and Applications', *Sociometry*, **11**, 31–58.

STEWART, J. Q. and WARNTZ, W., 1958 A, 'Macrogeography and Social Science', *Geog. Rev.*, **48**, 167–84.

— 1958 B, 'Physics of Population Distribution', *Journal of Regional Science*, **1**, 99–123.

— 1959, 'Some Parameters of the Geographical Distribution of Population', *Geog. Rev.*, **49**, 270–2.

THE SOCIAL AREA TYPOLOGY

BELL, W., 1953, 'The Social Areas of the San Francisco Bay Region', *American Sociological Review*, **18**, 39–47.

— 1958, 'The Utility of the Shevky Typology for the Design of Urban Sub-area Field Studies', *Journal of Social Psychology*, **47**, 71–83.

GREER, S., 1956, 'Urbanism Reconsidered: a Comparative Study of Local Areas in a Metropolis', *American Sociological Review*, **21**, 19–25.

HAWLEY, A. H. and DUNCAN, O. D., 1957, 'Social Area Analysis: a Critical Appraisal', *Land Economics*, **33**, 337–45.

SHEVKY, E. and BELL, W., 1955, *Social Area Analysis: Theory, Illustrative Application, and Computational Procedures* (Stanford).

SHEVKY, E. and WILLIAMS, M., 1948, *The Social Areas of Los Angeles: Analysis and Typology* (Berkeley).

VAN ARSDOL, M. D., CAMILLERI, S. F. and SCHMID, C. F., 1958 A, 'The Generality of Urban Social Area Indices', *American Sociological Review*, **23**, 277–84.

— 1958 B, 'An Application of the Shevky Social Area Indexes to a Model of Urban Society', *Social Forces*, **37**, 26–32.

THEODORSON, G. A., 1961, *Studies in Human Ecology* (Evanston, Ill.).

FACTOR ANALYSIS

BELL, W., 1955, 'Economic, Family, and Ethnic Status: Empirical Test', *American Sociological Review*, **20**, 45–52.

BERRY, B. J. L., 1961, 'A Method for Deriving Multi-factor Uniform Regions', *Przeglad Geograficzny*, t. 33, z. 2.

CATTELL, R. B., 1952, *Factor Analysis: an Introduction and Manual for the Psychologist and Social Scientist* (New York).

FRUCHTER, B., 1954, *Introduction to Factor Analysis* (New York).

HAGOOD, M. J., 1943, 'Statistical Methods for Delineation of Regions applied to Data on Agriculture and Population', *Social Forces*, **21**, 287–97.

— *et al.*, 1941, 'An Examination of the Use of Factor Analysis in the Problem of Sub-regional Delineation', *Rural Sociology*, **6**, 216–33.

HAGOOD, M. J. and PRICE, D. O., 1952, *Statistics for Sociologists* (New York).

HARMAN, H. H., 1960, *Modern Factor Analysis* (Chicago).

MOSER, C. A. and SCOTT, W., 1961, *British Towns: a Statistical Study of Their Social and Economic Differences* (London).

PRICE, D. O., 1942, 'Factor Analysis in the Study of Metropolitan Centres', *Social Forces*, **20**, 449–55.

SCHMID, C. F., 1950, 'Generalizations Concerning the Ecology of the American City', *American Sociological Review*, **15**, 264–81.

— 1960 A, 'Urban Crime Areas: Part I', *American Sociological Review*, **25**, 527–42.

— 1960 B, 'Urban Crime Areas: Part II', *American Sociological Review*, **25**, 655–78.

SCHMID, C. F., MACCANNELL, E. H. and VAN ARSDOL, M. D., 1958, 'The Ecology of the American City: Further Comparison and Validation of Generalizations', *American Sociological Review*, **23**, 392–401.

SWEETSER, F. L., 1965, 'Factorial Ecology: Helsinki, 1960', *Demography*, **2**, 372–85.

THURSTONE, L. L., 1947, *Multiple Factor Analysis: a Development and Expansion of the Vectors of Mind* (Chicago).

TRYON, R. C., 1939, *Cluster Analysis* (Ann Arbor).

— 1955, *Identification of Social Areas by Cluster Analysis: a General Method with an Application to the San Francisco Bay Area* (Berkeley).

SCALOGRAM ANALYSIS

FORD, R. N., 1950, 'A Rapid Scoring Procedure for Scaling Attitude Questions', *Public Opinion Quarterly*, **14**, 507–32.

GREEN, N. E., 1956, 'Scale Analysis of Urban Structures: a Study of Birmingham, Alabama', *American Sociological Review*, **21**, 8–13.

HAGOOD, M. J. and PRICE, D. O., 1952, *Statistics for Sociologists* (New York).

MOSER, C. A., 1958, *Survey Methods in Social Investigation* (London).

STOUFFER, S. A. *et. al.*, 1950, *Measurement and Prediction* (Princeton).

TIMMS, D. W. G., 1962, *The Distribution of Social Defectiveness in Two British Cities: a Study in Human Ecology* (Ph.D. dissertation, University of Cambridge).

CHAPTER THIRTEEN

Geographical Techniques in Physical Planning

E. C. WILLATTS

Principal Planner, Ministry of Housing and Local Government

A quarter of a century ago Britain had almost no professional geographers outside the field of teaching, but since then they have established themselves in various other fields and particularly that of physical, or town and country, planning. There was an early recognition of their contribution in carrying out and presenting surveys, not merely topographical, but social and economic. More slowly came the realization that their understanding of many of the complex factors which confront physical planners entitles them not only to make surveys on the basis of which others would prepare plans, but to analyse the problems which require to be solved and thus to indicate their solutions. It was his early appreciation of this which led Lord Justice Scott, a revered Honorary Vice-President of the Town Planning Institute, to write 'Town Planning is the Art of which Geography is the Science'.

This paper is concerned with a few examples of the way in which some of those geographers in central government who are concerned with physical planning on a national level have been applying their techniques with the whole country as their field of review, and with major problems which cannot be decided at a local level. Their first duty is to array the primary facts about the land; its configuration, its geological nature, vegetation and climate, its soils and its quality, for these are vital to the physical planner. These facts require to be readily available in map form, preferably in a co-ordinated series of maps. So do the changing facts about the use man makes of the land, for work and recreation, the way population is distributed over it, the changes in that distribution, and the many factors which influence those changes, such as the patterns of industry and employment, of power, offices, shops and of communications. It is an essential part of

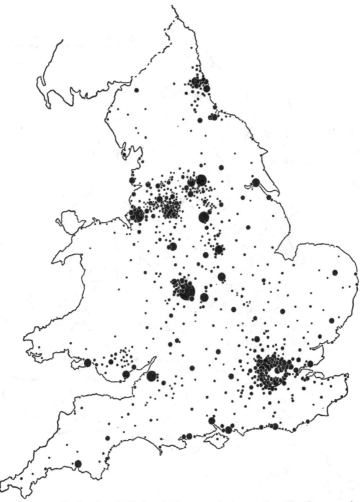

FIG. 13.1. *England and Wales: Urban population, 1961. The County of London is represented by a single open circle.*

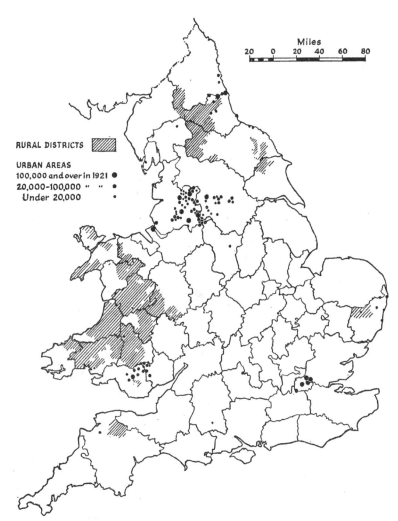

FIG. 13.2. *England and Wales: Areas of persistent population decrease,*
1921–31, 1931–9, and 1939–47.

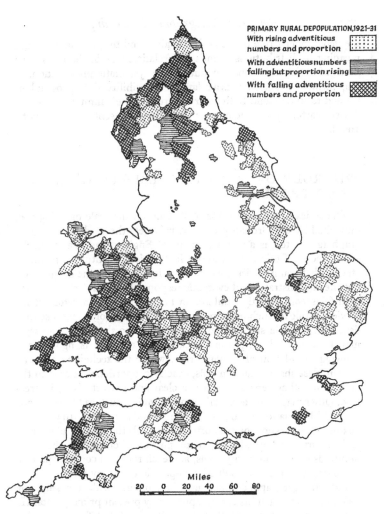

PRIMARY RURAL DEPOPULATION,1921-31
With rising adventitious
numbers and proportion
With adventitious numbers
falling but proportion rising
With falling adventitious
numbers and proportion

Miles
20   0   20   40   60   80

FIG. 13.3. *England and Wales: Different types of rural population decline,
1921–31.*

the geographer's role to recognize, analyse and to give character and meaning to distributions, trends and relationships in these complex fields of social and economic study. The systematic presentation, in map form, of all such data is a related responsibility of the geographer, trained to understand both their nature and their limitations and to devise cartographic techniques for their presentation (Willatts, 1963).

THE ROLE OF MAPS IN PRESENTING FACTS AND PROBLEMS

The series of national maps of Britain, on the scale of 1:625,000, published by the Ordnance Survey, is largely the work of such geographers. So, too, is a series of maps of England and Wales[1] on the smaller scale of thirty miles to one inch which constitutes a planning atlas of that country, designed to provide not merely the geographer, but the administrator and everyone responsible for advice and policy, with facts conveniently available in map form, and therefore at his service. While all such maps repay detailed study, some, because they deal with a single subject, can be presented so that their 'message' may be conveyed simply and clearly. An example is figure 13.1, showing Population of Urban Areas, where the technique of using the simplest shaped flat symbols, exactly proportional in area to the populations they represent, gives a clear and correct visual impression. Other maps are necessarily more complex, such as the 1/625,000 maps of Population Changes for the periods 1921–31, 1931–9 and 1939–47. These show by proportional symbols the amount of change, and by various coloured tints the proportion of change, in the many hundreds of local authority areas, and each reveals a complex pattern of varying change, the result of the operation of many factors.

But the geographer must not be content with devising the techniques by which his cartographers may present primary data. He must analyse that data so that the really significant facts concealed within it are clearly revealed, for facts do not speak for themselves until they are correctly marshalled. Trends in population change may be taken as an example. The detailed maps already referred to may each reveal much by close study, but further laboratory techniques must be applied to them before significant and persistent

[1] To be published by H.M.S.O. as 'Desk Atlas of Planning Maps of England and Wales'.

patterns emerge. Such techniques, adopted by workers with a trained curiosity, have resulted in the preparation of such a map as that shown in figure 13.2 (Willatts and Newson, 1953). This map is very simple, a ruling for rural areas and graded dots for the towns in England and Wales which were found to be persistently losing population in each of the three consecutive periods covered by the published series of maps. The map is a typical one prepared not primarily for geographers but for intelligent lay readers, a class of user to whom the essential truths need to be effectively and speedily conveyed. The simplicity of such a map is therefore its strength and merit. Its 'message' is clear to readers who only with the utmost difficulty might have discovered the facts from volumes of arid statistics and then would still not have seen them placed on a map where the patterns are so clear that they imprint a photographic impression on the mind. Having thus answered the question 'what?' and aroused curiosity by the presentation of significant patterns of distribution the geographer should face the obvious challenge and answer the inevitable question, 'why?'.

Not that a simple answer is usually possible, particularly with a subject so complex as population changes. Vince (1952) demonstrated this in a study of the structure and distribution of rural population in England and Wales, 1921–31, in which he divided the rural population into the *primary* population, directly dependent on the land,[1] the *secondary*, which serves the needs of the primary, and the *adventitious* which neither depends on the land nor serves those who do. He developed a detailed analysis by which he was able to show, among other things, not merely where both the total occupied population and the primary rural population had declined, but to recognize and delimit different types of decline, as shown on figure 13.3 which is reproduced from his paper. It reveals, in his words, that: 'While rural depopulation is almost everywhere a function of remoteness there is a significant difference between upland and lowland Britain. Even in lowland eastern England some primary depopulation occurred, but mostly it was accompanied by adventitious increases on a scale less than sufficient to maintain the total level. In Wales and the Highland zone generally, however, the adventitious numbers on the whole fell to a greater extent than the primary population.' His analysis of the three components of rural population has facilitated a clearer understanding of various aspects of the problem of rural population changes.

[1] In this context: agriculture, horticulture and forestry.

SYNOPTIC MAPS

Techniques which result in a complicated and laborious analysis of a subject being presented with graphic simplicity are suited to certain subjects, especially those where it is desirable that the results should be continually borne in the minds of planners and others. But there is another class of problem which calls, not for a simplified summary of the results of an analysis, but for a synoptic presentation of all the complex factors involved in a problem. The cartographic technique which has been evolved for such purposes, though commonly called a 'sieve' map, is, at its best, much more than that title implies. A good example of its use is in the selection of a suitable locality for some major project such as a new town. The method pre-supposes that the location factors are known and can be translated into terms of geographical data which can be marshalled. Its object is not to eliminate all areas where unfavourable factors operate, but to indicate the relative suitability of different parts of the 'search area' for the construction of a new town, and to do so in such a way that each of several factors operating in any locality may be clearly distinguished. In this way those concerned with taking decisions are enabled to appreciate the issues involved. These may be so many that they cannot be presented on one map without one or more overlays being constructed. A map of this type for the West Midlands of England, prepared in connexion with the selection of a site for a new town, covers 6,000 square miles and consists of a 1/250,000 topographic base map and two transparent overlays on which are shown areas subject to various disadvantages. The notations in which these are presented are carefully designed so that none is opaque and all allow two or more factors to be shown as operating in the same area.

The major factors shown on the base map are: physical; land too high, land too steep, and land which for other reasons (such as liability to flooding) is unsuitable for building; locational and accessibility factors, such as areas more than five miles from a major road or railway. Competing land uses are indicated, including areas used or allocated for mineral working; areas of the highest quality agricultural land; Green Belt land; land reserved for afforestation or for its natural beauty. Further information on competing uses is shown on a first overlay which shows land liable to subsidence arising from the underground working of minerals; other good agricultural land; areas of high landscape value; and land occupied by government uses.

The second overlay deals with other aspects of physical limitations, position and accessibility, such as the restrictions on the supply of underground water, the tracts which lie within certain arbitrary radii of the main conurbation and other large cities and towns, and areas more than two miles from a major road or railway.

The map and overlays together thus present a cumulative picture of the various difficulties to which the building of a new town at any location within the study area would be subject and obversely suggest areas which it would seem most profitable to examine in detail. The technique is applicable to geographical factors which can be expressed in relation to the land which they affect. It is not possible to use it to portray imponderable economic and social factors.

MAPS OF LOCATIONAL PROBLEMS

Similar techniques have been applied to other subjects, such as the location of nuclear power stations, projects which by their size and nature must make a significant impact on the landscape and life of the area in which they are placed. Once the necessary maps have been compiled they are often found to be of great assistance in the examination of the problems of siting other developments which must somehow be fitted into a crowded land in which there are many societies and individuals ready to object to the alleged ill-chosen siting of any new project. Such champions of natural beauty are often in as much, or more, doubt than technical experts as to the visible effect of some proposal. Artists and photographers have often combined their skills to show what some new building or other proposed erection, such as a radio or television mast, would look like in a town or country landscape. Their illustrations raise, but do not answer, the question 'from how much countryside would it be visible?' A geographical technique can be used to supply the answer.

On a close-contoured map, such as the 1:25,000, which has a contour interval of 25 feet, the position of the proposed project is marked and radial lines are drawn from this at intervals of, usually, 5 degrees with intermediate rays at greater distances. Along each of these a profile is plotted. The project is then marked on each as a vertical feature and from it sight lines are projected against the profile which is then marked to show from which parts of the surface there is unobstructed vision of the project. Due allowance must be made for trees and other features, which can often be studied, and their heights assessed, from air photos. The portions of each radial line from which the feature may be seen are then marked on the map and this is then studied to enable an interpolation to be made of the intervening areas from which sight lines would be uninterrupted and so a map is compiled to show the ground from which a view of the proposed erection would be obtained. In practice it has been found useful to divide this into two classes, the areas from which, say, more than 75 per cent and less than 25 per cent of it would be visible. In the case of a structure whose bulk and colour would render it obtrusive from a considerable distance, allowance must be made for refraction and the curvature of the earth. With minor adaptations the technique has been used to suggest locations where certain projects would be likely to give rise to much less objection on grounds of prominence than on the sites first proposed by their sponsors.

The problem of where to site the many new developments which are inevitable in a changing industrial economy does indeed provide a challenge to the geographer. He must not merely find the answer, but must demonstrate both the truth and the consequences of his conclusion. An example of such a problem is the location of oil refineries. In recent years radical changes have occurred with the building of refineries in consumer rather than producer countries, with the vast expansion in their capacity and the development of the so-called 'super-tanker' for importing the crude oil. The use of these vast vessels, of from 65,000 to 100,000 or more tons, remarkably reduces the cost of conveying oil. But they demand harbours with a draught of more than 45 ft, a sheltered anchorage and a quick turn-round. At the unloading point interference by wind, tidal streams and other vessels must be avoided, while nearness to markets is important. The ideal refinery site is one of 1,000 or more acres on flat land, affording good foundations, with ample fresh water and from which distribution can be easily and economically made, and it must be adjacent to a sheltered anchorage where super-tankers can unload quickly and safely.

A careful analysis of the chief factor, deep water in sheltered anchorages, has been a vital prerequisite to the discussion of the siting of new refineries. Once made, it can be expressed quite simply in map form on large or small scale, enabling the significance of the real scarcity of such localities to be appreciated. Figure 13.4 shows the close relationship of sheltered anchorages, suitable for large tankers, to oil refineries of which the eight largest have been built, or greatly enlarged, since the Second World War (James, Scott and Willatts, 1961).

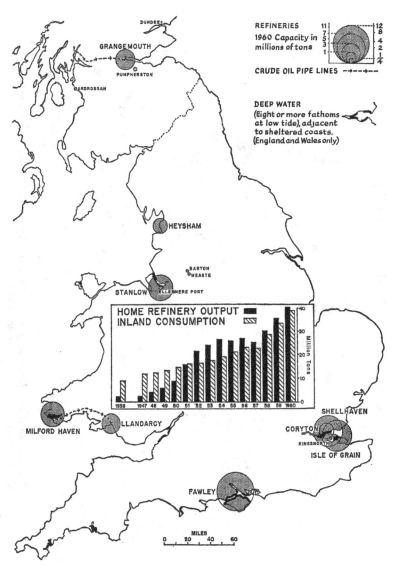

REFERENCES
1960 Capacity in millions of tons

11	12
7	8
5	4
3	2
1	¼

CRUDE OIL PIPE LINES →·→·→·

DEEP WATER
(Eight or more fathoms at low tide), adjacent to sheltered coasts. (England and Wales only)

DUNDEE

GRANGEMOUTH

PUMPHERSTON

ARDROSSAN

HEYSHAM

BARTON WEASTE

STANLOW ELLESMERE PORT

HOME REFINERY OUTPUT
INLAND CONSUMPTION

Million Tons

1938 1947 48 49 50 51 52 53 54 55 56 57 58 59 1960

MILFORD HAVEN

LLANDARCY

SHELLHAVEN

CORYTON

KINGSNORTH

ISLE OF GRAIN

FAWLEY

MILES
0   20   40   60

FIG. 13.4. *United Kingdom: Oil refineries, 1960, in relation to deep, sheltered anchorages. Another large refinery on Milford Haven has subsequently been completed.*

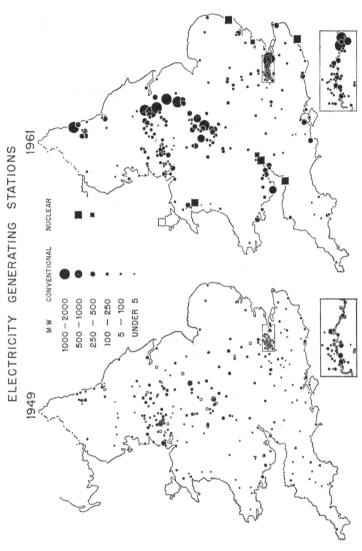

ELECTRICITY GENERATING STATIONS

1949   1961

M W   CONVENTIONAL   NUCLEAR

1000 – 2000
500 – 1000
250 – 500
100 – 250
5 – 100
UNDER 5

FIG. 13.5. *England and Wales: Electricity generating stations 1949 and 1961, completed or under construction. Projected stations in 1961 are in outline.*

Another power industry has provided a very different siting problem in which geographers have helped to evolve a solution. As shown in figure 13.5, the pattern of generation of electric power from coal-burning stations is changing from one of many small stations near the points of consumption to a few very large ones on or near the coalfields, whence the current is conveyed by high voltage cables. The greatest concentration of these in Britain is now in the Trent Valley where low grade small coal is readily available in proximity to the river from which cooling water is drawn.

Vast quantities of coal are burned (over 20,000 tons a day at one site) and from the group of power stations in this valley about three and a half million cubic yards of powdered fuel ash will soon be produced each year. This presents an enormous and growing problem of disposal, in spite of attempts to find economic uses for the material. However, the valley is also a very important source of gravel, worked from shallow wet pits, yielding 4½ million cubic yards per year. A careful study of the separate problems of the disposal of fuel ash, which tends to create a dust nuisance, and the phasing of gravel working, has led to arrangements whereby much of the ash is used to fill gravel pits, largely by pumping it as a slurry, while future gravel workings are planned in relation to the needs of the power stations for space in which to dispose of their ash. Thus it is proving possible to make the best and most economical use of land: the mineral is obtained and a waste product disposed of on the same sites, which can subsequently be restored to agriculture and additional land is not sterilized by being covered with heaps of waste ash. This solution was evolved by consultation and team work, in which geographers played a significant role, an important part, but only one of which, was expressing the facts quantitatively in map form so that the nature of the problem and the means of solving it were clearly evident to the policy makers. Figure 13.6 shows the relationship between the existing and proposed power stations in the middle Trent valley, their sources of fuel and the gravel pits to which some of the powdered fuel ash is conveyed. (The large station shown as 'projected' is not being constructed because the Ministry of Power subsequently rejected the application for consent to build it on that site.)

T

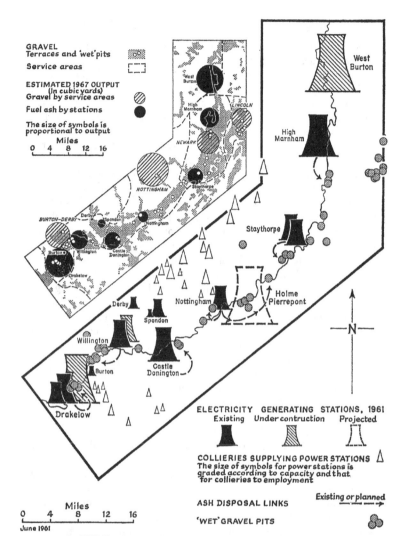

GRAVEL
Terraces and 'wet' pits
Service areas

ESTIMATED 1967 OUTPUT
(in cubic yards)
Gravel by service areas
Fuel ash by stations

The size of symbols is
proportional to output

Miles
0 4 8 12 16

West Burton
High Marnham
LINCOLN
NEWARK
Staythorpe
NOTTINGHAM
BURTON-DERBY
Derby Spondon
Nottingham
Burton Willington Castle Donington
Drakelow

West Burton

High Marnham

Staythorpe

Holme Pierrepont

Derby
Spondon
Nottingham

Willington
Burton
Castle Donington

Drakelow

—N—

ELECTRICITY GENERATING STATIONS, 1961
Existing   Under contruction   Projected

COLLIERIES SUPPLYING POWER STATIONS △
The size of symbols for power stations is
graded according to capacity and that
for collieries to employment

ASH DISPOSAL LINKS   Existing or planned

'WET' GRAVEL PITS

Miles
0 4 8 12 16
June 1961

FIG. 13.6. *Middle Trent Valley: The relationship between power stations,
fuel supplies and gravel pits, 1961.*

PROBLEMS IN LAND-USE COMPETITION

The extraction of minerals from underground or from the surface of our country, of which over 3,000 acres are annually made derelict by surface mineral working, gives rise to problems in the field of competing land uses which call for continuous study by both geographers and their colleagues, the geologists, and has been given close attention by S. H. Beaver and others (Beaver, 1944; Wooldridge and Beaver, 1950; Beaver, 1949). The problems raised can and do change with technical and social changes, but they still require to be studied. For example the development of modern earth-moving machinery has made it possible, and legislation has made it necessary, to restore to agricultural use land from which ironstone has been won by opencast workings. On the other hand, with the winning of gravel (now second only to coal in value and tonnage) the opposite trend has rapidly occurred. Only a few years ago it was generally thought to be very desirable to find means of filling the large lagoons left around parts of London and other large towns as a result of the extraction of river valley gravel. But changes in leisure habits have brought a rapid revolution. Lagoons or other stretches of water near large centres of population are now in great demand for recreational use particularly by sailing clubs, so that there has even been talk of a 'blue belt', as well as a green belt, around the metropolis.

The working of minerals which give rise to subsidence, particularly but not only, coal, gives rise to obvious problems in the siting of new developments in a country where 5,000,000 people still live on active coalfields. Large-scale development requires not only special structural precautions in building but in the detailed complementary phasing of mining and building. On the site of the New Town of Peterlee, Co. Durham, this was particularly important, for when its plans were first prepared there were about 30 million tons of coal beneath the site and although a cover of glacial drift tends to reduce the shock of subsidence, the fracturing of the underlying magnesian limestone causes serious surface distortion. Figure 13.7 shows, for the northern part of the town, how its development was phased in relation to the availability of land after the extraction of the underlying coal. A million tons had to be sterilized under the town centre, the early construction of which was imperative, but elsewhere detailed programmes were evolved so that building could take place, area by area, as the land became available after the coal had been mined.

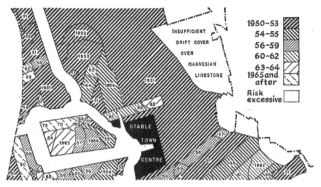

AVAILABILITY OF LAND FOR BUILDING

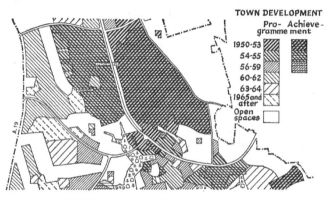

BUILDING PROGRAMME AND ACHIEVEMENT

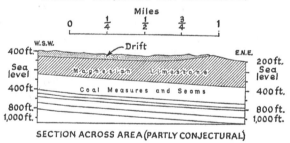

SECTION ACROSS AREA (PARTLY CONJECTURAL)

FIG. 13.7. *Peterlee, Co. Durham: Phasing of development in the northern part of the town in relation to the availability of land after the extraction of coal.*

MAPS OF URBAN RELATIONSHIPS

Many aspects of human geography fall to be investigated by geographers in connexion with planning. One which frequently occurs is the problem of 'journey to work', a study of which has many immediate practical applications. A general study of the numbers of persons who live in one place and work in another, as in the West Midlands conurbation, may reveal the very varying 'pull' of certain centres for employment. In such a complex area, a picture which simultaneously reveals the attraction exerted by a number of different towns is best achieved by the use of coloured symbols and the methods used in the Belgian *Atlas du Survey National* constitute an admirable technique. But there are many facets to the problem, and many false impressions which require correction by objective mapping. That a very large proportion of people working in the centres of large cities travel considerable distances to work is a common impression among the suburban 'commuters' to central London. But the facts revealed by the simple technique of showing the numbers who do make the journey from each locality suggests that popular impressions can be very misleading.

Figure 13.8 shows the pattern of commuting to central London as revealed by the 1951 census and emphasises that in general the largest numbers of workers travel relatively short distances. But, as Powell (1960) has pointed out, it also 'shows clearly that the daily influence of London in 1951 extended in some force as far afield as Brighton and Southend and in lesser degree to Luton and Reading'. It forms part of his study of the recent development of the London Region (bounded by the outer heavy line, the inner is the conurbation) which demonstrated that its planning, based on pre-war thinking, was outdated and that 'the expanding conurbation is the product of geographical and economic forces too powerful for man to reverse. He can only, within limits, direct them into convenient channels'.

The relationship between towns and the surrounding countryside which looks to the town for various services is a subject of great interest and importance to planners and many others, and is one in which geographical research and mapping techniques have been profitably applied to the recognition of centres, the delimitation of hinterlands and the estimation of their populations (Green, 1950). Such information is, for example, highly relevant to the development of central shopping and business districts of service centres. The subject now has a considerable literature.

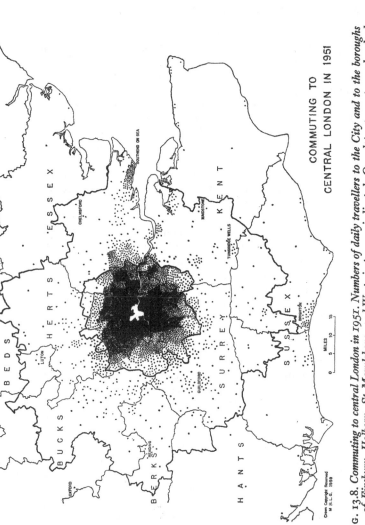

FIG. 13,8. *Commuting to central London in 1951. Numbers of daily travellers to the City and to the boroughs of Finsbury, Holborn, St Marylebone, and Westminster are indicated. One dot represents one hundred travellers, figures being rounded to the nearest hundred.*

COMMUTING TO
CENTRAL LONDON IN 1951

E S S E X

H E R T S

B E D S

B U C K S

B E R K S

H A N T S

S U R R E Y

S U S S E X

K E N T

SOUTHEND ON SEA

CHELMSFORD

MAIDSTONE

TUNBRIDGE WELLS

LUTON

GUILDFORD

READING

OXFORD

BRIGHTON

MILES
0   5   10   15

Crown Copyright Reserved
M.H.L.G. 1958

For the whole of Britain a map (Ordnance Survey, 1955) has been constructed and published showing the hinterlands of towns as determined by an analysis of bus services. The basic technique is to plot the complete bus route network for every centre, defined as a town having at least one regular service operating only to and from smaller places (which therefore look to it as a centre). By superimposing the diagrams on one another the hinterland boundaries are drawn in the same way that watershed boundaries may be drawn on a map showing the drainage pattern. Figure 13.9, reproduced from the

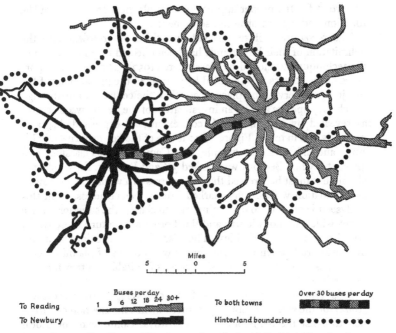

FIG. 13.9. *Reading and Newbury: A method of delimiting hinterlands.*

explanatory text published by the Ordnance Survey to accompany the ten-mile map, shows the method of delimiting the hinterland between Reading and Newbury. The population of each hinterland is calculated and on the final map, showing the boundaries of hinterlands, the population of each centre and its hinterland are shown by proportional concentric circles. The resultant map serves many purposes including the reconsideration of administrative boundaries, commercial sales organizations and the administration of medical, social and many other services.

It relates essentially to local accessibility, the relationship to their 'umland' of towns of local importance. These are towns of the 'Fourth Order' of importance, the single 'First Order' centre being the metropolis, the 'Second Order' the 'Provincial' cities such as Bristol, Birmingham and Leeds and the 'Third Order' the lesser towns of greatest regional importance such as Exeter and Lincoln. A realization, and understanding, of the hierarchy of service centres is important for the proper appreciation by planners and others of the function and pattern of urban settlements.

The recognition of the second and third order towns and the delimitation of their hinterlands has been effected by an extension of the technique, developed by Carruthers (1957). The relative importance of the centres was revealed by a study of the proportions of the bus journeys serving no place larger than the centre into or through which they operate and by an investigation of the ties between centres as revealed by bus and coach connexions, all expressed diagrammatically and in maps. Figure 13.10 is one of the key maps used in his study and indicates, for example, why Cambridge was classified as a 3A centre while Bedford was ranked 3B and Bury St. Edmunds 3C, although it must be stressed that the centres as a whole can be ranged in a gradation without any marked breaks. Nevertheless, from this analysis it is possible to determine, and to clarify, the superior service towns which are both fourth and third order and in turn those which are also second order cities. The classification provides a useful measuring rod which can assist in many problems of planning. One of the most recent is the consideration of suitable centres for new universities.

This type of technique was further developed by Carruthers (1962) for the study of suburban centres within the London conurbation to assist the deliberations of the Royal Commission on Local Government in London. Three separate analyses were made and presented on a series of maps. A study of the intensity of the equipment of banking, entertainment and selected shopping facilities in the central area of each 'centre' provides a 'status' classification. Figure 13.11 is a fragment of the map showing the status of the centres as determined by the provision of selected facilities. It shows how the varying grades of intensity of the six selected facilities graphically reveal the differences in status of these centres. The rateable value of

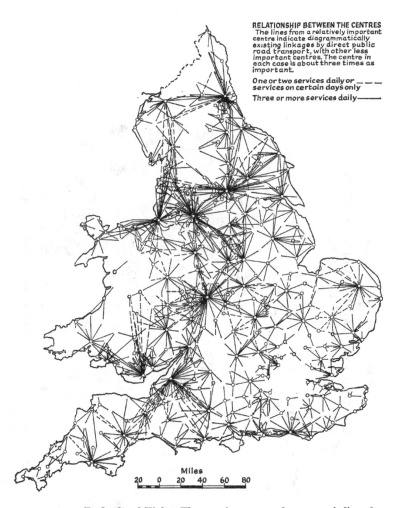

RELATIONSHIP BETWEEN THE CENTRES
The lines from a relatively important
centre indicate diagrammatically
existing linkages by direct public
road transport, with other less
important centres. The centre in
each case is about three times as
important.

One or two services daily or _ _ _
services on certain days only

Three or more services daily―――

Miles
20 0 20 40 60 80

FIG. 13.10. *England and Wales: The more important urban centres indicated by the public road transport links from other centres which are dependent on them.*

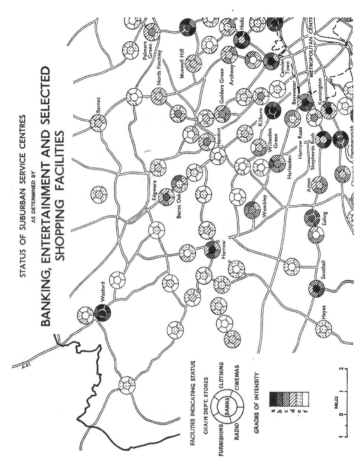

STATUS OF SUBURBAN SERVICE CENTRES

AS DETERMINED BY

BANKING, ENTERTAINMENT AND SELECTED
SHOPPING FACILITIES

FACILITIES INDICATING STATUS

CHAIN DEPT. STORES

FURNISHING / CLOTHING

RADIO / BANKS / CINEMAS

GRADES OF INTENSITY

a
b
c
d
e
f

MILES
0   1   2

FIG. 13.11. *North-west London: An extract from Carruthers' map showing the method of determining the status of suburban service centres.*

those shops was also recorded to give a qualitative evaluation and an analysis of the off-peak hour nodal bus journeys furnished another measure of the use of the centre.

A correlation of all three analyses furnished both a graded classification of centres and an indication of whether their provision and use are broadly equated, or whether they are under- or over-equipped, which is shown by whether the provision is low or high relative to value and nodality. As their author says, '. . . it is not sufficient to investigate only the provision and availability of services; an assessment of the intensity with which existing services are used offers, if anything, a more telling and sensitive index. This makes it possible to appreciate the true position of the centres in a constantly changing environment.'

MAPS OF REGIONAL RELATIONSHIPS

The recognition and analysis of distributions and relationships by geographers must take place at all levels from local to national, and studies on the latter scale can be particularly valuable. An example of a recent study of the whole of England and Wales is shown in figure 13.12, which shows the location and relative importance of almost all major post-war developments affecting land-use planning in England and Wales except increases in population and employment, which could not be effectively represented here in black and white.

The assembly of the facts is the first task in the preparation of such a map. But to present them so that their relative importance may be seen is less easy. For the purpose of this map the common factor of cost was used and a capital cost of £5 million was in general taken as qualifying for inclusion. The cost of certain projects, such as colleries, is published, but for others, e.g. manufacturing industry, an average cost per square foot for building and plant was applied. Each subject was normally graded into three sizes according to the relative magnitude of the projects in that subject and the size of the symbols used to represent the items was made exactly proportional to the average capital cost of the projects in the relevant grade of the particular subject. Thus the area of the symbols is a reasonably valid measure of the importance of the data they represent: the largest steel works, costing about £100 million, appears as ten times the area of the largest grade of colleries, which cost about £10 million.

Such a map reveals the extent to which many large new (or

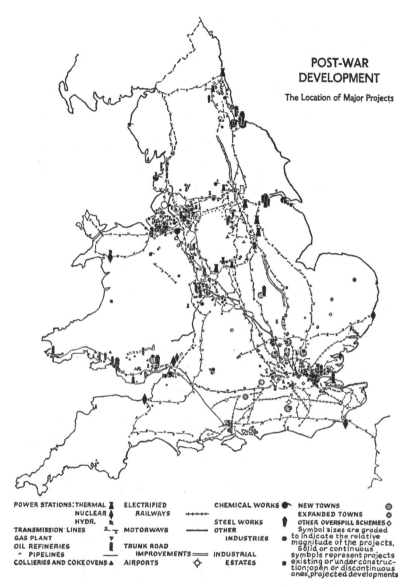

POST-WAR
DEVELOPMENT

The Location of Major Projects

POWER STATIONS: THERMAL	ELECTRIFIED RAILWAYS	CHEMICAL WORKS	NEW TOWNS
NUCLEAR			EXPANDED TOWNS
HYDR.	MOTORWAYS	STEEL WORKS	OTHER OVERSPILL SCHEMES
TRANSMISSION LINES		OTHER	Symbol sizes are graded
GAS PLANT	TRUNK ROAD	INDUSTRIES	to indicate the relative magnitude of the projects.
OIL REFINERIES	IMPROVEMENTS		Solid or continuous
" PIPELINES		INDUSTRIAL	symbols represent projects existing or under construc-
COLLIERIES AND COKE OVENS	AIRPORTS	ESTATES	tion; open or discontinuous ones, projected developments

FIG. 13.12. *England and Wales: Location of major post-war projects, 1961.*

expanded) projects have been concentrated in comparatively few localities linked by improved communications and power lines, although there have been some individual projects in more isolated localities (Willatts, 1962).

The uses to which such a map may be put are too numerous to discuss here but the point must be made that although the nature of the patterns on such a map may be very obvious, their significance is usually only to be understood after further study and analysis.

Figure 13.13 uses rating statistics to show the country-wide distribution of all industry. A symbol directly proportional to the value has been drawn in each local authority where the industrial rateable value exceeded £100,000. Thus, using what is perhaps the only useful comparable data, the map gives a quantitative impression of the general distribution of industry and comparison with the industrial symbols on figure 13.12 shows that there are significant differences between the total pattern of industry and that made by major post-war developments. This is even more true if the new projects of the latter are separated from the extensions of former developments. This has been done in figure 13.14 in which the new developments, whose nature can be read from figure 13.12, are shown to be making new patterns on the industrial map. Thus it may be seen that there is a very significant concentration, particularly by chemical, oil and steel plants, along the major estuaries and deep water havens such as the Humber, Southampton Water, Severn and Milford Haven, which hitherto had not been very attractive to industry. At the same time there has been an intensification of development on the Tees, Thames and Mersey. In the Greater London area, except in the New Towns and Luton, there have been few other new developments in the belt fifteen to forty miles from the centre. Otherwise in most of the major industrial axis from London to Lancashire and the West Riding most of the major post-war developments are expansions of older industrial plants. The reasons for these changes are complex and include the forces of geographical advantages, of government policy, the need for large new sites and the developments in the production and distribution of electric power which have emancipated industry from the bondage of the coalfields.

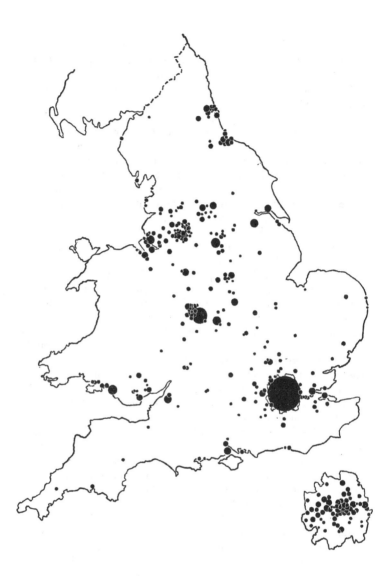

FIG. 13.13. *England and Wales: Distribution of industry as shown by rating statistics.*

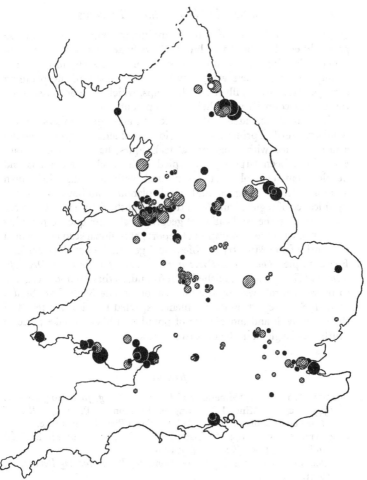

FIG. 13.14. *England and Wales: Major post-war industrial developments. New projects are shown by solid circles, extensions by shaded circles, and proposals by open circles, graded according to capital cost.*

CONCLUSION

This paper has been essentially concerned with examples of work carried out by a small, and changing, group of geographers employed in a headquarters section of the Ministry of Housing and Local Government. It has not attempted to deal with the work of geographers in the service of local planning authorities, although these probably employ about two hundred graduate geographers, both in general planning work and in research units, especially in the larger county councils. Their work may be seen in many of the planning surveys and analyses, illustrated by maps, which have been issued by the county councils with their Development Plans.

Nor has it considered the work of those geographers in the Ministry who deal primarily with regional problems and who, although also concerned with geographical techniques, have, with important exceptions, been largely preoccupied with development plans and development control 'case work' and with advising their own administrative colleagues and local planning authorities on the significance of regional and local trends. But the future is likely to see a new trend in the application of geographical techniques to planning problems, with the publication of reports of regional studies in which the work of the Ministry's professional geographers will be evident. For example, *The North-East: a Programme for Regional Development and Growth* was published in 1963 and, as this volume was going to press, was followed by the publication of *The South-East Study: 1961–1981*, a report on the problems expected to arise as a result of the big growth and movements of population likely to take place in south-east England in the next two decades.

References

BEAVER, S. H., 1944, 'Minerals and Planning', *Geog. Jour.*, **104**, 166–93.
— 1949, 'Surface Mineral Working in Relation to Planning', *Report. Town and Country Planning School*. Town Planning Institute.
CARRUTHERS, W. I., 1957, 'A Classification of Service Centres in England and Wales', *Geog. Jour.*, **123**, 371–85.
— 1962, 'Service Centres in Greater London', *Town Planning Review*, **33**, 5–31.
GREEN, F. H. W., 1950, 'Urban Hinterlands in England and Wales: an Analysis of Bus Services', *Geog. Jour.*, **116**, 64–81.
JAMES, J. R., SCOTT, S. F. and WILLATTS, E. C., 1961, 'Land Use and the Changing Power Industry of England and Wales', *Geog. Jour.*, **127**, 286–309.
ORDNANCE SURVEY, 1955, *Local Accessibility: the Hinterlands of Town and Other Centres as determined by an Analysis of Bus Services*, 1/625,000 (London).
POWELL, A. G., 1960, 'The Recent Development of Greater London', *Adv. Sci.*, **17**, 76–86.

VINCE, S. W. E., 1952, 'Reflections on the Structure and Distribution of Population in England and Wales, 1921–31', *Trans. Inst. Brit. Geog.*, Pub. No. 18, 53–76.

WILLATTS, E. C., 1962, 'Post-war Development: the Location of Major Projects in England and Wales', *Chartered Surveyor*, 94, 356–63.

— 1963, 'Some Principles and Problems of preparing Thematic Maps', *Proceedings. Conference of Commonwealth Survey Officers*, Cambridge.

WILLATTS, E. C. and NEWSON, M. G. C., 1953, 'The Geographical Pattern of Population Changes in England and Wales, 1921–51', *Geog. Jour.*, 119, 431–50.

WOOLDRIDGE, S. W. and BEAVER, S. H., 1950, 'The Working of Sand and Gravel in Britain: a Problem in Land Use', *Geog. Jour.*, 115, 42–57.

PART THREE

TEACHING

CHAPTER FOURTEEN

Geography in American High Schools

CLYDE F. KOHN

Professor of Geography, University of Iowa

The 1960's has been a decade of ferment in American education. Throughout the United States, there has been a growing tendency to undertake critical appraisals of teaching methods, to try out new approaches to learning, to incorporate new content into courses of study, to use new educational media, and to prepare experienced and prospective teachers to do a better job of educating American youth more efficiently and more effectively than ever before. In this movement, spurred by grants-in-aid from the federal government, scholars have joined with professional educators and classroom teachers in projects and studies designed to improve the quantity and quality of learning. These new developments are well reflected in the field of geography.

The assignment to develop an acceptable program in geographic education at the high school level is by no means an easy one. It involves a plan of action along at least five fronts: (1) a re-assessment of the role of geography in the secondary school curricula; (2) a development of new learning experiences for the student; (3) a reorganization of these learning experiences in some kind of orderly way so as to provide for the development of significant concepts and generalizations; (4) the preparation of new and interesting materials and teaching procedures for use in the classroom; and (5) the initiation of new programs of study for prospective teachers which will help them handle new concepts, methods of thinking, and materials as they appear. Before proceeding with a description of these new developments, it is wise to have in mind a brief outline of the developments of geography as an area of study in American high schools.

DEVELOPMENTS DURING THE PAST CENTURY

Geography is not now, nor ever has been, a well-intrenched subject in American high schools. During the past century, however, it has experienced a series of advances and retreats. Its content has also changed over the years, responding in large measure to changing emphases in geographic research and to social demands.

During the early part of the nineteenth century, attention centered on the study of individual countries. Students memorized answers to set questions concerning the location and character of places. Such a geography, then as now, has little scholarly appeal, and interest in the subject was never high.

A more systematic type of geography was introduced into a number of American high schools about the middle of the nineteenth century. These new courses undertook to study the distribution of natural features from a world point of view, and were based on the appearance of Fitch's *Outlines of Physical Geography*.

The study of physical geography was further advanced during the 1890's with the publication of the 'Report of the Committee of Ten of the National Education Association', headed by William Morris Davis of Harvard. Resulting courses were built around the principles of physical geography, rather than the memorization of descriptive facts. They included a study of the relation of the natural environment to man.

Courses based on the work of the Committee of Ten were well-planned, but proved difficult to teach. Most students found them dull and uninteresting, especially when taught by teachers who were poorly trained in physical geography as a laboratory science. They were unable to handle the subject matter of the course properly. Before long, interest in these courses waned, and by 1910 the study of physical geography no longer played an important role in the high school curriculum. Much of the content of the former course is now found, however, in courses called Earth Science.

At the turn of the twentieth century, the United States began to look outward to distribute its growing surplus of goods, and to invest its surplus capital. This development had repercussions in the educational world. High schools of commerce, and with them, the study of commercial geography, were introduced into the American educational system. Courses in commercial, or economic geography, were generally offered to non-college-bound students, and became intrenched in vocationally-oriented programs of study. In fact, during the 1930's, commercial geography was the only kind of geography to be represented in most American high schools; such courses are common even today.

Courses in commercial geography usually deal with major industries, their basic resources, and the leading commercial nations of the world. Throughout such courses, the all-important question of making a living appears again and again. The natural conditions and resources on which industries depend are kept constantly in the foreground, and students are expected to gain a working knowledge of the outstanding commercial nations of the world in terms of their production and consumption. The international significance of regions is developed by the ever-recurring theme of trade and commerce, a theme, which like that of making a living, runs throughout the course. A well-known textbook for such a course was prepared in the 1940's by Charles Colby and Alice Foster of the University of Chicago, entitled, *Economic Geography*. More recently written textbooks are also available.

There was a rebirth of interest in geography as an academic study with the beginning of World War II. The biggest impetus to the acceptance of geography as a *bona fide* subject of study for college-bound students came, however, during the 1950's. These courses were commonly organized as 'world geography'. They seek to place in perspective the economic, social, and polical conditions and problems of different parts of the world. They have their basis in the geographer's interest, expressed during the 1930's and 1940's, in the study of differences from place to place, a theme clearly enunciated in Hartshorne's *The Nature of Geography*. The same theme is further developed at the college level in regional courses so prevalent until recent times.

The least interesting of these courses are organized around national units, and present a description of one nation after another, much in the manner that the very early nineteenth-century courses were organized.

Fortunately, two more academically acceptable alternatives are available. One approach organizes the knowledge of the earth as the home of man in terms of physical, principally climate, regions. Such courses provide the general characteristics of particular parts of the earth's surface (drylands, semi-arid lands, hot humid lands, and so forth) and trace the ways people with different objectives, customs, and technical abilities live under similar climatic conditions. As a first course in world geography, its adherents claim that it makes pupils aware of the close relationship of man to his natural environment, a theme earlier introduced into the physical geography courses of the 1890's. Students learn that the houses people build, the crops they grow, the occupations they follow, the games they play, and the clothes they wear are all influenced by the climate of the region in which they live, depending on their culture and stage of technological development.

The second approach introduced in the late 1950's divides the world in terms of 'culture regions'. Each region is characterized by a set of common economic, social, and political conditions and problems. Within each there is a particular assortment of resources and habitat conditions, unique social institutions, and a way of life and problems that justify its definition as a major world region. Equally important, the entire area is experiencing similar forces for change and innovation, and has a particular relation to all other culture regions. A student who has had an opportunity to pursue such a course will not confuse Latin America with Africa south of the Sahara, nor the Soviet Union with South Asia where different sets of conditions and problems are to be observed. He will, instead, obtain a knowledge and understanding of the basic differences, natural and cultural, from place to place on the earth's surface.

RECENT DEVELOPMENTS IN AMERICAN HIGH SCHOOL GEOGRAPHY

As indicated earlier, the 1960's has been a decade of fermentation in American education. There is much evidence that extensive and profound revisions of the secondary school curriculum are taking place, and at an accelerating rate. The nationwide movement in curriculum reform and teaching practices emphasizes the understanding of principles rather than the acquisition of information. Thus, in the study of geography, instead of being asked to memorize facts about one culture region after another in today's world, students are being taught *how to* observe and interpret the character of places so that they might develop the skills and abilities to study regions by and for themselves later on in life. In the words of John W. Gardner, former U.S. Secretary of Health, Education, and Welfare: 'We are moving away from teaching things that readily become outmoded, and towards things that will have the greatest long-term effect on the young person's capacity to understand and perform. Increasing emphasis is being given to instruction in methods of analysis and synthesis, and modes of attack on problems.' Thus, teaching today is moving from what might be termed *expository teaching* to *guided discovery*. In the former, the teacher communicates to the students verbally. He, or the textbook, gives them the data, the concepts, and the generalizations developed by scholars as products to be memorized. In guided discovery, the teacher presents data – readings, statistical information, audio-visual materials, and so forth – and asks a series of questions in order to challenge the student to formulate hypotheses and make generalizations from them. The extent of the teacher's guidance, that is, the clues or questions offered, and the frequency and nature of rewards necessary to sustain inquiry, vary with the sophistication of the student, the skill of the teacher, and the particular objectives of a day's instruction. The materials and learning activities of the High School Geography Project have been structured according to an inquiry or discovery method, and encourage innovative teaching strategies. They develop skills in the use of data to arrive at generalizations applicable to a wide variety of circumstances. New materials being developed by commercial publishing firms also attempt to create in students a fuller understanding of the conceptual structure of the discipline, and methods of inquiry that are geographic in nature. Thus, both the High School Geography Project and commercial publishing firms are providing students with more interesting and lively data, and are asking them to deal directly with the kinds of questions that geographers ask.

The set of units being developed by the current High School Geography Project differ from the units now currently being taught in American high schools in regional geography courses not so much in terms of their orientation towards the inquiry method as in the nature of the overall purpose of the course. The present High School Geography Project course helps students learn how to develop generalizations, laws, models, or theories and to apply them to man's economic, social, or political behavior. By understanding certain generalizations or regularities in the location and distribution of phenomena over the earth's surface, the student is better able to answer the question asked by all men, 'What do I, or we, do next?' The hypotheses, or generalizations, developed in the course are looked upon as analytical tools for the solution of everyday problems. Thus, the course attempts to develop analytic concepts and the analytic mode of thought, and to apply these concepts in decision-making experiences.

In contrast, the regional courses now taught in most American high schools, although incorporating the inquiry method of learning, focus on integrative concepts and the integrative mode of thought. They lead to an understanding of places rather than to policy-making activities.

To understand this fundamental difference between what is currently being taught in high school regional geography courses, and the new units of study being developed by the High School Geography Project, it is necessary to consider these two modes of thought, the analytic and the integrative, at greater length.

PRESENT REGIONAL GEOGRAPHY COURSES: EXPERIENCES IN THE INTEGRATIVE MODE OF THOUGHT

The study of natural or culture regions, like the study of 'periods' in history, calls for the development of a particular mode of thought, commonly referred to as the integrative mode. Learning depends on the development of four inquiry processes: observation, analysis, holistic integration, and application.

Observations in the integrative mode are as comprehensive as possible, and attempt to include all significantly knowable facts and features of the area under consideration. The observations are made with an eye to the holistic integration which takes place later in the inquiry process. They center on the distribution of economic, social, political, and physical phenomena that give the region its unique character. In the classroom, these observations are gained through such media as maps, documents, still or motion pictures, statistics, reports of interviews, or textual materials.

Analysis consists mainly of noting similarities and differences from place to place resulting from the unequal distribution of physical and cultural phenomena over the earth's surface. In the integrative mode there is also a second kind of analysis, the comparison of regions with some known culture region like one's own. It is through this kind of comparison that the inquirer generates and tests *believability*, the over-riding standard for assessing the validity of inquiry in integrative thought processes.

Thought in the integrative mode attempts to examine a single area, and to describe and interpret the nature of that region. To do so, it employs a wide variety of contributions from analytical thought, as well as from cultural knowledge and intuition, in order to provide the richest and most varied set of perspectives from which the area under study can be viewed and understood. Hence, the integrative mood is more subjective than objective. It is not within the scope of the scientific method. Rather, holistic integration focusses on the unique qualities of particular places at particular times, trying to understand these places as a whole. In terms of culture regions, for example, holistic integration results from treating a given cultural area as an entity whose parts or processes are mutually supporting or reinforcing. The integration may involve focussing upon one or more institutions or human processes that seem to permeate life in a particular society. For example, James in *One World Divided* helps the student interpret a culture region in terms of its stage of economic and political development. Kohn and Drummond in their textbook, *The World Today: Its Patterns and Cultures*, help students interpret a region by noting its development in terms of the transition theory of population growth, its stage of industrialization and urbanization, its ability to provide food for its peoples, and the nature of its political and social systems.

Conclusions drawn from an integrative study of a culture region can be applied to different places only with special caution. For example, conclusions drawn from an integrative study of Latin America can help the inquirer to ask questions about the Orient, but he must be extremely careful not to overgeneralize from his study of Latin America to that of the Orient. The danger of overgeneralization in integrative inquiry is that conclusions from the study of one culture region might be misapplied to a quite different region.

THE HIGH SCHOOL GEOGRAPHY PROJECT: AN EXPERIENCE IN THE ANALYTIC MODE OF THOUGHT

Geography in an Urban Age, the course prepared by the High School Geography Project, differs substantially from the regionally-oriented courses now being taught in the American high school. This new one-year course for students in grades 9–12 focusses on a 'settlement' theme. There are seven units in the course, very loosely related to one another. The course is designed to rely on student inquiry and analysis in determining why cities, industries, and institutions are located where they are. A variety of audio-visual materials, educational games, pupil activities, and map experiences are employed to develop the major skills and concepts of the course.

Unit I deals directly with cities. Emphasis is placed upon the selection of city sites, the factors which influence city growth, urban land-use patterns, and inter-relationships of urban places. The central feature of the unit is the construction of a hypothetical city, Portsville, using historic data from Seattle, Washington.

Unit II centers on economic geography concentrating on manufacturing and agriculture as related to urban settlements. The importance of farm and factory to man, how they affect his landscape, and how they are affected by location are some of the issues presented. The two major activities in this unit involve decision-making and role-playing games which deal with the location of a factory and the risk involved in American farming.

Unit III deals with cultural relativity by examining various ways different cultures view cattle. The concept of cultural diffusion is developed through game activities, one dealing with sports and the other with the expansion of Islam. Unit III also develops the notion of culture regions, to which both the Islam activity and one on regions in Canada contribute. A summary activity introduces students to the concept of cultural uniformity through modernization.

Unit IV carries students through the geography of political processes by developing such basic concepts as territoriality and political hierarchies, and moving to problems of legislation and boundaries. There is a role-playing game in which students, representing a variety of political interests, are faced with the need to compromise.

Unit V is designed to acquaint students with the elements of man's habitat. Water and energy provide the basis for considering patterns of human interaction with the habitat. Attention is given to problems of resource utilization, and a variety of educational media is used in teaching the unit.

Unit VI on Japan is the one regionally-oriented unit in the course. Building upon the previous units, it focusses attention upon a country that is rapidly moving through the processes of modernization, urbanization, and industrialization. Emphasis is on student use of graphed and mapped data to check hypotheses.

Unit VII involves students in research on topics considered to be on the frontiers of geographic research. This relatively short unit is designed for the more able students.

Except for Unit VI (Japan), which calls for the application of the integrative mode of thought, the units designed by the High School

Geography Project depend on the student's ability to handle the learning processes involved in the analytic mode of thought.

To initiate the analytic mode, students need to be confronted with some kind of problem. In geographic studies, the problem is either to account for the location and distribution of selected phenomena, or of the spatial interaction between two or more places; or it may be a behaviorally-oriented problem that asks the student to make a decision, such as to locate a firm or decide what to plant, and so on.

Observations are made in terms of the statement of problem. They may be *direct*, if field work is involved, or *mediated*, that is, obtained through such media as documents, photographs, statistics, maps, films, and so on, much as in the case of integrative studies. It should be noted that to collect the necessary data involves a number of abilities – the ability to observe carefully in the out-of-doors and to record one's observations; the ability to read tables and charts; the ability to obtain facts from maps and globes; the ability to read pictures of all kinds including aerial photographs. It is the duty of the teacher to help students gain these abilities in order to make the proper observations.

Customarily, analysis proceeds by setting up certain hypotheses that might account for the distribution or spatial interaction being studied. This process requires the selection and definition of one or more independent variables, which in their turn, must be observed and classified before analysis can continue.

Once the dependent and independent variables have been established, the spatial identities and contrasts among these categories can be studied. In other words, the hypotheses can be tested. If the hypotheses are found to be valid, they are accepted as statements of relationships, or generalizations. If they are found not to be valid, they are rejected and the search is continued with a different set of hypotheses. It should be noted that this type of analysis differs from that employed in the integrative mode. The latter does not attempt to set up hypotheses about the distribution of phenomena at one time, nor of their redistribution over time. In contrast, though in the integrative mode attempts to examine a single area, and to describe and interpret the holistic nature of that area.

The products of inquiry in the analytic mode (hypotheses, generalizations, models, or theories) may be used for further inquiry in both the analytic and integrative processes of thought. For example, a generalization, such as 'variations from place to place in the density of rural farm population are directly related to the variations in average annual precipitation' can be used in developing further generalizations about population distribution (analytic mode), or in understanding population distribution in a particular society (integrative mode).

Application of generalizations resulting from the analytic method of inquiry are based on the assumption that acceptable patterns of behavior in reality conform to logical relationships. If repeated observations fail to confirm the applicability of the generalization to reality, then the generalization should be modified or discarded as being useless. Thus, the outcome of work in the analytic mode can always be evaluated, according to a very important standard: its *scientific fit*. Does it contribute to the development of a logically consistent conceptual frame that is useful in understanding reality, and has predictive and heuristic value? Analytic concepts and generalizations developed in the units of study of the High School Geography Project will be educationally acceptable only if they enable the student to make logically valid decisions about the location of phenomena within cities, about the location of a firm, about the kind of crops to grow given certain constraints, and so on.

THE NEED FOR TEACHING BOTH MODES OF THOUGHT

To summarize, thought in the integrative mode attempts to examine only a single region, such as 'India', 'Australia', or 'The Arab World'. It does so by employing a wide variety of contributions from analytic thought, from cultural knowledge, and from intuition. From these sources, the student is able to develop a rich and varied set of perspectives with which the region under study can be viewed and understood.

Thought in the analytic mode focusses upon an examination of a small set of closely defined distinctions as they are embodied in a wide range of social, economic, political, and physical phenomena. An example is the analysis of the spread of the Islam culture in order to develop generalizations about the innovation and diffusion of an idea.

To understand reality from a geographer's point of view, both analytic and integrative modes of thought must be developed. Insights can be derived from a close analysis of the separate activities in which mankind is engaged regardless of time and place; other insights can be derived from an integrative view of man in a specific culture region.

Also, since no analysis is ever finished and unchangeable, the search for new dimensions and more explicit distinctions is a continuous one. With each new analysis there comes a need for a new integration. Also, the job of integration is an unending one. The concepts and generalizations resulting from the analytic mode of thought are viable only at a given point in time. They do not provide a sufficient basis to assure understanding of all future problems.

Thus geographic instruction in American high schools is in the process of changing from a factually-oriented course of study, to one based on inquiry methods. The units of study in the High School Geography Project, save for the unit on Japan, are designed to help students develop greater analytic powers. Regionally-oriented courses are being revised to develop integrative modes of thought.

Perhaps the future will see the development of courses that help students develop both modes of thought. Such a course might, for example, attempt first to develop analytic concepts and generalizations concerning the distribution and growth of population, increase in the world's supply of food, industrialization and urbanization, and the political systems by which men are governed; and then help students understand the major culture regions in terms of these analytic concepts and generalizations.

At this particular time in the history of American educational reform, it is not clear just what the role of geography in the American high schools will be, nor what kinds of learning experiences in geography high school students will be afforded in the immediate future. How widespread the acceptance of analytically-oriented units of study developed by the High School Geography Project will be is still an unknown. The acceptance of a regionally-oriented course based on the development of the integrative mode of thought has not been so widespread as had been anticipated early in the 1960's. Much remains to be done to impress upon high school administrators and supervisors the importance of geography education at the high school level if many of the problems now confronting Americans, both at the national and international levels, are to be solved in a satisfactory manner.

References

BRIGHAM, A. P. and DODGE, E. E., 1933, 'Nineteenth-Century Textbooks of Geography', *The Thirty-Second Yearbook of the National*

Society for the Study of Education (Public School Publishing Co., Bloomington, Indiana).

COHEN, S. (Ed.), 1967, *Problems and Trends in American Geography* (Basic Books, Inc., New York), pp. 251–63.

DAVIS, W. M., 1954, 'The Progress of Geography in the Schools', in *Geographical Essays* (Dover Reprint, New York), pp. 23–69.

HELBURN, N., 1968, 'The Educational Objectives of High School Geography', *Jour. Geog.*, **67** (5), 274–81.

MAYO, W. L., 1965, *The Development and Status of Secondary School Geography in the United States and Canada* (University Publishers, Ann Arbor, Michigan).

CHAPTER FIFTEEN

Recent Trends in Undergraduate Geographic Training in American Universities and Colleges

PLACIDO LAVALLE

Professor of Geography, University of Windsor, Ontario

Within the last two decades, American geographers have been involved in a serious re-evaluation of their undergraduate college and university curricula. The motivation for this re-evaluation of geographic training may be attributed to the following factors: (1) the increased demand for geographers in education, government, and industry; (2) the increasing concern on the part of many professional geographers to firmly establish geography in its rightful place among the sciences; and (3) the revolutionary changes in geographic technology.

American geographers are attempting to restructure their geography programs to meet the demands of the times. Evidence of this concern for the improvement of undergraduate training may be found in the recent publications of the Association of American Geographers' Commission on College Geography (see Lounsbury, 1965; Hart, 1968). As a consequence of this concern for improved undergraduate geographic training, several institutions have initiated a series of major changes in their geography curricula which are being closely monitored by the American Geographic Profession. Within the last decade several new geography courses have been incorporated into the undergraduate offerings of many universities and colleges. Recently, several new approaches to the teaching of traditional geography courses have been published in the Association of American Geographers' Commission on College Geography reports (see Randall, 1967; Cohen, 1967). Many other institutions have been involved in a gradual modification of already existing programs. In

W

this report an attempt will be made to describe some of these innovations and some of the salient features of American undergraduate geography curricula.

American universities and colleges are characterized by a great diversity of organizational structures and programs (Schwendeman, 1968). Over 1,440 universities and colleges offer geography courses, but only 351 institutions have a geography department. Graduate work leading to either a master's degree or a doctorate is offered in only 120 American institutions.

When examining the differences in geography curricula which exist among American universities and colleges, it is most useful to differentiate those institutions which do offer graduate work in geography from those which offer an undergraduate program only. Universities having graduate geography programs tend to have sufficient personnel and resources to offer a versatile program. The larger faculties of these departments also increase the probable number of communication links with the rest of the profession which can be used to monitor more effectively current trends in geographic methodology. In general, those departments offering graduate programs tend to be the first departments to offer new programs or courses, due to their having the resources necessary to initiate such experiments.

Smaller departments not offering graduate work are often restricted in the variety of courses which they may offer. Such departments usually concentrate on teaching only the basic courses in cultural, physical, and regional geography. However, there are some notable exceptions to this generalization which will be discussed later in this paper.

It is important to make the distinction between the privately endowed institutions on one hand and the publicly supported institutions on the other. Within the United States, one will find that geography tends to occupy a more prominent position in those colleges and universities which are supported by state and local government agencies. In spite of the fact that geography was first taught in the private colleges and universities, one finds that a smaller proportion of American privately endowed colleges and universities offer an undergraduate major in geography, and even a smaller proportion of the privately supported American institutions offer any form of a graduate geography program. One possible explanation of this situation is that many of the publicly supported institutions put a strong emphasis on teacher training, and traditionally the primary employer of geography graduates has been educational institutions. However, it is also necessary to examine the history of geography in American university education in order to obtain a greater insight into this situation.

THE HISTORY OF GEOGRAPHY IN AMERICAN COLLEGES AND UNIVERSITIES

Throughout the history of American education, geography has alternatively been included and excluded from university curricula. Perhaps the first appearance of geography on the American academic scene occurred shortly after the establishment of Harvard College in the middle of the seventeenth century. Early records indicate that a knowledge of geography was a prerequisite to entrance into early colonial colleges, and geographic training was a part of colonial college curricula. However, after 1816 geography was eliminated from most university programs, and geography did not return to the American university scene until 1854 when Arnold Guyot was appointed Professor of Physical Geography at Princeton. From 1860 through 1900 geography courses were slowly introduced into a number of new universities and colleges; and during this era, geography courses were re-introduced to a number of eastern universities. The first American graduate department of geography was established in 1903 at the University of Chicago. After 1903 geography expanded rapidly into a large number of state institutions. However, it did not enjoy the same rate of growth in the privately endowed institutions and was eliminated from many curricula after World War I.

Up until World War I, American university geography was strongly oriented in the direction of physical geography and cartography, but after World War I emphasis on physical geography declined while greater emphasis was placed on cultural, economic, and regional geography. During the era between the two World Wars, American geographers had a tendency to concentrate their efforts on microgeographic studies and large scale regional synthesis. Thus geography programs which were developed during this era tended to be characterized by a large number of regional geography courses and an increased number of courses in cultural, historical, and economic geography. Also during this period, courses in geomorphology, climatology, soils, and plant geography tended to lodge in geology, meteorology, agronomy, and botany departments, although recently some of these trends have been reversed.

After World War II, there was a gradual shift away from the emphasis on regional geography to a stronger emphasis on systematic geography. Several new geography courses gradually became firmly established in the programs of various departments including courses in population geography, urban geography, and air photo interpretation. During the postwar era, there was a stronger emphasis on economic geography which eventually led to the emergence of mathematical statistics as a major geographic research technique. In part, this trend may be related to the increased demand for geographers by government and industry.

With the advent of modern high speed computers and the development of sophisticated mathematical techniques of spatial analysis, geographers have become more concerned with providing geography students with an adequate background in mathematical and statistical methodology. The increased demand for geographers in government and industry has encouraged the training of geography graduates who are capable of carrying out the more sophisticated forms of spatial analyses. In order to provide young geographers with an adequate background in quantitative spatial analysis, some geography departments have inaugurated special courses in quantitative techniques, while other departments have encouraged their students to take additional quantitative work in statistics or mathematics departments. However, the introduction of quantitative techniques into American geographic programs was met with considerable opposition from the more conservative members of the geographic profession (see Spate, 1960; Robinson, 1962; Burton, 1963). Numerous arguments raged over the suitability of the quantitative approach as it was applied to geography. Some of these arguments are still not resolved, but for the most part quantification has become an established part of American geographic methodology. Therefore, one of the most significant and recent developments in contemporary geographic curricula has been the inclusion of quantitative methodology in some of the traditional geography courses and the addition of specialized quantitative techniques courses to many geography programs.

Coincident with the increased emphasis on quantification have been the efforts to amplify the theoretical aspects of geography which may have been motivated by the desire to place the subject on a more solid scientific footing. Many of the recent revisions of standard elementary courses in cultural, economic, and physical geography have emphasized theory and the development of theoretical concepts. This represents a trend away from the more traditional emphasis on the description of areas, but this phenomenon seems to be currently restricted to a small number of American institutions.

GENERAL ASPECTS OF AMERICAN UNDER-GRADUATE TRAINING

Compared to British universities, American universities tend to require their undergraduates to take more general educational survey and enrichment courses while a smaller proportion of the undergraduate's time is allocated to courses in his field of specialization. In British universities the undergraduate concentrates in his chosen speciality throughout his undergraduate program, but in American institutions the student in many cases may not start work in his ultimate major field until his third or fourth year. The British undergraduate usually takes courses in his major field of interest or in closely allied cognate fields. The American student is exposed to a great variety of courses in a large number of departments during his undergraduate career. Some of these courses represent university general education requirements while others are purely electives. Thus, training in American institutions tends to cover a broader mixture of subject matter. Also the American undergraduate program in geography tends to be less concentrated than its British counterpart.

Usually, the American undergraduate is required to take a certain number of courses in English composition, the physical sciences, the biological sciences, mathematics, the social sciences, physical education, foreign languages, and the humanities. These requirements vary from university to university, but they usually make up from two-thirds to three-quarters of the undergraduate's course load. The geography major in American universities usually consists of twenty-four to thirty-six semester hours in geography courses. Many departments also encourage their students to take additional work in mathematics, statistics, geology, economics, sociology, anthropology, history, political science and biology. Frequently these additional courses can be used by the student to meet university general education requirements. It is interesting to note that mathematics or statistics courses were most frequently cited by departmental chairmen as being recommended to undergraduate majors. However, few departments require their majors to take work in mathematics or statistics as part of the geography program.

AMERICAN GEOGRAPHY DEPARTMENT DIFFERENCES

An important aspect of American undergraduate geographic training is that there are tremendous variations in the structure of the geography major program which exist among the various universities and colleges. Some of these variations may be of a regional character while others are related to the methodological orientation of the personnel found in the various departments. Some of the variations in the structure of the geography program may be traced to the educational orientation of the institution offering undergraduate courses in geography, while other sources of curricula variation may be traced to the relative size of the individual geography departments.

In order to cope with the great variety of undergraduate geography curricula found in the United States, this survey will divide the American institutions into the following classes: (1) those universities offering bachelor's, master's and doctoral degrees in geography; (2) those privately endowed colleges offering only a bachelor's or master's degree in geography; and (3) those public institutions offering either a bachelor's or a master's degree in geography. The institutions offering doctorates in geography as well as bachelor's degree are grouped together, for these are the institutions which supply most of the faculty personnel to the other American colleges and universities offering degree programs in geography. The privately endowed colleges are separated from the remaining institutions, because geography in these institutions generally does not have the same status that geography has in the tax supported colleges and universities.

Geography curricula in those departments which offer a doctoral program tend to be more diversified than those found in other institutions, and often the geography curricula of departments which do not offer the doctorate are strongly influenced by the curricula of nearby doctoral departments.

GEOGRAPHY CURRICULA IN DOCTORAL DEPARTMENTS

Generally, the undergraduate major course in a large geography department which offers a doctorate consists of a core program made up of courses in cultural geography, economic geography, physical geography, cartography, and research methods. This is often supplemented by additional regional geography, systematic geography, and research methods courses. In addition, many departments require of their majors a senior seminar or colloquium in some aspect of geography. The variety of undergraduate course offerings is often conditioned by the size and methodological orientation of the department.

Usually a small department which offers a doctoral program will structure the undergraduate curricula around a small core program of introductory courses in physical geography, cultural geography, economic geography and some form of research techniques. These courses will be augmented by a restricted variety of regional courses and systematic specialties which are determined by the research orientation of departmental personnel. If the small department has a strong traditional orientation, one will find a fair variety of regional offerings, but these departments will tend to offer a limited program of systematic and research methods courses. One will often find few offerings, if any, in quantitative methods, and less than half of them will offer undergraduate field techniques courses. Usually the only techniques courses found in these departments are cartography courses.

The University of Texas is a good example of a relatively small department with a more traditional methodological orientation which offers undergraduate courses in physical world regional, and economic geography. Although this department offers two courses in cartography and several regional courses, it does not offer a wide variety of undergraduate courses in the systematic fields of geography or in geography techniques.

A second trend in relatively small geography departments which offer doctoral work, can be found in those departments with a particularly strong orientation toward one aspect of systematic geography, for example, economic geography. These departments will usually offer a larger variety of techniques courses, such as quantitative methods, than departments which have a regional geography orientation. Some examples of departments in this category are those at

Northwestern University, the University of Iowa, and the University of Cincinnati.

At the University of Iowa, the undergraduate program starts with an Introduction to Geography course followed by a core program of basic courses in social, political, economic, urban, and physical geography. These courses are then supplemented by others in cartography, field techniques, and quantitative methods. The department also offers intermediate and advanced level courses in physical, social, economic, political, and urban geography. There is a strong emphasis on research methods and the development of geographic theory. Perhaps one of the most unique aspects of the Iowa program is that the introductory course is focussed on the development of geographic concepts and theory rather than a descriptive world survey of human activities or natural landscape patterns. Most American undergraduate programs start with an introductory course in physical, human, or world regional geography, and these courses do not emphasize the theoretical aspects of geography to the same extent as the Iowa course. Thus Iowa's introductory course represents one of the most significant innovations in the undergraduate training of American geographers. Already several other institutions have incorporated similar courses into their curricula.

Undergraduate geography in the larger departments offering doctoral programs tends to be characterized by a wide spectrum of course offerings, but undergraduate majors in these departments usually start their geographic studies with introductory courses in physical geography, cultural geography, and map reading or an introduction to geographic methods. After the introductory sequence, the undergraduate is then faced with a wide choice of courses in regional geography, systematic geography, and in geographic techniques. Usually the department will specify that the student should choose a certain number of courses from a regional specialty, systematic specialty, and from the geographic techniques group. A few of these larger institutions will also require that all students take a course or courses in either cartography, field techniques, or quantitative methods. The University of California at Berkeley, University of California at Los Angeles and the Universities of Georgia, Minnesota, Wisconsin, Illinois, Michigan, Kansas and Michigan State University fall into this category.

At the University of California, Los Angeles, the undergraduate starts with a three-course sequence in physical geography, cultural geography, and an introduction to geographical analysis which is then followed by intermediate level courses in physical geography, cultural geography, field techniques, and other elective courses in geography. The student chooses his electives from a wide variety of courses but he must choose a certain number of courses from the regional geography group, physical geography group, and economic geography group. In such a department, a student has the opportunity to sample a large variety of specialties.

GEOGRAPHY IN PRIVATE COLLEGES AND UNIVERSITIES

Geography in the privately endowed institutions, which do not offer a doctoral program, tends to be restricted to a narrow range of course offerings with a strong emphasis on human and regional geography. Undergraduate majors in such departments are required to take introductory courses in human, physical geography and world regional geography. This is often followed by numerous regional geography courses plus a restricted number of systematic specialty courses. Usually one does not encounter a strong emphasis on research techniques courses in these departments (see Table 18). Of the twenty-five privately endowed institutions sampled, four offered courses in field methods, and only 70 per cent offered cartography; physical geography does not tend to be emphasized as much as cultural geography. Both of these trends may be related to the fact that these departments tend to be small. Another factor which may be responsible for these trends is that geography in these departments frequently originated as a series of social science service courses for history, political science, or economics programs. Based on our sample, it was found that such departments tended to offer only the more traditional courses in regional and systematic geography; thus their programs do not seem to have a strong emphasis on research techniques and the development of geographic theory.

GEOGRAPHY IN PUBLIC NON-DOCTORAL INSTITUTIONS

The largest group of American geography departments is found in publically supported institutions where a Ph.D. in geography is not offered. This group includes departments in state universities, state teachers colleges, city colleges and other public four-year institutions.

Curricula and program trends in these institutions tend to be patterned after those in institutions offering teacher training programs which has indirectly supported their emphasis on regional geography. Although their geography curricula are modeled after those in the doctoral departments, new course innovations tend to be adopted several years after they are established in major doctoral departments. Evidence of this trend can be found in Table 18 where it can be seen that a smaller proportion of the non-doctoral public university and college departments offer specialized courses in quantitative methods and air photo interpretation. However, these departments tend to offer more traditional cartography and field techniques courses in roughly the same proportions as the doctoral departments. Another trend which can be found in the non-doctoral government supported departments is that the variety and nature of course offerings tends to be conditioned by department size and methodological orientation. The larger departments will offer an almost complete list of systematic regional, and research methods courses. It is in these larger departments that one is most likely to encounter courses in quantitative methods and air photo interpretation. The smaller departments tend to offer a much smaller variety of courses, and they tend to concentrate on more traditional course offerings in regional geography, conservation, economic geography and physical geography.

Some small geography departments, such as that in the University of California at Irvine, have been involved in the development of some exciting new approaches to the teaching of geography.

At Irvine, undergraduates start with an introduction to geography course that is similar in nature to the introduction to geography course given at the University of Iowa, but it is then followed by a three-course intermediate level sequence in geographic analysis, which is supplemented by courses in urban geography, conservation, methods of regional analysis, and other electives in research methods or systematic geography. This is followed by a three-course senior level seminar sequence which emphasizes the solution of geographic problems. In the first course of this sequence the student concentrates on a review of the literature pertinent to a particular problem, and in the second course he concentrates on study design and data collection, which is then followed by a third course where the student makes an analysis of his problem and completes his thesis. This curriculum is basically one in human geography, and it is part of an interdisciplinary program leading to a bachelors degree in social science. Very little emphasis is placed on physical geography or on

Table 18. Frequency of Undergraduate Course Offerings by Institutional Type*

	Four-year University with doctoral program	Four-year Private institution without doctoral program	Four-year Public institution without doctoral program
Number Sampled	35	25	70
Quantitative methods	23	0	12
Air photos & Remote sensors	21	3	23
Field techniques	24	4	36
History & Geographic Thought	17	2	20
Cartography	34	17	58
Population geography	10	1	12
Cultural geography	25	8	45
Political geography	30	15	56
Economic geography	34	15	62
Industrial geography	24	6	19
Agricultural geography	10	0	13
Urban geography	33	14	52
Conservation	23	10	44
Geomorphology	25	5	38
Climatology	28	12	58
Soils	7	1	10
Biogeography	12	0	7
Research methods	14	5	23

* Compiled from University and College catalogues, calendars and brochures.

the traditional regional geography courses. In addition, a strong emphasis is placed on training in mathematics which is rare in small departments. Some geographers feel that this approach represents the wave of the future, but this has been contested by geographic conservatives.

Otherwise the small non-doctoral state and local university under-graduate geography programs usually consist of a core program and a small list of electives. The core program will usually consist of courses in physical geography, cultural geography, world regional geography, and sometimes map reading. In the more traditionally oriented departments this will be followed by several courses in regional geography plus one or two systematic courses. In other departments the emphasis on regional geography courses will give way to an emphasis in course work in geographic methods, physical geography, cultural geography, economic geography, conservation or planning.

In general the emphasis on methodology courses at all non-doctoral public institutions tends to be much less than that which is found at the doctoral granting departments. Again, the major exception to this rule is found at the University of California, Irvine. In general, one finds that there is a smaller chance of finding quantitative methods, air photo interpretation of courses on the history and philosophy of geography in the non-doctoral departments (see Table 18). A smaller percentage of these departments either requires or recommends work in either mathematics or statistics. However, they do tend to offer courses in field techniques and cartography more frequently than their doctoral-granting counterparts. Thus the methodology emphasis in the non-doctoral departments tends to be more traditional in character.

CONCLUSIONS

Within the last twenty years there have been some major changes in undergraduate geography curricula, which have been concentrated mainly in those geography departments offering a Ph.D. program and other publicly supported four-year institutions. There has been a marked trend towards emphasizing the systematic aspects of geography in the public institutions, and these institutions have also made an attempt to broaden their course offerings in research techniques. One aspect of these trends is that there has been a greater emphasis on training in quantitative techniques in the major doctoral departments and a lesser emphasis on quantitative techniques training by the non-doctoral public institutions. In the non-doctoral geography departments in the private colleges, the geography curricula has remained relatively unchanged and these departments continue to emphasize regional geography. This pattern may also be encountered in some of the smaller tradition-bound public colleges. Today geography curricula are moving toward a greater emphasis on the development of geographic theory and problem solving with a consequent de-emphasis on area description. The stronger emphasis on theoretical geography can be seen in the new introduction to geographic behavior courses which are very gradually replacing the old introduction to world geography courses in some geography departments. Whether or not these trends will prevail depends on a host of factors, but they are a significant aspect of the changing character of American undergraduate geography.

References

BURTON, I., 1963, 'The Quantitative Revolution and Theoretical Geography', *Canadian Geog.*, 7, 151–62.

COHEN, S. B., 1967, 'New Approaches in Introductory College Geography Courses', *Commission on College Geography, Association of American Geographers*, Pub. 4.

HART, J. F., 1966, 'Geographic Manpower – A Report on Manpower in American Geography', *Commission on College Geography, Association of American Geographers*, Pub. 3.

HART, J. F., 1968, 'Undergraduate Major Programs in American Geography', *Commission on College Geography, Association of American Geographers*, Pub. 6.

JONES, C. F., 1961, *Status and Trends of Geography in the United States*, Report by The Committee of the Association of American Geographers.

LOUNSBURY, J. F., 1965, 'Geography in Undergraduate Liberal Education', *Commission on College Geography, Association of American Geographers*, Pub. 1.

RANDALL, J. R., 1967, 'Introductory Geography-Viewpoints and Themes', *Commission on College Geography, Association of American Geographers*, Pub. 5.

ROBINSON, A. H., 1962, 'On Perks and Pokes', *Econ. Geog.* 37, 181–3.

SCHWENDEMAN, J. R., 1968, *Directory of College Geography of the United States*, The Geographical Studies and Research Center, Association of American Geographers, Vol. XIX.

SPATE, O. H. K., 1960, 'Quantity and Quality in Geography', *Ann. Ass. Amer. Geog.*, 50, 377–94.

CHAPTER SIXTEEN

Geography in Great Britain

PART 1: GEOGRAPHY IN BRITISH SCHOOLS

P. BRYAN

Senior Geography Master, Cambridgeshire High School for Boys

In the period after the Second World War, geography came of age in the secondary schools of Great Britain. Before the War, a relatively small band of teachers had struggled to gain recognition for the subject with varying success, but by and large it was regarded as a minor subject, taught to few really able children and often thought of as a refuge for the intellectually weak. Since the War there has been a marked reversal of fortunes. The number of well trained graduate and non-graduate teachers has increased rapidly. In a growing number of schools good teaching and successful examination results have begun to attract an ever increasing flow of able pupils. Educationalists have realized that the subject has something to say which runs beyond mere general interest in the world, and thus it has gained academic respectability within the general teaching world. We no longer have to defend our existence. On a rising tide of popularity the subject has swept forward in the schools, and if numbers taking the subject in public examinations are any guide, it has been one of the three most rapidly growing disciplines. But within this success story it could be said that the seeds of stagnation lay dormant. Perhaps complacency was induced by success. Be that as it may, it is clear that in recent years the torch of inquiry which concerns itself with the teaching methods and content of school subjects has passed into other hands, notably those of mathematics and science. It was in these subjects that inquiries were set in hand which sought to look critically into content, and to investigate new methods of promoting and testing knowledge and understanding. Belatedly this movement has reached geography, and over the past few years some energetic but decidedly ill-co-ordinated work has been done on the content and methods of teaching school geography. From this has sprung what has become known, inappropriately in most cases, as the 'New Geography'. It is not an easy task to make an assessment of the impact of this movement while it is barely out of its infancy, but it is important to do so because there are few countries where geography is taught so widely or to such a high level in schools. If the nature and content of school geography is to change radically here, then it is bound to have repercussions outside our own frontiers.

How much of a change has taken place in geography teaching in this country? So far very little, if one measures quantitatively. In the vast majority of secondary schools the syllabuses are basically little changed from those which prevailed in the 1950's. The majority still accept that a school course should be based on the regional approach, aiming to cover virtually the entire world from both the physical and human points of view. This view is clearly and fairly set out in *The Teaching of Geography in Secondary Schools* (1967). But there is a strong movement in some quarters, particularly among younger teachers, which indicates an important change in attitude towards the traditional approach. They accept the need for radical change, even if they are not yet agreed on what to change or how to change it. It is certainly no longer accepted that there is a single corpus of geographical knowledge which should form the basis of all school syllabuses, and there is a strong demand at all levels that the teaching of the subject should be freed from the excessive factual burden which it seems obliged to demand of its students. Young graduates entering the schools seem often to find themselves greatly out of sympathy with the type of geography they are asked to teach. Indeed, there are those who are actually unable in some measure to teach it, and this is largely due to the type of degree course they have taken. It is of course no part of a university's task to provide instruction suitable for intending teachers, but there is cause for reflection when the two get so far out of step.

In many ways the most pleasing feature of the movement for change is that it has come very largely from practising teachers. It was not the higher academic bodies who initiated the pressure, although it is true that such bodies have given their encouragement at a later stage. The Geographical Association now has a committee studying 'The Role of Models and Quantitative Techniques in Geographical Teaching,' initiated because it became plain that there was a great need for such a body to co-ordinate some of the work being done. Another very gratifying feature is that much of the dynamism has been generated by the co-operation of School and University Geographers. A few years ago it could be said with some justice that there was a measure of the ivory tower attitude in the universities' attitude to school geography. Now it is clear that there are a considerable number of University Geographers who are actively aware of, and interested in, conditions in schools. This co-operation has been invaluable, partly because of the cross fertilization of ideas, and partly because it has lent to the teachers a voice of greater weight and authority in organizations where these things matter, such as the Schools Council. It is clear then that there is a movement for change, both in the content and the methods of teaching school geography. It has so far made limited progress, mainly among younger teachers. It has the active support of a body of university teachers, and the blessing of official organizations. But how far has it altered the nature of the subject in schools?

In this country the content of a school course has a dual origin. Since nearly all secondary schools sit some kind of public examination as a leaving certificate, courses have to be geared in part to the syllabuses of these examining bodies. They are autonomous bodies, although strongly influenced by the voices of elected teachers who represent various teachers' organizations. Until recently, the syllabuses of these examining bodies have shown no significant variation in content, and virtually none in aim. Generally speaking their Ordinary Level examination was framed in the context of a two-year course, to be taught in the two years preceding the taking of the examination at the age of 15 or 16. But it is also normal to find that the terms in which examination questions are posed make it plain that a considerable amount of geographical background is being tested so that the examination is really a test of a whole school course. As such it spreads its dictates over more than the two years suggested, and indeed anyone who sat this examination purely on the basis of two years' work would find himself at a considerable disadvantage, as any mature student working on his own via correspondence courses will testify. The other element in the framing of a school course is the wishes of the teacher himself. Ostensibly he has the first three years of a secondary school course in which he can teach what he likes. This freedom is to some extent illusory, for he must be mindful of the long-term aim of preparation for the final examination, but it is none the less true to say that he has a very fair latitude of method and approach in these three years. With regard to the Advanced or Higher Level examination, the situation is much simpler.

This always has been, and still is, regarded as an examination designed primarily for university entrance qualifications. This, incidentally, is one of the main sources of unrest in the schools, because at this time of rapidly expanding sixth forms it is quite clear that the majority of senior pupils are not aiming at university entrance, and are therefore being forced through a hoop which is not in any way designed for their needs. Be that as it may, the Advanced Level syllabuses have always dictated the whole of the two-year sixth-form course for 16–18-year-olds, and have left the teacher no freedom with regard to the content.

This dual control of syllabuses has not caused any stress until recently, because there was broad agreement between the teachers and the examining boards about the content of school geography. It is probably true to say that there is still a large body of school geographers who are in agreement, but it is obviously pertinent to ask why there is now an element of discontent. Probably the most important single reason is dissatisfaction with the traditional course, and especially with the role played by regional geography within it. No single issue seems to cause more controversy than whether it is necessary to attempt to cover the regional geography of the whole of the world in a school course. A fundamental dichotomy exists between those who maintain this viewpoint, and those who believe that the task of school geography lies more in the analysis of problems, hypotheses and correlations which can and must be illustrated by reference to regional examples. This has been construed by some as an attack on the central place of regional geography in school syllabuses. This is not the place to argue the merits and demerits of regional geography in schools; its merits have been very ably argued in a recent book by Roberson and Long (1966). It could be submitted that the real question is not whether regional geography is of central importance in the philosophy of the subject, but whether there is a need for so much of it. Some also think that it is not the only intellectual framework within which the subject can be taught in schools. It may well prove true that no other approach can offer the same coherence, but equally one cannot hold with certainty that the dicta of such great teachers as Fairgrieve or Cons, to mention only two, represent the last word on such a rapidly changing subject.

There is no doubt that the greatest bone of contention has been the Ordinary Level syllabuses. These have become grossly overloaded with facts, and teachers have increasingly come to feel that they are

x

being forced into the role of retailers of information. They are asked to cover so much ground in so much detail that only stereotyped methods which impart information quickly and accurately enable the job to be done. This is why the first moves in the rethinking of school geography have been the modification of external syllabuses, because until these are reduced in content there is little room for the individual teacher to reconsider his position. In the light of the development of systematic studies at University level, some teachers have also come to believe that the regional synthesis does not always give as clear and detailed a picture of the working of human phenomena as is needed today. A good example of this lies in the development of urban studies. In a country like Britain, with its very high urban percentages, some teachers have believed it necessary to attempt a more detailed exposition of the factors operating in urban geography than is normally offered in school regional geography. To them such matters as urban hierarchies, rank size correlations and central place theory seem more vital to a child's understanding of his environment than the customary lists of facts about the site, position and functions of individual towns the world over. It is not a question of one being better than the other, but of one being better suited to a particular purpose. Land use, industrial location and transport studies offer similar examples. This trend, of course, merely mirrors a much stronger current within the universities themselves, where systematic studies have reduced the traditional regional geography to a mere shadow of its former self. It is hardly surprising if a similar trend makes itself felt in the schools.

In line with this new trend of thought has come the pressure for new syllabuses to mirror its aims, nearly all of which aim at a reduction in the amount of regional geography taught. The impetus has come first at Advanced Level. This is because it has proved much easier to devise a new syllabus for these students, who are generally among the more gifted and who can be said to have some modicum of professional interest in the subject. New syllabuses have been produced by the Cambridge Local Examinations Syndicate, the Oxford and Cambridge Local Examinations Board, and the Southern Universities Examination Board. All of these are based on systematic geography, but none departs entirely from the application of the work to regional geography, for there is an insistence that the candidate must be able to illustrate his studies from regional examples. For instance, the Southern Universities syllabus states that 'in both the Physical and Systematic papers the candidate must be able to illustrate the principles of physical and human geography with reference to the British Isles and three other regions chosen from:

A (North America or the Common Market Countries of Western Europe).
B (U.S.S.R. or China and Japan or India and Pakistan).
C (Middle East and Africa North of the Equator or Africa South of the Sahara or South America).'

It is obvious that the examiners have been very conscious that systematic geography could at this level lend itself very easily to vague and facile generalizations, and that the insistence on regional application is a deliberate attempt to meet this danger. Unlike the other two boards, the Cambridge Local Examinations Syndicate's revision of its syllabus has not abandoned the traditional regional syllabus, which is being maintained in its original form as an alternative to the new paper. This is a useful principle of choice which will no doubt be applauded by those teachers who have no wish to change the form of their Advanced Level teaching. The new Oxford and Cambridge paper goes further than the other two in that, in its paper entitled 'Techniques and Human Geography', it makes specific reference to statistical work of the newer kind, including the methods and problems of data collection, methods of mapping the information gained, and the problems of assessing the reliability of information by statistical tests. All three boards also make it plain that a great emphasis will be laid on the use and application of the candidate's own field work and local studies. It is also clear that it is greatly desired that much of the candidate's work should be based on the study of original materials, such as large-scale Ordnance Survey maps, aerial photographs, statistical materials, photographs, and local reports from such bodies as Planning Offices, Town Halls and Civil Service Offices. The general drift of this would seem to be that in future candidates will have much less chance to rely purely on the memorizing of facts, and much more chance to show their ability to think originally and to assess solutions to simple problems. One further aspect of this approach is that it calls into question the status and value of the paper set by many boards at Advanced Level called the 'Practical Paper'. In most cases this consists of mapwork, the representation of statistics, weather maps, surveying and, in some cases, map projections. It has become increasingly obvious that many of these papers bear some resemblance to simple parlour games whereby it is possible to score good marks providing you know the rules of the game well enough. Candidates who have never held a surveying instrument in their hands answer questions on surveying. Questions are asked on Ordnance Survey maps, the answers to which must partially involve inspired guesswork. For instance, questions are set about the pattern of settlement in an area, the answers to which cannot be other than speculative or deterministic without a knowledge of the history of settlement. Many candidates can answer questions on weather maps, but are quite unable to relate them intelligently to the weather and climate of the British Isles, because they have been studied as an entirely separate 'mapwork exercises'. It is thought, therefore, that all this material could be translated much more intelligently to the physical and systematic papers, where the knowledge could be set in a more meaningful context; or at the very least it should be remodelled to do away with the arbitrary element and to bring it more into line with modern thinking.

Thus there is already clear evidence that the bonds of the traditional Advanced Level syllabus are loosening. There can be no doubt that this mirrors the changes taking place in the universities, but this is hardly surprising when the setting of Advanced Level papers is largely in the hands of university teachers. It represents a fairly considerable change in the traditional content of Advanced Level work, but not necessarily a total rejection of it. It is evolution rather than revolution.

Reform of Ordinary Level syllabuses presents a much more complex problem. The Ordinary Level is by tradition a form of leaving certificate. A subject like geography is taken by thousands of children of a very wide range of ability, most of whom have no special or lasting interest in the subject from a professional point of view. As an examination it represents a test of what is thought to be a reasonable content of work during the five-year school period, and of the pupils' ability to digest and organize this material. Although the amount of work prescribed for the examination may be notionally based on the period of 2 years' work, it inevitably tests the geographical understanding built up over the five-year period. This is indeed a fundamental issue: Is school geography to be like English Language, a test of cumulative knowledge gained over the whole of a school course, or like English Literature, a test of specific material studied over a two-year period? Because of its more varied aims, and the much greater range of opinion about its purpose, the Ordinary Level examination is much more difficult to revise. It is not difficult to produce new syllabuses which differ from the existing ones in content. However, many believe that a more fundamental change in the whole approach to the subject, rather than just to its content is needed at this level. Unfortunately even those teachers who wish to experiment are trapped in a vicious circle. Existing syllabuses leave little or no time for radical experiment in new approaches. There is little or nothing in print in the way of new textbooks to point the way for those who wish to experiment. In the absence of proven new material there is much less demand for the revision of syllabuses.

Thus it is hardly surprising that there is at present much less to show publicly in this field. Some teachers have concentrated their efforts first in the Advanced Level field. Other and bolder spirits have tackled their problem from the other end, starting with their first year intakes and experimenting with their school syllabuses. In a movement which dates back little more than five years, their work has barely passed through a complete generation, and few of them as yet feel sure enough of their ground to want to promulgate new syllabuses. Only the Southern Universities Examination Board has so far produced anything greatly different from the standard syllabus, for its new papers will have no orthodox regional geography at all. There will be two papers, one on Map Reading, the other on the Principles of Geography, which will include both physical geography and the distribution of man and his major economic activities in relation to environment. Other Boards, are also conscious of the need to bring their papers more into line with current thinking, and the Cambridge Local Examinations Syndicate has established a committee which is looking into the possibility of introducing new alternative syllabuses at Ordinary Level.

The protagonists of change are not united in their aims; many would frankly admit that they are groping their way towards rather ill-defined objectives, and few would claim that their experiments have so far produced a syllabus as coherent as that of the established school course. Their common element is, to quote Chorley and Haggett (1967), 'some measureable contrast between their approaches to Geography, various as they are, and those which characterize the greater part of the established geographical patterns of thinking as evidenced in existing syllabuses and textbooks.'

It remains to assess how far there have been changes in method as well as in content. Here one moves from the realm of fact into that of opinion. The interest in new methods has been stimulated by two movements, one domestic, the other imported. The first was the New Mathematics developing in the primary and secondary schools in this country. The second the knowledge of the experimental work connected with the American High School Project. Like many other innovations in learning, the development of the link with the New Mathematics seems to have worked down from university level, no doubt occasioned by the increasingly mathematical nature of much university research. It has become obvious that many of the concepts of the New Mathematics such as sets, matrices and topology are applicable to geographical phenomena, and that they can provide a fresh and exciting approach to the teaching of the subject. It must be admitted that initially some of the protagonists of the link may have been over-enthusiastic about its charms; but many were equally put off a fair evaluation by the demands which seemed to be made on their limited and often rusty mathematical skills. The use of mathematics in school geography has to be judged, not by its logic and beauty, but by whether it does the job better than any other method. To paraphrase Yeates in his Preface to *An Introduction to Quantitative Analysis in Economic Geography* (1968) – 'it is not the techniques by themselves which matter. The basic concern is how they may be used in Geography.' As far as the question of individual skill in mathematics is concerned, most teachers were over-apprehensive. The range of knowledge demanded of a university researcher is not needed by a classroom teacher, and it can be said from experience that anyone possessing Ordinary Level Mathematics and the willingness to submit himself to a short intensive study of the techniques, has little to fear. Unfortunately a great many geography teachers fear and dislike this development because they are as wary of mathematics as are many of their pupils, and perhaps because they see in it something which may destroy their stock in trade. Whilst one can sympathize greatly with their fears, it is hardly an attitude which can commend itself to any academic discipline. Experience already shows that it is possible to devise exercises based on simple mathematical techniques for all levels, including the very young. We do not as yet have very much concrete evaluation of this kind of work, but we suspect it produces more thought on the part of the pupil. Obviously in some fields it produces a more precise answer, if this is what you want. Will it produce better geographers? This so much depends on what you think a geographer ought to be that the question automatically begs itself. Some see the new mathematical approach as something which could utterly destroy geography as it has been traditionally taught and known; some as something which has given the subject its biggest leap forward in decades; some as simply another tool in the armoury of geographers. The value of the work is already proved in the sphere of the universities, but this does not automatically make it suitable for school geography. In the schools it will be many years before we see whether we have produced a method which fundamentally revolutionizes the teaching of the subject, or merely an additional technique to underpin existing methods. Meanwhile, one would have thought that the subject needed a fair hearing, and the encouragement of skilled objective research into its merits.

To many teachers the experimental approach of the American High School Geography Project makes a much more immediate appeal, because it falls into line with the trend away from the mass accumulation of factual knowledge by moving towards a technique of individual work by the pupils. More and more teachers see their job not as retailers of information, but as providers of material from which, with guidance, knowledge and understanding can be gained. It moves in spirit towards the individual and group work which has been so much in evidence in our own primary schools in recent years. The High School Project contains much that is not suited to our approach to school geography, but the method is exciting, and it has stimulated many teachers in this country to strive for a similar project based on our own resources. Many references have been made to the desirability of a Nuffield Project on the teaching of geography, similar to that now being evolved for the Classics. If this is not possible, it would seem an admirable project for sponsorship by an Institute of Education, or perhaps a publishing group.

Quite certainly the role of models in geography teaching would fall within the ambit of such a research project. Many teachers would claim to have been using models in their teaching for years, and in a limited way this is true. In the realm of physical geography few have not resorted to the models of world wind and pressure patterns, or of depression structure, but for most these were but isolated uses of the method, carried out without conscious awareness of its philosophical basis. Obviously numbers of teachers have also used hardware models of their own devising, instinctively recognizing their great value as a teaching method in their ability to simplify and provide a structure for a more complicated reality. But the research done in this country in recent years, especially by Chorley and Haggett (1965), has added a new dimension to the work, for it has added a philosophical basis which turns the use of models from an occasional aid into a network within which teaching can be organized. The traditional idiographic approach to school geography does not at first sight sit happily with the approach of model building, but as Haggett and

Chorley (1965) point out 'the apparent gulf must either be bridged or must lead to the dismemberment of the subject', a fate which some obviously fear. The dilemma is a terrifyingly real one. The descriptive and analytical approach of regional geography can hardly be cast out completely, especially for the younger children, for to do so would be to reduce the subject to such an arid state that few would want either to teach or learn it. At the same time it must be recognized that at least up to Ordinary Level, the vast majority of children are fed on an unstructured diet of descriptive information, accompanied by elementary and often deterministic explanations. One of the most valuable contributions of the model theory has been the introduction of the stochastic element into geographical thinking as a corrective to our traditional explanations. Children do not seem to mind greatly the traditional kind of teaching, because the world is interesting, relatively undemanding to learn about, and because the subject lends itself to more lively methods of teaching. It is incidentally, interesting to speculate how many of us are confusing interesting lessons with lively lessons, the two not necessarily being synonymous. But this 'one gear' kind of teaching, which changes in the period of a school course only by adding ever more details, is increasingly subject to criticism, not least from the children themselves. It is a not un-common complaint at the end of a Ordinary Level course that by this time so many facts about so many regions have been thrown at the pupils that even the most intelligent have lost their power to dis-tinguish between one area and another in any real sense. The whole thing has become a mechanical exercise.

Are the two approaches so incompatible? The approach of model building seems to offer a solution to their apparent opposition. It provides the answer to the 'one gear' concept, for the number and variety of models provides the scope for the intellectual changes of gear which our teaching so often lacks at the moment. Their ability to select and structure important information tackles effectively the problem of the large amount of 'random noise' (insignificant regional detail) in our lessons, although it must be admitted that the opponents of the concept are hardly likely to admit that the noise is random or unimportant. They can help us in the problem of 'a lack of *a priori* rationale in the selection of regions, or of the gains we expect to make from studying them' (Chorley and Haggett, 1965); again an argument not likely to be accepted by those who view whole-world coverage as the norm of school geography. Again and again one comes back to this point: once it is allowed that a more selective approach to the study of the world is possible and proper for school geography, an integration of the idographic and nomothetic views present no difficulties which cannot be removed by experiment and time.

One cannot claim that school geography in Britain at this moment has been revolutionized. In the vast majority of schools one would see no appreciable change in methods or content; nevertheless, the movement is there, although it represents perhaps no more than a ground-swell at the moment. But it is a ground-swell fostered from below, originating in the schools and in the hands of younger teachers, for which reason it can be expected to grow. Already there are visible signs of this growth in the form of new syllabuses and methods. Below the surface lies much work which can be expected to come to fruition in the next few years. To some these developments are anathema, because they challenge a valid, successful and much loved philosophical tradition of teaching the subject. It is probably true that the experimenters cannot as yet offer as coherent an alternative. But no subject stands still, in university or school. If it does, it runs the risk of being pushed aside, a fate which seems very near to befalling the Classics. However unwelcome these developments are to some, one would think that such signs of intellectual vigour would be a source of great pleasure to most geographers.

References

CHORLEY, R. J. and HAGGETT, P., 1965, *Frontiers in Geographical Teaching* (Methuen, London), 379 pp.

— 1967, *Models in Geography* (Methuen, London), 816 pp.

INCORPORATED ASSOCIATION OF ASSISTANT MASTERS IN SECONDARY SCHOOLS, 1967, *Teaching of Geography in Secondary Schools* (Cambridge), 5th Edn., 396 pp.

LONG, M. and ROBERSON, V. S., 1966, *Teaching Geography* (Heinemann, London), 416 pp.

YEATES, M. H., 1968, *An Introduction to Quantitative Analysis in Economic Geography* (McGraw-Hill, New York).

PART 2: FIRST DEGREE COURSES IN GEOGRAPHY AT BRITISH UNIVERSITIES[1]

ALAN R. H. BAKER

Lecturer in Geography, University of Cambridge

More than thirty British universities together offer over seventy first degree courses in geography from which an increasing number of students graduate annually. In 1967, well over 1,000 students graduated in geography with single honours and more than 150 graduated with combined honours in geography and one other subject. An inspection of calendars and prospectuses for these universities in the late 1960's reveals the broad characteristics of university instruction in geography in Britain. These characteristics will be discussed here first in terms of elements of the traditional pattern and secondly in terms of some changes discernible within that pattern.

Given the many problems of generalization inherent in a discussion of so many degree courses, it is possible to distinguish at least four principal common and traditional elements. The first is the length of the course, for at most universities it lasts three years. Only at Aberdeen (M.A.), Dundee, Edinburgh, Glasgow, Keele, St. Andrews and Sussex (B.A. in the School of European Studies) do the courses always last four years. The courses at Aberdeen (B.Sc.), Belfast, Liverpool (B.Sc.), Newcastle (B.Sc.) and Strathclyde may also take four years, although it is possible for students to obtain exemption from the first year by achieving a high standard in their Advanced Levels of the General Certificate of Education. Secondly, virtually all the courses begin with one and often two years' study of the basic groundwork, divided into the familiar physical and human elements together with regional geography and cartography. This is intended to give a broad view of the subject and to enable the options chosen later to be seen in perspective. Thirdly, specialization takes place in the final one or two years in the form of compulsory and optional subjects. The compulsory subjects are relatively limited in scope and

[1] For further information, see *Geography. CRAC Course Guide* (Careers Research and Advisory Centre, Cambridge: 1968). This Guide was prepared in collaboration with the present author, who wrote the present chapter in 1968.

in general emphasis is given to regional geography and methodology; the optional subjects vary widely but the most common are climatology, economic geography, geomorphology and historical geography with only relatively few departments offering options in, for example, hydrology, population geography and surveying. An indication of the principal specialized studies (i.e. those more narrowly-based courses usually taken during the final one or two years of the degree course) taught in British universities, is provided in Table 19. This Table is not comprehensive. It does not, for example, record the courses in polar studies available at a few universities. Nor could the Table accurately record the newly developed and developing courses in 'locational', 'regional' and 'spatial' analysis. As the number of options made available by individual university departments has considerably increased since 1945, so the amount of choice open to an Honours student has correspondingly expanded.

Fourthly, there remains a strong emphasis on all forms of field and practical work and an assessment of competence in them is usually made for examination purposes. Most universities require attendance at an annual field course, generally lasting a week. On such courses, undergraduates usually participate in exercises in data collection and in the description and analysis of landscapes and of the processes forming them. The exact role of field work in the education of geography undergraduates varies among university departments but a detailed exposition of the part which it plays in one department, that of University College London, has recently been published (Mead, 1967). Field work in that department is seen as the purposeful appreciation of a piece of country, or more selectively, of certain elements or processes within it. The accent is always on personal experience – the translation of the printed word or cartographic symbol into the reality of the countryside, or conversely the translation of the personal experience into the written word or the mapped symbol. In each of their first two years, undergraduates participate in a nine-day Easter field excursion principally carrying out exercises in physical geography (such as soil and slope, drainage and vegetation surveys), in economic geography (such as surveys of field boundaries, of the shifting edge of cultivation and of building materials) and in historical geography (such as studies based on earlier editions of the Ordnance Survey maps, with a view to identifying changes in the human landscape). In their third year, undergraduates either attend a similar excursion in a mainland European setting or carry out independent systematic surveys according to the student's field

Table 19: Some Specialized Studies available within Geography Courses

Key: C = compulsory   O = optional   CO = some study compulsory, further optional study possible

University or College	Oceanography	Hydrology	Geomorphology	Climatology	Biogeography	Quaternary studies	Pedology	Cartography	Mathematical geography/surveying	Economic geography	Political geography	Historical geography	Urban geography	Settlement geography	Social geography	History of geography	Geographical methodology	Applied geography/planning	Population geography	Photogrammetry	Transport geography	Regional geography
Aberdeen (M.A.)		CO	CO	C	C	O	C	C	C	C	O	CO	O			C	CO					COl
Aberdeen (B.Sc.)		C	CO	C	C	O	C	C	C	C	O	CO	O			C	CO					COl
Belfast (B.A.)			O	O	O					O	O	O	O	O			C					c
Belfast (B.Sc.)			O	O	O				C	O	O	O	O				C					c
Birmingham (B.A.)			CO	CO	CO					O		CO	O				C		O	O		c
Birmingham (B.Sc.)			CO	CO	CO					O		CO	O				C		O	O		c
Bristol (B.Sc. in Social Science)		O	O	O	O		O			C	C	C	C	C	C	C						c
Bristol (B.Sc.)		C	C	C	C	C				O	O	O	O	O	O	O	C		O		O	
Cambridge (B.A.)		O	O					O	O			O	O			C	C	O	O			O
Dundee (M.A.)			O			O				O	O	O	O			C	O					c
Durham (B.A.)	O	O	O	O	O	O	O	O		O	C	O	O						O	O		O
Durham (B.Sc.)	O	O	O	O	O	O	O	O		O	C	O	O						O	O		O
East Anglia (B.Sc.)	O	O	O	O		O			C	O			O	O	O				O	O		
Edinburgh (M.A.)		CO	CO	CO				CO	CO	CO	CO	CO	CO	CO		C	C		CO	CO		c
Edinburgh (B.Sc.)		CO	CO	CO				CO	CO	CO			CO			C	C		CO			c
Exeter (B.A.)		CO	O						C	C	O	O	O	O								COl
Exeter (B.A. in Social Science)									C	C	C	O	O	O								COl
Exeter (B.Sc.)		CO	O						C	C	O	O	O	O								COl
Glasgow (M.A.)		O	O	O					C	O		O	O									c
Glasgow (B.Sc.)		O	O	O					CO	O		O	O							O		
Hull (B.A.)	O	O	O							O			O					O	C			c
Hull (B.Sc.)	O	O	O							O			O					O	C			c
Leeds (B.A.)			CO	CO	C	O		CO		O	O	O	C						O	O	O	c
Leeds (B.Sc.)	CO	CO	CO	CO		CO		CO		CO			CO	C		C	O					c
Leicester (B.A.)			O					O				O	C	O	O		C				O	
Leicester (B.Sc.)	O	O	O	O	O	O	O	O			C						C					
Liverpool (B.A.)	O	O	O	O	O	O	O	O		O			O			C	C	O	O		O	C
Liverpool (B.Sc.)	O	O	O	O	O	O	O	O		O			O			C	C	O	O		O	C

London:

University or College	Oceanography	Hydrology	Geomorphology	Climatology	Biogeography	Quaternary studies	Pedology	Cartography	Mathematical geography/surveying	Economic geography	Political geography	Historical geography	Urban geography	Settlement geography	Social geography	History of geography	Geographical methodology	Applied geography/planning	Population geography	Photogrammetry	Transport geography	Regional geography
Bedford (B.A.)			O		O	O	O	O	O	O			O									O
Bedford (B.Sc.)			O		O	O	O	O	O	O			O									O
King's (B.A.)		CO	CO	CO				CO	C	CO	O	O	CO	O	O	O	O	O	O	CO	CO	COl
King's (B.Sc.)		CO	CO	CO				CO	C	CO	O	O	CO	O	O	O	O	O	O	CO	CO	COl
L.S.E. (B.A.)		CO	CO	CO				CO	CO	O	O	O	CO	O	O	O	O	C	O	CO	O	COl
L.S.E. (B.Sc.)		CO	CO	CO				CO	CO	O	O	O	CO	O	O	O	O	C	O	CO	O	COl
L.S.E. (B.Sc.Econ.)		CO	CO	CO					C	C	O	O	O	O	O	C		C	O		O	CO
Queen Mary (B.Sc.)		CO	CO	CO					C	CO	O		CO	O	O	CO	C	C	O			COl

Key: C = compulsory   O = optional   CO = some study compulsory, further optional study possible

University or College	Oceanography	Hydrology	Geomorphology	Climatology	Biogeography	Quaternary studies / Pedology	Cartography	Mathematical geography/surveying	Economic geography	Political geography	Historical geography	Urban geography	Settlement geography	Social geography	History of geography	Geographical methodology	Applied geography/planning	Population geography	Photogrammetry	Transport geography	Regional geography
Queen Mary (B.Sc.Econ.)	DETAILS NOT AVAILABLE																				
University (B.Sc.)		o	o	o					o	o		o	o	o	o		o				o
Manchester (B.A.)		co	co	co	o	c	o		o	o	co	o			o	o			o		c
Manchester (B.Sc.)		co	co	co	c	c	o												o		c
Newcastle (B.A.)		co		o					o	co	o	o	o						o		
Newcastle (B.Sc.)		co		o					co	co	o	o	o						o		
Nottingham (B.A.)	o	co	co	co		c	c		co	co	co	o		c	c	c					co
Nottingham (B.Sc.)	o	co	co	co		c	c		co	co	co	o		c	c	c					co
Oxford (B.A.)		o	o						o	o	o	o			o		c		o		c
Reading (B.A.)									o	o	o	o				c	o			o	o
Reading (B.Sc.)									o		o					c	o			o	o
St. Andrews (M.A.)		co	co			co	co	c	c		c	c	c		c		co				c
Sheffield (B.A.)		o	o	o					o	c	o	o				c					co
Sheffield (B.Sc.)		o	o	o					o	c	o	o				c					co
Southampton (B.A.)		co	co	co	c	c			co	co	co	co	c	c		c	co				co
Southampton (B.Sc.)		co	co	co	c	c			co	co	co	co	c	c		c	c	c			co
Strathclyde (B.A.)	o	o	o	o		o			o	o	o	o		o							c
Sussex (B.A., School of African and Asian Studies)									co	o	o	co	c	co	c	o	o				c
Sussex (B.A., School of European Studies)									co	o	o	co	c	co	c	o					c
Sussex (B.A., School of Education Studies)									c			c	c	c	c						c
Sussex (B.A., School of Social Studies)									co	o	o	co	c	co	c	o	o				o
Sussex (B.Sc., School of Biological Studies)	c	co	c	c											c	c		o			c
Ulster (B.Sc.)		o	co	co	c	o	co		co	c		c	c	c	c		co	c			o
Wales:																					
Aberystwyth (B.A.)									o	o	o	o	o	o	o						
Aberystwyth (B.Sc.)	o	o	o	o		o	o												o		
Aberystwyth (B.Sc.Econ.)									c	c		c		c							
Swansea (B.A.)	o	o	co	co	c		o		co	co	co		co	o	c		co	c		co	c
Swansea (B.Sc.)	o	o	co	co	c		o		co	co	co		co	o	c		co	c		co	c
Swansea (B.Sc.Econ.)									o	o	o		o	o			o	o			o

of specialization. At the undergraduate level, such field activities are also seen as exercises in methodology. It is the nature of these exercises, Mead argues, that they should be largely exercised for their own sake. But to indicate the applicability of any exercise is to lend a new dimension to reality and to introduce the practical potentiality of the subject (Mead, 1967, pp. 1–2).

In addition to annual field courses, most universities require students to submit a dissertation of between 5,000 and 15,000 words, normally in their final year: the exceptions are Exeter and Leeds, while at London an additional examination paper is available as an alternative to a dissertation. The dissertation always involves personal research, but its subject varies. At most universities, it can be on any approved subject. At Leicester, Manchester, Newcastle, Oxford, St. Andrews and Sheffield, on the other hand, it takes the form of a detailed study of one or more aspects of a region not exceeding 150 square miles in size. At Glasgow a dissertation related to the third year optional subject is required as well as a regional essay, while Hull requires both an 8,000 word essay on an area of about 100 square miles in the second year and a 5,000 word essay on any approved topic in the final year. There is a long tradition of studies of this kind in Britain. Oxford and the London School of Economics were among the pioneers of this form of test and their studies provided some of the models for the regional surveys of the 1920's (Board, 1965, p. 298). Greater significance ought to be attached to these dissertations as an assessment of research potential than is the case at present. As a criterion for the selection of graduates for research, performance in written examinations receives over-riding importance. A good Upper Second degree class is almost automatically accepted today as a measure of a student's ability to do research. A recent survey has thrown doubt on the validity of this selection procedure. J. W. R. Whitehand (1966–7) took a sample of 85 students (all the graduates in single honours geography at Newcastle upon Tyne during the three sessions 1963–6) and made a comparison between the marks attained for dissertations and marks attained by the same students in their final examination. This revealed no significant correlation between the two sets of marks and of those students with the highest dissertation marks (i.e. within the α bracket) considerably less than half attained a final examination mark of an Upper Second or First Class. In other words, a majority of the writers of outstanding dissertations had no opportunity to proceed to a higher degree by thesis. If research ability at undergraduate level is a reliable indication of research ability at postgraduate level then it would seem that there are serious limitations in the existing method of selection. There is, however, one saving feature in the existing system: the outstanding examination candidates, although they may not produce the best dissertations, generally achieve a competent dissertation mark. The danger therefore, as Whitehand concluded, is not so much that the existing system promotes incompetent research students but that it precludes a majority of the students who, at least from their undergraduate dissertations, appear to be potentially good research students (Whitehand, 1966–7, p. 46).

Straddling as geography does, both in content and in method, the boundaries between the observational sciences such as geology and the critical humanities such as history, it is hardly surprising that most universities award both B.Sc. and B.A. degrees in geography.[1] Where this is so, the degree available to a student normally depends on the A-level subjects which he has passed but within the degree courses themselves this distinction usually becomes blurred, in particular because the optional subjects which are available in the final year are usually basically the same for both B.A. and B.Sc. students. A few universities offer only a B.A. degree and in their case the degree title may not imply any particular arts affiliation at all: this is the situation at Cambridge, Keele, Oxford and Strathclyde. Conversely, there is no B.A. in geography at Bristol and it is possible there to take the B.Sc. either in human and physical geography or in physical or human geography alone, while at London (Queen Mary and University Colleges) all students are at present admitted to a B.Sc. course which combines both science and art subjects. There is no B.A. in geography at East Anglia or at Ulster; the B.Sc. at East Anglia does have a clear bias towards the scientific side of geography, while the B.Sc. at Ulster allows specialization in either physical or human aspects of geography.

In addition, it is becoming increasingly possible to study geography within a social science framework. Nearly a dozen universities now offer such courses, which usually lead to a B.A. or a B.Sc. in social science. In most of these courses a broadly-based first-year includes the study of such subjects as economics, history, politics and sociology and there is a body of opinion which regards this as a more appropriate setting for human geography than 'pure arts' subjects. Nevertheless, the actual geographical content of these courses remain

[1] Only general remarks are made here. Notes and articles on geography courses at individual universities are listed at the end of this chapter.

broad, including both physical and human aspects: the main exception is the B.Sc. in social science at Bristol in which students may choose to study human geography alone.

There is also discernible a developing trend towards systematic studies in geography and away from regional geography which has long been the core and culmination of the subject. The actual ratio of systematic to regional studies varies quite widely from department to department. It is difficult to make an accurate evaluation of their relative importance within individual universities, in particular because teaching in a regional course may be systematically organised. Nevertheless, in broad terms the courses fall into two main groups: Bristol, East Anglia, Reading, Southampton, Sussex and Ulster are primarily concerned with systematic analysis; all of the other departments offer a middle-of-the-road approach, combining systematic analysis and regional studies. Symptomatic of the trend towards systematic studies is the increasingly mathematical treatment of the subject and it is clear that in the future a greater degree of competence in mathematics (and more specifically in statistics) will be required of those wishing to read geography at university. Already an Ordinary Level pass in mathematics is a necessary qualification for entrance to some geography courses at Belfast, Birmingham, Bristol, East Anglia, Exeter, Leeds, Liverpool, Manchester, Reading, Southampton, Ulster and Wales (Aberystwyth), and most other universities would now regard it as a desirable qualification.

Unfortunately, little seems to be known about the general level of numeracy of students admitted to first degree courses in geography at British universities but it may be reliably assumed to be not particularly high. For example, of those students who entered Cambridge in 1966 to read geography, 15 per cent possessed an Advanced Level pass in mathematics or statistics. In 1967, the figure was 17 per cent. Not much more is known about the general level of training in statistical techniques given to geography undergraduates. From replies to a questionnaire circulated to Departments of Geography during the spring of 1968, it would seem that some elementary statistical training is given at all universities with the exception of Oxford and Durham, where the matter was under discussion. Characteristically, training in quantitative techniques involves 20–50 hours of combined lectures and practicals in each of the first two years of the course and covers such topics as deviation and variability, frequency distributions, sampling, correlation and regression. Most courses comprise the statistical analysis and often the mapping of economic, historical, climatological or geomorphological data. In the third year, training in quantitative techniques is more varied and optional, being linked to the specialized studies undertaken by final year students. Only as yet at a few universities – such as Bristol, London (King's), Nottingham, St. Andrews and Wales (Aberystwyth) – do geography undergraduates undergo training in the use of computers.

To some extent, statistical competence is replacing fluency in a foreign language as an integral part of first degree courses in geography at British universities. Only Glasgow and the School of European Studies at Sussex require an A-level foreign language as an entrance qualification. Most other universities require an O-level foreign language but this is merely desirable rather than necessary for courses at Leicester, Bedford College (London), King's College (London) and Newcastle. While at university, all geography undergraduates need to be able to read some foreign books and articles but only a few universities require a reading knowledge of a foreign language as an integral part of the degree course. Students at Birmingham, Leeds and Manchester (B.A.) are required to sit a translation test in one modern foreign language. At Glasgow students must take a translation test in French, German, Russian, Spanish or Portuguese. At Newcastle students must satisfy the Head of Department concerning their knowledge of modern languages and may be required to take a course in German if they have not studied this language previously. At Oxford students must take a translation test in French, German or Russian. At Sussex (School of European Studies) students must make some study of literature and take translation classes in French, German, Greek, Italian, Latin or Russian and sit examinations in these subjects. At Wales (Swansea) students are required to sit a translation test in German.

Inevitably and unfortunately, an inspection of the calendars and prospectuses provides an incomplete picture of current university instruction in geography in Britain. The frontier in teaching lies in the lecture-hall and the seminar room; the official syllabuses represent long-settled territory, all too often containing many relict features of an erstwhile frontier. This is quite characteristic of the development of university studies in Britain, where changes in the teaching programme are much more easy to effect than changes in the formal structure of examinations. Since such changes are almost always dependent upon the policies and agreement of other departments included in the faculty (whether Arts, Science or Social Science),

Y

they generally lag behind changes in the character of geography teaching and research within departments. But the winds of change detectable in the calendars are only the official manifestations of the gales blowing in the classroom.

References

BALCHIN, W. V. G. B., 1961, 'University News. University College of Swansea', *Geog. Jour.*, **127**, pp. 131–3.

BOARD, C., 1965, 'Geography in the Older Universities' in R. J. Chorley and P. Haggett (eds.), *Frontiers in Geographical Teaching* (Methuen, London), 297–302.

BROWN, E. H., 1963, 'University News. University College London', *Geog. Jour.*, **129**, pp. 242–3.

CAREERS RESEARCH AND ADVISORY CENTRE, 1968, *CRAC Degree Course Guide: Geography* (Cambridge), 39 pp.

ELKINS, T. H., 1965a, 'Geography in a New University' in R. J. Chorley and P. Haggett (eds.), *Frontiers in Geographical Teaching* (Methuen, London), pp. 303–8.

— 1965b, 'University News. University of Sussex', *Geog. Jour.*, **131**, pp. 575–6.

— 1965–6, 'Geography Today: Geography in a New University', *Geog. Mag.*, **38**, pp. 812–15.

FISHER, W. B., 1967, 'University News. University of Durham', *Geog. Jour.*, **130**, pp. 124–6.

JACKSON, C. I., 1966, *Degrees in Geography at British Universities.*

KIDSON, C. and BOWEN, E. G., 1967–8, 'Geography Today: Geography on the Western Seaboard', *Geog. Mag.*, **40**, pp. 1460–3.

LEWIS, W. V., 1961, 'University News. University of Cambridge', *Geog. Jour.* **127**, pp. 130–1.

MEAD, W. R., 1966–7, 'Geography Today: University College London', *Geog. Mag.*, **39**, pp. 19–22.

— 1967, 'Field Work in Geography', *Terra*, **79**, pp. 1–6.

MILLER, R., 1966–67, 'Geography Today: Geography in Glasgow', *Geog. Mag.*, **39**, pp. 910–3.

PEEL, R. F., 1966–7, 'Geography Today: Bristol Fashion', *Geog. Mag.*, **39**, pp. 830–3.

SMAILES, A. E., 1964, 'University News. University of London, Queen Mary College', *Geog. Jour.*, **130**, pp. 437–9.

STEEL, R. W., 1965–6, 'Geography Today: Liverpool Studies the Tropics', *Geog. Mag.*, **38**, pp. 322–5.

— 1967, 'Geography at the University of Liverpool' in R. W. Steel and R. Lawton (eds.), *Liverpool Essays in Geography*, pp. 1–23.

WHITE, H. P., 1967, 'University and College News. University of Salford', *Geog. Jour.*, **130**, pp. 274–5.

WHITEHAND, J. W. R., 1966–7, 'The Selection of University Students', *Universities Quarterly*, **21**, pp. 44–7.

WILKINSON, H. R., 1964, 'University News. University of Hull', *Geog. Jour.*, **130**, pp. 580–1.

CHAPTER SEVENTEEN

Teaching the New Africa

R. J. HARRISON CHURCH

Professor of Geography, London School of Economics

Few if any parts of the world are changing so fast as Africa. The 'wind of change' has become both a cliché and a typical British understatement, for a veritable whirlwind is roaring through Africa, bringing formidable changes in the economy and the political scene. These demand our attention, and cannot and should not be ignored; to do so would be to turn our backs on an outstanding aspect of the twentieth century. In any case, Africa occupies between one-fifth and one-quarter of the inhabited area of the world, the African states compose one-third the membership of the United Nations, and the Afro-Asian group about one-half.

POLITICAL EVOLUTION

The colonial era has almost ended and most of Africa became independent with little or no bloodshed between 1956 and 1961, except for the long and bloody prelude to independence in Algeria, the similar aftermath to it in the Congo, and the uncertain future for Portuguese Africa. The federations of French West and Equatorial Africa and that of Rhodesia and Nyasaland have broken up, and Africa is now the second most politically-divided continent. In part this is a natural reaction to excessive centralization in the ex-French lands, and to the failure to develop true partnership in the formative years of the Rhodesian Federation, but there are other reasons. Separate independence means a voice at the United Nations with which to seek aid, and ministries and embassies with posts for those who have been prominent in the independence movement. Africa appears to be following in the footsteps of South America; indeed, Africa is even more divided and its nations the successors of more varied colonizers than those of most of South America, while racial divisions and hatreds are far stronger in Africa.

FIG. 17.1. *Political divisions of Africa.*

On the other hand the trends to co-operation and integration in Africa are not inconsiderable. Pan-Africanism is a potent force which has brought the African nations together at many conferences, and unites them in their hatred of *Apartheid* and Colonialism. Part of the former United Kingdom Trusteeship of the Cameroons has joined the ex-French Trusteeship in the Cameroon Federation, while ex-British Somaliland has joined the ex-Italian Trusteeship of Somaliland in the Somali Republic. An East African Federation is a possibility, and the Gambia may eventually join Senegal in some kind of federation or association.

Certain countries, while remaining independent, have come into new associations. In 1958 Guinea alone answered 'No' to De Gaulle's referendum and all French aid, personnel and equipment were withdrawn as a reprisal. Ghana was the only non-communist country to offer help to Guinea in her distress, granted her a loan of £10 million, and a 'union' of the countries was signed at the end of 1958. In 1960, when Senegal left the Mali Federation formed only in 1959, the Mali Republic joined the union, then renamed the Union of African States in the hope that other states might join. Meanwhile, the union unified nothing, but was merely an association of radicals. It demonstrated the difficulties of unifying former colonies of two or more powers with different official languages, currencies, constitutions and administrative methods – the more so as they are not all contiguous. The union, never effective, is defunct. Yet these countries have the potential for economic collaboration since they produce different commodities and Guinea's alumina might be used in Ghana's future aluminium industry (Figure 17.1).

The Benin–Sahel Entente formed in 1959 and now grouping the Ivory Coast, the Upper Volta, Dahomey, Niger and Togo has less ambitious aims but has achieved more because all four countries are ex-French administered and so have the same official language, currency, similar constitutions and election dates, and were already joined in a customs union. Even more significant is the fact that the first two and the next two already had close economic ties. The Ivory Coast is the inlet and outlet for the overseas trade of the Upper Volta, although this very poor and overpopulated country has little to offer except its labour. Men from the Upper Volta customarily work for several years on Ivory Coast coffee, cocoa and banana farms and plantations, on timber cutting and sawing, and in diamond and manganese mines. Dahomey is normally the land of transit for much of Niger's trade, although some of this also passes through Nigeria and more would do so were it not for economic nationalism which usually diverts trade through Dahomey. Both the Upper Volta and the Niger produce groundnuts, cattle, hides and skins, but the Niger more than the Upper Volta. Dahomey is mainly a producer of palm oil and kernels. The countries adjoin each other and are fairly well linked by roads, railways and air services. Collectively they have a population of 16 million, and with this and their resources are together a stronger counter than in isolation to the economically richer Ghana with 7 million or Nigeria with some 55 million people.

To both the Ghana–Guinea–Mali Union of African States and to the Benin–Sahel Entente the Upper Volta occupies a strategic position. Although poor and a member of the Entente, the country had quite as many interests with the Union, for her labourers go in even greater numbers to Ghana than to the Ivory Coast, and from Ghana the Upper Volta receives Sterling Area imports that are often cheaper and more varied than those from the Franc Zone of which she is a member. Furthermore, substantial transit trade between Mali and Ghana passes across the Upper Volta, especially kola nuts and Sterling Area imports from Ghana in exchange for cattle, hides and skins, and dried Niger River fish from Mali. Lastly, had the Upper Volta left the Entente and joined the Union, the members of the latter would have been contiguous and would have isolated the Ivory Coast from its partners, putting the Ivory Coast in the situation that Ghana now has relative to Mali and Guinea. Ghana and the Upper Volta once declared their boundary customs-free, but so long as these countries belong to different currency zones the full effect of free trade is lacking.

The Union and the Entente represent contrasted approaches to African integration, the political approach of the Union, and the economic one of the Entente. From 1961 until 1963 the former was the view of the 'Casablanca countries' which included the three of the Union, Morocco and Egypt; while the latter was the view of the 'Monrovia countries' which included almost all the other nations of Africa. At the Conference of African Heads of State in Addis Ababa in May 1963 the Monrovia view was accepted, and the rivalry of approach ended officially.

Whatever its other demerits or evils, colonialism restrained tribalism, which every now and then reappears as a disrupting influence in modern Africa. Most of its leaders are against tribal traditions and chieftaincy but these were elements in the Congo (Kinshasa) troubles, the estrangement of Togo and Ghana, the separation of the small new Mid-Western Region from the Western Region of Nigeria in 1963, in the difficult discussions prior to independence in Kenya, and the Nigerian disturbances of 1966.

Given the need for national unity in independence and the formidable problems of government and economic development, it is not surprising that African governments have become progressively more authoritarian. African societies have been accustomed to strong rule by chiefs balanced by very free discussion in councils, and 'One Party Rule' in states is often referred to as a natural evolution. Steps towards national unity where it is, perhaps, as yet not very strong, have been the incorporation of Eritrea into a unitary Ethiopia in late

1962 (Eritrea was previously in a federal relationship with Ethiopia), and the elimination of the three provinces of Libya (Fezzan, Tripolitania and Cyrenaica) in early 1963.

The colonial map of Africa was often a reflection of the 'Scramble for Africa' after 1885. Many boundaries were the limits of penetration by a European country, such as the more northerly extension of Nigeria compared with that of Ghana, or the north-westward thrust of Mozambique which reflects the farthest penetration of the Portuguese up the Zambezi Valley. Other boundaries are the result of exchange, such as African Zanzibar for European Heligoland, or of colonial administrative convenience when independence was never envisaged, such as the excision of the Hodh from the then French Sudan in 1944 and its addition to Mauritania, so causing independent Mali to have a narrow waist-line in the centre and a shape like that of giant butterfly wings. Several other African states have extraordinary shapes, e.g. Zambia. African boundary disputes are likely to trouble the world, for the boundaries are not only exceptionally long but are the work of non-Africans, and they are usually utterly unrelated to ethnic or economic patterns. Africa is likely to give more trouble in this respect even than South America where, despite the numerous disputes, most of the peoples are of European descent.

There is tremendous variation in the size of African states as, indeed, there is in the World generally, but in Africa this may be the more serious given the poverty of the countries and the paucity of skilled and dedicated administrators. Many states include much desert, notably Mauritania, Algeria, Libya, Egypt, Sudan, Mali, Niger and Chad, although it has brought good fortune with iron to the first and much oil to the next two. In several states there are major problems of integration because of size, natural and human diversity or length of communications, especially in the Republic of South Africa, the Congo (Kinshasa), Ethiopia, the Sudan and Libya.

Some states are very small, especially Gambia (4,008 square miles), Rwanda (10,166), Burundi (10,744) and Lesotho (11,716). Many have small total populations, such as Gambia (315,000), Gabon (449,000), Botswana (560,000), Congo Brazzaville (790,000), Mauritania (1,000,000 and mostly nomadic), Liberia (1,016,000), Lesotho (1,025,000 including migrants), the Central African Republic (1,203,000), Libya (1,564,000 and many nomadic), Togo (1,630,000), the Somali Republic (2,250,000 – mostly nomadic), Dahomey (2,260,000), Sierra Leone (2,180,000) and Guinea (2,900,000), to mention only those under the very low figure of 3 million which Dr

Nkrumah once gave as the lowest desirable minimum population for a state. And while most of Africa is very thinly peopled, mainly because of the difficulties of the environment and the consequences of the slave trade, Malawi and the Upper Volta are over-peopled and large numbers of their men must seek work abroad for long periods.

Most of the land-locked states and territories of the world are in Africa (14 out of 25), and they comprise 25 per cent of the area and have 15 per cent of the population of Africa (Hamdan, 1963). In addition, Mauritania, Ethiopia and the Congo (Kinshasa) while not land-locked have had many of the problems of such states. The short coastline of Africa relative to its area is well known, and the small extent is made worse by its inhospitable character with few deep inlets or substantial peninsulas, surf and coral (east coast). As Africa is so overwhelmingly dependent upon overseas trade this is the more serious, and the construction of ports has been a major engineering feat in Africa, notably at Casablanca, Monrovia, Abidjan, Takoradi, Tema and Cape Town.

So many of the recently independent states of Africa rely on foreign aid that it might be objected that consideration of shape, area and population are academic points of no importance, yet it is difficult to be sure that external aid can or will continue indefinitely on the present scale, and while it continues aid tends to be unduly mixed up with the politics of the 'cold war'. Colonialism may be replaced by Aid-Dependence.

AGRICULTURAL DEVELOPMENTS

Change is no less rapid or profound in the economic field than in the political one. Africa is mainly, and will long continue to be, a producer of vegetable and mineral raw materials. High or at least good prices for agricultural raw materials for many years after the Second World War put money into the hands of villagers, and made prosperity fairly widespread. Cocoa production has been increased by new plantings on the eastern fringes of the Nigerian cocoa belt, and on the western side of the Ghana and Ivory Coast areas. Improved pest and disease control have also helped, so that African countries (mainly the West African ones) now produce some three-quarters of the world's cocoa (Table 20); Ghana alone produces over one-third, and had record harvests in the early 'sixties. Almost all the West African crop comes from small farms.

Groundnut production is about 900,000 tons annually in Senegal and 1,400,000 tons in Northern Nigeria, almost all again from small African farms, such amounts being secured by the use of improved seed, artificial and animal fertilizer. Substantial quantities of groundnuts are used as food in Africa or crushed for oil. Turning to the oil palm, there are extensive foreign-owned plantations in both Congos, but Nigeria produces as much (nearly one-third of world exports) from small African-owned farms, and the quality of its oil has been vastly improved since the Second World War by extraction in small oil mills, by initial bonus payments for high grade oil, and by clarification and bulking in large holders at the ports. West Africa also supplies about four-fifths of the palm-kernel exports of the world.

Table 20.[1] *Production of Selected Crops in Africa in Thousands of Metric Tons*

Crop	Average 1934-8	Per cent of World Production[2]	1963-4	Per cent of World Production
Cocoa	484	66	900	74
Groundnuts	1,500	17	4,670	31
Cotton Lint	590	11	880	8
Sugar Cane	800	4	32,010	7
Coffee	140	6	1,000	25

[1] Sources: *Production Yearbook*, 1951 and 1964, F.A.O., Rome.
[2] Excluding the U.S.S.R.

Another crop which has expanded greatly in some countries is cotton, especially in the Sudan, Tanzania, Mozambique, Chad, Cameroon, Nigeria, Egypt and the Republic of South Africa. Most of the increased production is for use in African mills, of which there are some very modern examples in several countries, e.g. in most of the above-mentioned states, the Congo (Kinshasa), Angola, Rhodesia and Ethiopia. Cotton production has, however, declined in the Congo (Kinshasa), while remaining stable in the rest of Africa, so that the percentage of African cotton production in world output has declined (Table 20). Sugar is also being grown much more widely in Africa to satisfy local needs, which increase rapidly with improved standards of living. Important areas using irrigation are in and near the Lundi Valley (Rhodesia), near Moshi and Kilombero in Tanzania, and at Wonji in the Rift Valley south of Addis Ababa in Ethiopia, while the crop is developing in the Niger Valley in Nigeria, and in the Inland Niger Delta in Mali.

The most outstanding crop development has, however, been coffee. Whereas in 1934–8 the continent accounted for only 6 per cent of world production, in 1964 the percentage was 25. This rise came about after the Second World War as the result of a guaranteed market at very high prices in France for coffee from the then French colonies, of the founding of co-operatives for African growers in Tanzania, of the demand for soluble coffees which use the robusta variety much grown in Africa, of prosperity and improved living standards in coffee-consuming countries, and of increasing demand for coffee in countries like the United Kingdom which are traditionally greater consumers of tea. The Ivory Coast is now the third or fourth world producer of coffee after Brazil and Colombia, and surpassing Angola. Uganda and Ethiopia are other important producers, so that Africa now exports a quarter of the world's coffee.

Table 21.[1] *Exports of Selected Crops from African Countries in Thousands of Metric Tons*

Crop	Average 1934–8	Per cent of World Exports	1963	Per cent of World Exports
Cocoa	460	67	738	76
Groundnuts[2]	770	42	1,180	83
Palm-oil[3]	230	50	303	56
Palm Kernels	670	92	532	99
Cotton Lint	560	18	740	22
Raw Sugar[4]	720	7	1,647	5
Coffee	130	8	684	25

[1] Sources: *Trade Yearbook*, 1951 and 1964, F.A.O., Rome.
[2] Total in shelled equivalent.
[3] 1962 figures.
[4] Africa is also a considerable importer (470,000 tons average in 1934–8 and 1,043,000 tons in 1963).

The exports of Africa are almost everywhere being further processed locally before export. Thus more than one-half (470,000 tons) of Senegal's average groundnut harvest is crushed locally and the oil exported by tanker, while some 245,000 tons of the Northern

Nigeria harvest are crushed in four Kano mills. Together with oil from small Niger Republic mills it is taken south in specially-designed saddle-back railway tankers, the interior tank being used to carry mineral oil northwards and the outer 'saddle' tanks taking groundnut oil southwards to Lagos. Likewise, far more varied forest trees are being felled than in the past, and whereas the fewer species of the past were exported entirely as logs, the now more-varied species are also exported sawn, as veneer or plywood from ultra-modern factories at Abidjan (Ivory Coast), Samreboi (Ghana), Sapele (Nigeria) and Port Gentil (Gabon), the latter claiming to be the world's largest plywood works. Nevertheless, Africa produces only some 5 per cent of the world's timber, although its value is much higher.

New methods of agricultural production are being tried. Whereas the British had rarely permitted plantations in West Africa (although freely in East, Central and South Africa), African governments are developing them as means of increasing output and improving the quality of crops. Nigeria is experimenting with Farm Settlements where farmers and their families are thoroughly instructed in new techniques over a number of years. Under Nkrumah Ghana had over a hundred State Farms on the Russian model, run with the help of Russian experts. Each farm was rather under a thousand acres in area, but most were to be larger when more mechanical equipment was available, and the aim was to produce a standard crop for processing in local factories. There were also farms run by the Workers Brigade, where the aim was to provide work for urban unemployed who had recently come from rural areas, but the farms were quite considerably mechanized. The Ghana Army was to undertake farming to provide its own food, as the Ivory Coast one is already trying to do with second-year conscripts. Ghana also has some Young Farmers Settlements for school-leavers but they are fewer and less developed than in Nigeria. Most countries have for long tried to develop co-operatives for ordinary farmers but progress is slow, partly because of the scarcity of honest and efficient clerks. In Kenya Africans are acquiring land in the former European areas of the highlands, while there has long been a vigorous effort to improve methods on African highland farms, mainly by introducing livestock and modern methods.

Many African governments are anxious to extend irrigated agriculture. The most spectacular success has been in the Sudan where the government has added 800,000 acres of irrigated land in the Managil Extension, on the north-west of the one million acres of the Gezira Scheme,[1] and where cotton is the vital crop. The Roseires Dam has been built to supply future water-needs of the Managil, which were at first met from the Sennar Dam. The height of this has been raised, and now produces hydro-electric power which is gridded away as far as Khartoum. Another remarkable scheme is for the comprehensive development of the Medjerda Valley in north-eastern Tunisia. The valley has suffered severely from soil erosion, silting, flooding, gross disparity of land-holding, and insufficiently productive farming. All these matters are now being tackled and some 27,000 acres have been irrigated for the intensive production of vegetables. Over 1,350 families have been settled on sub-divided farms. A third very notable achievement is on the Limpopo in Mozambique. A dam at Guija, which also carries the railway opened in 1956 between Bannockburn (Rhodesia) and Lourenço Marques, provides water for Portuguese and African peasant farmers who each have between 10 and 25 acres of irrigated land and 62 acres for dry farming. The main crops are rice, wheat, cotton, maize, vegetables (especially tomatoes, onions, beans, and potatoes), kenaff and alfalfa, and only here and in Angola can European and African settlers be seen living and working side by side under identical conditions for similar rewards. There have also been many irrigation developments with mechanical cultivation and paid labour, as at Richard-Toll in Senegal for rice (but at high cost), by private companies as in Ethiopia at Wonji for sugar and at Tendaho on the Awash in the Danakil Plain for cotton, and for sugar in Tanzania and Rhodesia.

CHANGES IN MINERAL PRODUCTION

The most dramatic change has been the rapid development in much less than a decade of the great oilfields of Algeria, Libya and Nigeria, the laying of oil and gas pipelines to the Algerian, Tunisian and Libyan coasts (Figure 17.2), and of an oil pipeline from the Nigerian fields to the sea at Bonny. A new world region of petroleum and gas production has been created. Oil is now by far the leading export of Algeria, Libya and Nigeria. Although North Africa produces the now

[1] There is an excellent film about this entitled *White Gold* obtainable free on loan from the Sudan Government Publicity Officer, Cleveland Row, London, S.W.1.

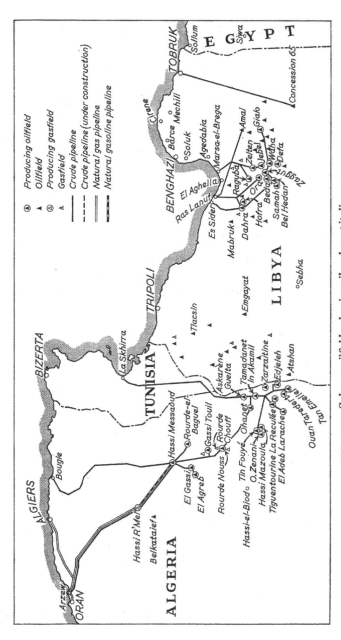

FIG. 17.2. *Saharan oilfields, showing oil and gas pipelines.*

less-needed high-octane products these fields are nearer to European markets than the Middle East or Venezuela fields. Moreover, natural gas is being brought to Europe by tanker, and this with considerable natural gas developments in Europe (e.g. in the North Sea, Netherlands, France and Italy) may also help that part of the Old World to secure some of the enormous benefits the United States has enjoyed from vast resources of natural gas.

Mineral oil has also been found in recent years on the coasts of Gabon near Port Gentil, of the Congo (Brazzaville) at Pointe Noire, and of Angola near Luanda. These are of much less significance. Very intensive prospecting is going on there and in some other producing and non-producing countries, but African oil production is still no more than some 4 per cent of World output.

Many oil refineries have been or are being built by companies in the hope of larger local markets and to stake a claim in such expanded markets (Hoyle, 1963). Refineries are sought by African governments as part of their drive for industrialization and as prestige symbols. Apart from the fairly numerous older-established ones in the more developed extremities of the continent, refineries have been built or are in progress at Luanda (Angola), Matola (opposite Lourenço Marques), Tema (Ghana), Port Harcourt (Nigeria), Mombasa (Kenya), Port Sudan, Abidjan (Ivory Coast), Dakar (Senegal), Dar es Salaam (Tanganyika), Assab (Ethiopia), Monrovia (Liberia) and Tamatave (Madagascar). Another refinery at Umtali (Rhodesia) is an African example of the world trend to establish refineries of imported crude oil nearer to inland markets. The pipe-line normally carries oil from Beira (Mozambique) across mountainous country along the international boundary.

Among the solid minerals, Africa is becoming an important supplier of iron ore. Before the Second World War only Algeria, Morocco, Tunisia, Sierra Leone and South Africa were significant producers of rich iron ore and the latter alone had an iron and steel industry. Since 1951 numerous other deposits, mostly of rich haematite ore, have been developed. In that year an American firm began operations in the Bomi Hills 50 miles north of Monrovia, and in 1962 it opened another reserve on the Mano River on the Liberian–Sierra Leone boundary (Figure 17.3). Both are linked by a mineral railway to special loading piers in the deep-water port of Monrovia, and iron-ore has displaced rubber as the leading Liberian export. A richer deposit on Mt. Nimba, at the meeting point of the Liberian, Guinea and Ivory Coast boundaries, has been developed by another

FIG. 17.3. *Liberia.*

company. This ore is taken out by a railway to a new port at Buchanan. A fourth Liberian reserve is being exploited by a third company in the Bong Hills north-east of Monrovia, and is linked with the latter by Liberia's third mineral railway.

After the development in Liberia in 1951 there followed one near Conakry of laterized ore of ferruginous magnetite. It has chrome, nickel, alumina and much moisture (which delayed its use), but it is easily quarried and near the port. In the later 'fifties rich ferromanganese and titaniferous iron ores were developed on the Luanda Railway in northern Angola, and on or near the Benguela and Moçâmedes lines. In late 1962 commercial production began of the rich haematite deposits near Fort Gouraud in Mauritania (Figure 17.4). These are joined to the deep-water port of Port Etienne by a 400-mile mineral line through desert and keeping just within Mauritania. Ore carriers of 45,000 tons are taking the ore to participating

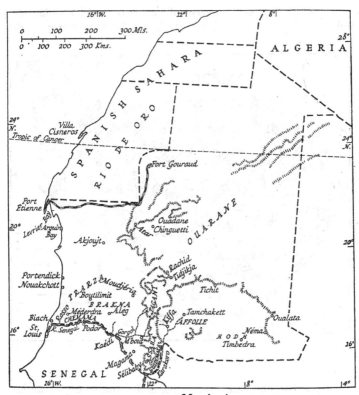

FIG. 17.4. *Mauritania.*

firms in France, Britain, Germany and Italy; in the first case to the
new coastal integrated steel works at Dunkirk. Royalties on this
ore are making Mauritania into a viable state. The same is happening
to Swaziland, as iron ore reserves are opened up in the Bomva Ridge
close to the boundary with the Republic of South Africa. To make
this possible Swaziland's first railway has been built across the
country to link the deposits with Lourenço Marques, in Mozambique, from where the ore goes to Japanese works. The distance from
Japan well illustrates the way that modern steel industries in countries
with costly fuel will go far to secure the richer ores that countries
in Africa are increasingly producing. More such reserves are likely
to be developed, e.g. in Gabon, given good markets for steel.

z

Meanwhile, Gabon is the scene of another remarkable mining and transport development (Figure 17.5). In the west-centre of the

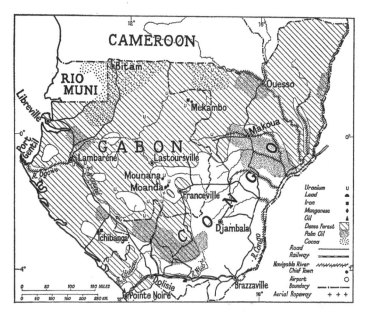

FIG. 17.5. *Gabon and Congo (Brazzaville).*

country at Moanda, near Franceville, is the world's largest worked deposit of manganese, with a content of 48 per cent. To export it from this remote area of difficult terrain the ore is carried first by a fifty-three-mile aerial ropeway, then by 2,000-ton trains of forty wagons on a 178-mile long mineral railway to the Congo-Ocean Railway, along which the ore travels a further 124 miles to Pointe Noire in the Congo (Brazzaville). All this came into full operation in 1962 and is another engineering marvel of modern Africa, made worth while by the richness of the manganese ore. Gabon quickly became the fourth world producer of manganese.

THE DEVELOPMENT OF HYDRO-ELECTRIC POWER

Africa's hydro-electric power potential has been described as equal to the amount of developed power in the world; that on the lower Congo is alone equal to that already developed in the United States. The difficulties of developing Africa's power potential are enormous – especially the greater costs of any such development in the tropics because of the seasonal rainfall, high evaporation, the need for greater insulation, and higher costs of transmission. All this is made more difficult by the smallness of the general market and the usual need to have one or more large industrial users of the power. So far Africa has only just over 2 per cent. of the world's hydro-electric power, and that has been developed mainly for mining needs – particularly in the Congo (Kinshasa). The Kariba development is in line with this, since it was conceived principally to serve the power needs of the Zambian Copperbelt; nevertheless, it has also served the industries and other needs of Rhodesia. In this latter respect it is aiding industrialization in Africa, and this was the purpose of the Owen Falls Scheme which has supplied power to new cement, textile and copper-refining industries, as well as for domestic needs in Uganda and western Kenya. The Mabubas, Biopio and Matala plants in Angola supply power for numerous industries and domestic needs in Luanda, Lobito, Benguela and Matala, while the Edea dam in Cameroon provides power for Africa's first aluminium smelter close by, as well as for industrial and domestic needs in Douala and Yaoundé. The great Volta River Dam at Akosombo on the lower Volta in Ghana (Figure 17.6) is likewise designed mainly for the provision of power to an aluminium smelter at the new port and planned town of Tema, east of Accra, although power is also available for Ghana's existing mines and southern towns. At first imported alumina will be used but Ghana bauxite may be used after the first eight or ten years. The Kainji Dam on the Niger above Jebba in Nigeria will produce power for use generally in Nigeria.

QUICKENING INDUSTRIALIZATION

Although the Republic of South Africa is the only African country with a considerable range and diversity of industry, and with substantial industrial exports, Rhodesia and, perhaps, Egypt, are within sight of the same goal, while the Maghreb lands and the Congo (Kinshasa) have a fair range. Industrialization is sought as a

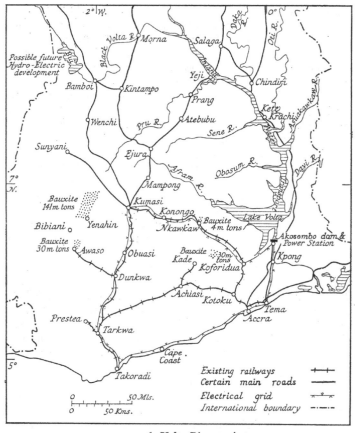

FIG. 17.6. *Volta River project.*

means of avoiding the price fluctuations of the raw materials now mainly exported, of securing further income from such materials by processing them locally (e.g. the extraction of groundnut oil described above), of broadening the wage structure and enhancing labour skills, and of providing employment in towns to which there is such a formidable exodus from the African countryside. To all these is added the prestige factor of industrialization, the belief that without an iron and steel works, oil refinery and aluminium or other smelter a country is primitive. It is not surprising, therefore, that the same industries have been set up in adjacent countries, even in the several regions of federal Nigeria, so that markets are being severely restricted to each national area or region. This problem is made more acute by the small and poor populations of many countries.

Apart from industries processing exported vegetable or mineral produce, a common type are those manufacturing simple yet common consumer goods such as textiles, soap and drinks, especially when these are costly and bulky imports. Some industries in these categories will start by using mainly imported goods, replacing them gradually by local material, e.g. tobacco and leather. There are cycle, scooter and car assembly works at ports putting together articles more costly to import in their finished form, as well as flour mills at ports to which a necessary import must come. Ports are thus significant industrial areas for this reason, and because they are often capitals they have local markets and are the termini of transport systems. Inland towns with industries are usually large centres of population, such as Ibadan with 627,379 (1963) or Kinshasa with 402,492 (1959). Industries are commonly grouped on industrial estates, such as the Trans-Amadi Estate at Port Harcourt, which is also supplied with natural gas from the Nigerian oilfields.

TRANSPORT

Outstanding characteristics of modern Africa are the importance of roads and road transport. The latter is usually far more important than rail traffic, so that atlases which emphasize railways give a false impression. Michelin, Shell and other tyre or petroleum companies issue road maps of African countries, and these are helpful in showing the number and importance of roads. They are constantly being added to or improved, and road transport is a common African enterprise. Air services are still for the relatively few, but most African countries have their own airline and there are numerous internal services as well as international ones in almost all countries. Airlines have done more than anything else to enable Africans to know and meet each other; the aeroplane is often regarded as the pacemaker of Pan-Africanism.

Railways are still vital for long-distance hauls of bulky produce, and new mineral lines in Mauritania, Liberia, Gabon and Swaziland have been mentioned. A general purpose line has been built in Nigeria from Jos to Maiduguri, the Luanda and Moçâmedes lines are being extended in Angola, where branches are also being made to the Benguela line, and the line from Nacala in Mozambique is nearing Lake Malawi. Rhodesia has since 1956 had a second outlet through Mozambique to Lourenço Marques. Among important new or re-developed ports are Port Etienne (Mauritania), Monrovia (Liberia), Tema (Ghana), Cotonou (Dahomey), Nacala (Mozambique), Assab (Ethiopia) and Bougie (Algeria), the latter for export of Saharan oil.

CONCLUSION

Africa is entering another great epoch during which it will not only be transformed but will impinge increasingly upon the rest of the world. Africa cannot be ignored, and the fascination of the continent and the needs of its peoples demand that it be studied adequately. There are already a number of detailed studies of large parts of the continent (Cole, 1966; Wellington, 1955; Harrison Church, 1966 and 1963) which are available for teachers and for reference by senior pupils. Three modern A-level studies of the continent (Jarrett, 1966; Harrison Church, Clarke, Clarke and Henderson, 1967; Mountjoy and Embleton, 1965) have appeared. Articles on Africa are listed regularly in *Geography*. *The Geographical Digest*, Philip, annual, details the main developments. The Central Office of Information, Horseferry Road, London, S.W.1., publishes *The Changing Map of Africa* and leaflets on Commonwealth countries, as does the Commonwealth Institute, W.8. The Petroleum Information Bureau, 4 Brook Street, London, W.1. has useful handouts on *Oil in Africa, Oil in the Commonwealth, Saharan Oil in the French Economy*, etc. The British Iron and Steel Federation, Steel House, Tothill Street, S.W.1., provides similar material on iron ore in Africa. Both organizations have excellent films, as have Unilever, Blackfriars, E.C.4., the United Africa Company, United Africa House, S.E.1., and Shell International, Shell Centre, Waterloo, S.E.1. Several of the African Embassies or High Commissions publish gratis monthly or quarterly bulletins, notably Liberia, Ghana, Nigeria, Rhodesia, and the Republic of South Africa.

References

COLE, M. M., 1966 (2nd edit.), *South Africa* (London).

HAMDAN, G., 1963, 'The Political Map of the New Africa', *Geog. Rev.*, **129**, 418–39.

HARRISON CHURCH, R. J., 1966 (5th edit.), *West Africa* (London).

— 1963, *Environment and Policies in West Africa* (London).

HARRISON CHURCH, R. J., CLARKE, J. I., CLARKE, P. J. H. and HENDERSON, H. J. R., 1967 (2nd edit.), *Africa and the Islands* (London).

HOYLE, B. S., 1963, 'New Oil Refinery Construction in Africa', *Geog.*, **48**, 190–4.

MOUNTJOY, A. B., and EMBLETON, C., 1965, *Africa, A Geographical Study* (London).

JARRETT, H. J., 1966 (2nd edit.), *Africa* (London).

WELLINGTON, J. H., 1955, *Southern Africa*, 2 Vols. (Cambridge).

CHAPTER EIGHTEEN

Frontier Movements and the Geographical Tradition

P. HAGGETT and R. J. CHORLEY[1]

Lecturers in Geography, University of Cambridge

RETROSPECT

In preparing this volume for press we have been forced to take a retrospective view back over the months since the inception of the Madingley Symposia and are surprised by even the small measure of order which seems to have emerged from chaotic beginnings. The make-up of the contributors to the First Symposium, from which this volume was developed, reflects a combination of least effort and of randomness that would have delighted Zipf (1949). Contributors were gathered both from personal contacts and colleagues who could be persuaded to leave the safety of their prepared geographical positions and 'go over the top' into the exposed conflict of pre-university education. However, there was no guarantee that, once over the top, everyone would identify the same enemy and charge in the same direction and, indeed, this volume is in no way a concerted and radical attack by angry young men. Neither does this volume represent any special 'Cambridge approach' (whatever this means) to the problems of geographical teaching for, although the myth of distinctive 'schools' dies hard, we are certain of receiving as critical a reception in Cambridge as in any other centre of scholarship.

Perhaps the only common features linking the fifteen contributors are that they share, in Quaker terminology, a 'concern' for geography; that they are actively engaged in some aspect of geographical work in the 1960's; and that none of them believe that the best geography, like the best wine, must be necessarily both French and long-matured.

---

[1] The views expressed in this chapter are those of the two authors and do not necessarily reflect the opinions of the other contributors to this volume.

Indeed, a recognition of the need for a complete and radical re-evaluation of the traditional approaches both to geography and to geographical teaching in Britain characterizes many of the contribu-ions (e.g. Wrigley, Chapter 1; Chorley, Chapter 2; Haggett, Chapter 6; Board, Chapter 10). If the average view of the contribu-tors is radical, however, it is an average with a recognizably large standard deviation.

Indeed, we should not expect any well-defined common philosophy to emerge from these essays – nor does it. There has been no con-scious attempt to dictate viewpoints, neither to integrate nor recon-cile opposing ones. In some contributions sharp contrasts emerge, for example, regarding the significance of regional method (see Wrigley, Chapter 1, and Timms, Chapter 12). There are, none the less, themes which appear and reappear with variations throughout the volume and, in so far as this work presents any unified attitude towards certain key aspects of geographical teaching, it is in these themes that the coherence rests. So individually modulated are these themes, however, that to present any one of them in a simple general-ized form seems to destroy much of its singularity. It is possible, however, to isolate some of them which appeared both in the written contributions and in the productive discussions with the teachers attending the Madingley Symposia. The present clash between the historical and the functional approaches to geographical matters (Wrigley, Chapter 1; Chorley, Chapters 2 and 8; Smith, Chapter 7) was commented on, and certain of the contributors pointed up the shortcomings of the largely historical treatments of physical (Chorley, Chapters 2 and 8) and human geography (Wrigley, Chapter 4; Collins, Chapter 11). The need for geographers to be at least aware of attitudes and techniques in the associated social (Pahl, Chapter 5; Haggett, Chapter 6) and physical sciences (Beckinsale, Chapter 3; Chorley, Chapter 8) in general, was allied with a concern for an increase of quantification in geography (Haggett, Chapters 6 and 9; Chorley, Chapter 8; Board, Chapter 10; Timms, Chapter 12), and an opposition to the idiographic character traditionally assumed by British geography. The recognition of 'man in society' as a focus of geographical interest emerged strongly (Wrigley, Chapter 4; Pahl, Chapter 5), with especial reference to the character of urban centres (Collins, Chapter 11; Timms, Chapter 12), in contrast with the strong current emphasis on primitive and agricultural societies fostered by the traditional man/land approaches to the subject. Attitudes to mapping also showed interesting features, not the least significant of which was the increasing regard for the map as a framework within which data can be organized as a springboard to higher and more sophisticated analysis (Haggett, Chapter 9; Board, Chapter 10; Willatts, Chapter 13), rather than as an end-product of geographical labour. This last attitude was also part of a wider concern regarding the place of geography in national planning (Pahl, Chapter 5; Collins, Chapter 11; Willatts, Chapter 13), where the contribution of geographers should be more than merely to map the information collected by others and then to surrender these maps for interpretation and analysis. At such a symposium on geographical teaching it was natural that the problems and opportunities of teaching on many different levels were also discussed (Chapters 14, 15 and 16), together with the need to keep both material and, especially, teaching attitudes up-to-date (Harrison Church, Chapter 17), and this last need gave rise to many criticisms of the existing British geographical publications as the means of providing the vital 'academic retooling' necessary for practising teachers of a rapidly developing subject in a rapidly changing world.

Three other matters of interest seemed to us of such especial significance that we have attempted to develop them in the following sections. They are the use of model teaching in geography, the integration of such models with conventional regional treatments, and the need to clarify the model of geography itself which we hold.

MODEL THEORY IN GEOGRAPHY

Although it is unnecessary here to recapitulate the significance of the construction of theoretical model structures in geographical teaching and research (see Chorley, Chapter 2, and Chorley, 1964), it became readily apparent from the contributions to these essays, and particularly from discussions with teachers, that such structures have an especial value in the understanding and presentation of information of geographical interest. The regional model (Wrigley, Chapter 1), the Davisian physiographic model (Chorley, Chapter 2), climatological models (Beckinsale, Chapter 3), the spatial locational models of economic geography (Haggett, Chapter 6), historical models (Smith, Chapter 7), urban models (Timms, Chapter 12) and those others less sharply outlined in other chapters seem to provide specially appropriate frameworks for both research and teaching. We cannot but recognize the importance of the construction of theoretical models, wherein aspects of 'geographical reality' are presented together in some organic structural relationship, the juxtaposition of which leads one to comprehend, at least, more than might appear from the information presented piecemeal and, at most, to apprehend general principles which may have much wider application than merely to the information from which they were derived. Geographical teaching has been markedly barren of such models, partly as a result of the interest which has centred largely on the unique and special qualities of geographical phenomena, and geographers have been loath to make use of these powerful frameworks, despite the teaching successes of, for example, the Davisian cyclical model. This reticence stems largely, one suspects, from a misconception of the nature of model thinking, wherein such frameworks are expected to be 'true', or 'real', or to possess other equally equivocal qualities. Models are subjective frameworks, constructed for specific purposes, relating to a limited range of reality and only possessing relevance within well-defined levels of information content, sophistication and time. On certain levels the Davisian system has a large measure of 'reality' and 'truth', on other levels it has a smaller measure. Model frameworks are like discardable cartons, very important and productive receptacles for advantageously presenting selected aspects of reality, but no one model should be expected to accommodate many aspects of reality, at different levels of information content, sophistication and time.

Figure 18.1 indicates the manner in which model thinking has a bearing on the structure of teaching. In figure 18.1A a loose framework is adopted at an early stage (perhaps some regional or simple spatial one), and through time attempts are made to incorporate progressively larger amounts of information and, to a much more restricted extent, to increase the level of sophistication of the framework. The lack of sophistication of the initial basic structure of the framework, together with the restrictive properties which any one structure most possess, however, imply that attempts to adapt it to encompass more and more information will result in further losses of internal form and cohesion, and in its degeneration into an amorphous mass of loosely-related information. Attempts to increase the level of sophistication with the passage of time by *ad hoc* tinkering with the framework in the hope of making it serviceable throughout a wide range of educational experience usually have a similar effect. It is such 'one-gear' teaching which is largely responsible for obtaining geography its poor reputation as a scholastic academic discipline.

Sometimes some measure of structural coherence of knowledge is obtained by the teacher adopting a *classification* type of approach (Figure 18.1B). Classifications have the advantages of well-defined structure (i.e. the relationships between different elements of classified information are usually clear), of the ability to subsume all the apparently relevant information in a more or less satisfactory manner at any one time, and, most significantly, of interchangeability in that an embarrassing increase in the amount of information can be met by discarding some outmoded classification and leaping to a new and more sophisticated one. This last quality is attractive to teachers in that geographical material can thus be presented to different scholastic levels in contrasting and intellectually-appropriate frameworks. Classifications, however, have obvious shortcomings as teaching vehicles in that their construction involves the dissection and categorization of information such that, instead of associating related information intimately together in a suggestive and productive manner, it is disassociated and usually incapable of promoting the student towards novel speculations as to *how reality operates*. Despite these shortcomings, classifications are susceptible of much sophistication and, on higher levels, the so-called genetic classifications merge into models. Teaching by means of *model frameworks* is illustrated diagrammatically in figure 18.1C. Each model framework is tightly knit such that different pieces of information are set in provoking juxtaposition, and the whole is capable of considerable exploitation through time as experience with the model leads to more and more sophisticated handling of the information within it. One has only to recognize the difference of levels on which the Davisian cyclic scheme can operate educationally to recognize this property of exploitability as representing one of the most striking contrasts between models and classifications.

However, as has been recognized, every theoretical model framework becomes strained and less appropriate as the amount of information required to be built into it and the level of sophistication on which the model is required to operate increase, and it is therefore necessary within any vital teaching programme to make imaginative leaps from time to time from less to more sophisticated models. These leaps (which might, for example, involve transitions from simple form-space observations for young students of physiography, to the cyclic model of Davis, and to the process-form-equilibrium Gilbert model for the older students) are most difficult to accomplish, largely because of the common desire to retain a previously-successful

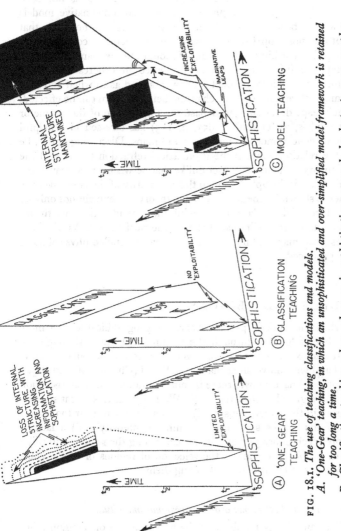

FIG. 18.1. The use of teaching classifications and models.

A. 'One-Gear' teaching, in which an unsophisticated and over-simplified model framework is retained for too long a time.

B. Classification teaching, where advances in sophistication are made by leaps to more complex classificatory schemes, although within each scheme no real advance in sophistication is possible.

C. Model teaching, where different levels of teaching are met by imaginative leaps to progressively more sophisticated models each capable of exploitation without the destruction of its internal cohesion.

model long after the time of its optimum utility. This has been particularly true in geography where first-rate imaginative models have always been at a premium. We would suggest, however, that one of the most hopeful and pressing requirements of contemporary geographical teaching is the giving of thought to the manner in which geographical information can be presented in progressively more and more sophisticated model frameworks, and as to when and how the imaginative leaps or 'gear changes' between them can be effected.

Such leaps do nothing to debase the importance of the simple or early models, as Toulmin (1953, p. 115) has stressed in a helpful analogy between model frameworks and maps. Discussing optics, he sees the relation between the geometrical refraction theory and the wave theory of light as '. . . not unlike that between a road map and a detailed physical map'. Although the superiority of the latter over the former is shown by the fact that wave theory can explain not only all the features of the refraction model but more besides, the wave theory has not necessarily eliminated the geometrical theory. As Toulmin argues, road maps did not go out of use when detailed physical maps were produced.

MODELS AND REGIONS

Perhaps no working problem vexes the geographical teacher more than that of 'getting through the regional syllabus'. Although the nature and extent of the regions covered varies with school examining board and with university syllabus, as Board (Chapter 14) and Bryan (Chapter 16) have shown, there is some common basic dissatisfaction with the regional part of the work. We find that this dissatisfaction stems from three main causes: (i) an attempt to cover far too large a part of the earth's surface at a 'uniform intensity'; (ii) a lack of accepted techniques or concepts for examining the region; and (iii) a lack of *a priori* rationale in the selection of regions or of the gains that we expect to make from studying them.

Scale in regional selection: orders of regional magnitude

There are clearly no absolute limits to the size of a region other than the limits of the earth itself, with its total land and sea area of 196,836,000 square miles. Below this the region contracts continuously towards the very small. Attempts to demarcate regions on the basis of size have been reviewed elsewhere in this book (Haggett, Chapter 9) and it is worth while here to try to draw together some of these schemes into a single regional-functional model in which the magnitude of the region is related to the functional use to which the region is to be put.

Table 22 presents a tentative order of magnitude for areal studies based on successive logarithmic subdivision of the earth's surface. It uses an index, the G-scale (Haggett, Chorley and Stoddart, 1965), in which G is a dimensionless ratio given by the formula:

$$G = log\ (G_a/R_a)$$

where G_a is the area of the earth's surface (i.e. 1.986×10^8 square miles) and R_a is the area of a regional subdivision. Large areas have small G values (e.g. U.S.A. $= 1.82$) while small areas have large values (e.g. Rutland $= 6.11$), and the scale may be continued down to the smallest areas of geographical investigation. The general advantages and disadvantages of the scale and rapid methods of computation have been given at length elsewhere (Haggett *et al.*, 1965), but it seems appropriate here to examine some of the implications of a logarithmic view of regions for geographical teaching. Eighteen regional works of very different characteristics have been plotted on Table 22 in relation to both their regional magnitude (*y*-axis) and their subject matter (*x*-axis) to show something of the range of interest on both axes. At best these studies are a 'grab-sample' of the relevant literature and show no distinctive pattern. Work in hand, however, suggests that the great bulk of 'classic' regional geography (represented on Table 22 by Vidal de la Blache's *La France de L'Est*) may cluster significantly around the central part of the diagram; the volume of work by geographers falling off steeply both upwards towards the world level, and downwards towards the site level. Whether this represents a fundamental characteristic of geographic writing or a convenient 'ecological niche' within the academic climax is not yet clear. Certainly, for most disciplines concerned with field study (e.g. geology, botany, agriculture, etc.) the 'centre of gravity' of their work falls well below that of geography on any table of areal magnitude.

Individual geographical monographs may show, in their internal structure, distinctive approaches to the problem of generalizing statements over a large area: e.g Platt in his *Latin America: Country-sides and United Regions* (1942) uses small sample studies ranging down to the $G = 10-11$ level, whereas James' *Latin America* (1959)

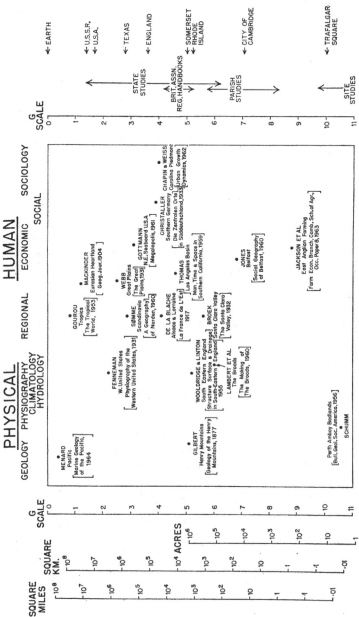

TABLE 22. *Areal studies referred to the G-scale.*

adopts a two-stage hierarchy of political and physical units ranging down to about the $G=6-7$ level. To be accurate, therefore, geographical works should be plotted not as points but as either graph networks or zones. Some tentative work along these lines with an added time-dimension (z-axis) has been begun, and we hope that Table 22 may be useful in arousing or annoying its readers into making their own experiments with the areal structure of works with which they are themselves familiar.

Although the system may seem over-elaborate it has two major virtues. First, it supplies a simple reference scale for regional magnitude. Secondly, it allows ready comparison between regions (the area

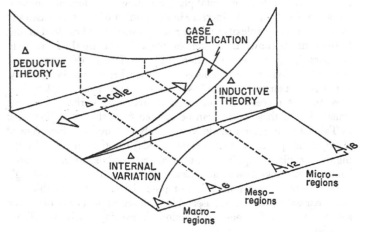

FIG. 18.2. *Implications of regional scale changes for teaching models.*

of regions in adjacent classes will vary on average by 1:10 and never by more than 1:100) on a ratio scale with the reference 'benchmark' being the whole planetary surface. Whether differences in regional magnitude have any deeper significance is a matter for debate, but it is perhaps noteworthy that the importance of dimensional differences is of key importance in classical physics where changes in one dimension (e.g. length) may be associated with disproportionate changes in area, mass, viscosity and so on. These problems in 'similitude' also have crucial importance in biology where D'Arcy Thompson (1917) devotes a considerable part of his *On Growth and Form* to a consideration of magnitude in zoological and botanical design. As geographers draw increasingly on physical models and their biological derivatives, we shall need to be increasingly alive to the dangers of spatial or dimensional 'anachronisms', if we may call them such. Measurements of distance inputs (length), boundaries (perimeters), populations (masses) in the 'gravitational' models of economic geography (Isard, 1960, pp. 493–568) may need to be successively re-cast at different areal levels if we are to retain their principles of similitude. Likewise, the models of airflow, or erosion, or migration, or regional development may not necessarily hold equally well at all magnitude levels of regional application.

Some of these implications of changes in magnitude are shown in Figure 18.2. On the horizontal plane are shown two logical concomitants of scale change: the increase in the number of potential cases and the decrease in complexity as the regions get smaller. These lead in turn to the changes shown in the vertical plane. These are the increase in comparability, in case-replication, and therefore in the significance levels of findings as the regions get smaller, whereas, on the contrary, as regions get larger there are fewer cases to compare and explanations have to rest increasingly on external analogies. Perhaps a simple illustration of these points can be derived from the cases illustrated in Table 22. In geomorphology, the study of Gondwanaland (one of a very small population of ancient shields) has been marked by the highly speculative application of analogue models, together with a considerable range in academic views; conversely, the study of shingle spits, where a large population of potential cases is available, has been marked by the careful assembly of observational evidence and a smaller range of academic opinions regarding their significant features.

Whether our reliance on external theory for explanation of the few macro-regional features of the earth is a built-in characteristic of geography is an interesting but unresolved point. Certainly in the economic units in Table 22 we rely heavily on international trade theory to explain major differences (e.g. developed versus under-developed areas), while at the lower levels of the meso- and micro-region both field observation and field-based models (like that of Christaller (1933)) play a more important part. Geography at the moment certainly appears to be more self-supporting in theory at the lower levels of regional magnitude.

Fusion of regions and models: the search for the modular unit

The separation of regional and systematic geography may not be as great in teaching practice as is often suggested in methodological reviews (e.g. Hartshorne, 1959, Chapter 9), but it might be largely resolved if the ideas of 'models' and 'regions' could be fused. If, for example, we select the Great Plains of North America (the *region*) to examine Turner's frontier concept (*the model*), or the North Atlantic region to examine the model of sub-tropical high-pressure systems, the lower Mekong region to examine models of river-basin development, the London region to examine alternative models of urban growth, or South Germany to examine settlement-spacing models, we are intuitively fusing our regions and our models. The advantage of such an approach is twofold. Regional delimitation is based on the 'modular unit' (the pressure system, river basin, migration field, hinterland, etc.) and the resulting regional treatment, far from being a mystical amalgam of 'landforms and life', is geared to the specific demands of the systematic model.

One disadvantage we see in such a fusion is some loss of the 'regional integration' of material irrelevant to the chosen systematic model. However, the feature of focal interest (e.g. settlement spacing in South Germany) must obviously be studied in relation to local variations in history, terrain and the conventional range of regional variables. Through such correlation the 'vertical integration' which characterized the classic French regional geographies will be preserved. Equally important in our view is the possibility for inter-regional or 'horizontal integration' through comparison of the region under study with other regions in the same 'set'. (See later in this chapter.) Thus we should expect a study of the Turnerian concept in the Great Plains to raise important questions of comparison in relation to the other major mid-latitude grasslands (i.e. the Pampas, Steppes, Veld, Murray-Darling plains, etc.)

To object to a fused region-model system on the grounds that it failed to cover the whole world at a uniform intensity we regard as a trivial criticism. There are few more discouraging teaching experiences than attempting to teach the 'flat' regional geography of an area which has no outstanding systematic problem or which has played no part in the development of any systematic geographical models, and we doubt the wisdom of selecting such regions for syllabus and examination purposes. Certainly Wooldridge's castigation on 'the eyes of the fool . . .' (Board, Chapter 10) is well justified if we direct our prime efforts towards remote and little-studied areas rather than towards equally appropriate nearby examples having also strong systematic significance. To ask for facts and nothing but the facts is, as Wittgenstein clearly saw, to demand the impossible; regions without theories are like maps without scale or projection.

If we compare geography with such sister disciplines as botany, geology, chemistry and history, we see with Bunge (1962, pp. 14–26) that regional geography may be regarded as a type of classification, a 'taxonomy of the earth's surface'. If this analogy is correct then an examination of these parallel taxonomies suggests the following implications:

1. No unique taxonomy is likely. The history of classification in each of the subjects we cite is a history of continuous reappraisal (i.e. the 'leaps' previously referred to), and indeed of conflicting reappraisal. Classifications merely represent useful working frameworks at a given state of information.

2. No attempt is made in teaching such taxonomies to 'go right through the card'. From the zoologist to the student of English literature the basis of taxonomic teaching is rational sampling. Whether the object is *Chlamydomonas* or *Othello* it is studied in part for itself and in part as a sample of a larger population.

3. The most efficient classifications are often those with model tendencies or affiliations.

Finally, one assumption which is worth challenging in such a reappraisal is that conventional regions for study should be 'nodular areas'. 'Linear' regions such as that along the 100° W meridian in the United States, the fringe of sub-Arctic settlement, the moorland edge in Britain, the London green belt, also come to mind as topics appropriate to regional study.

A SYNOPTIC MODEL OF GEOGRAPHY

In each of the preceding chapters the views argued have been moulded by the general view of geography held by the author. In some chapters, notably those by Wrigley (Chapter 1), Smith (Chapter 7), Board (Chapter 10), and Timms (Chapter 12), and in all the contributions to Part III, these views have been explicit and in others implicit. What kind of picture do these views give of geography? What light do they throw on such thorny problems as 'VI Arts' or 'VI Science', in which our educational administrators are continually faced with the perplexing problem of 'locating' geography within some broader academic framework? In short, what sort of subject is geography?

The Problem of Classification

One initial problem is quickly resolved. To ask whether geography is or is not a science is like asking whether sports are games. Geography can, like any other subject, be studied scientifically or aesthetically, and occasionally both. The study of birds can yield either sonnets or genetics, human suffering may provoke a Goethe or a Pasteur, rocks may inspire an Epstein or a Lyell. Our subject matter is passive and it is we who make the decisions as to how we shall study it or indeed if we shall study it at all. A more appropriate question is then: Can geography be studied as a science? It is in response to this question that we hope this book will show that in part it can, and, more important, that to an increasing extent it is being studied scientifically.

To say that geography can be studied scientifically is to say that it shares a trend common to many academic disciplines, a trend from which even history is not entirely exempt (Postan, 1962), but what kind of a science is it? Here we have two alternative approaches to an answer. One is to follow Hartshorne who, in his classic study on *The Nature of Geography* (1939), urged us to comprehend the nature of geography by studying what it had shown itself to be from the Greek geographers on, such that '. . . we must first look back of us to see in what direction that track has led' (Hartshorne, 1939, p. 31). A second source is to deduce the nature of geography from as few basic assumptions as possible, constructing a deductive-logical skeleton for what geography should be – even if it isn't. Bunge (1962) in his provoking *Theoretical Geography* argues this view and suggests we should be wary of our great forbears '. . . because the great men of the past might now, in view of more recent events, hold opinions different from those they then held' (Bunge, 1962, p. 1). In fact, our views of geography stem from both sources. These views must be, to a greater extent than we perhaps realize, influenced by the manner in which geography has evolved – partly because we ourselves have been involved in its evolution. At the same time outside influences enable us to see this solution in some perspective and to envisage the kind of changes which we would like to achieve.

We suggest that confused evaluations of geography stem not from any evasiveness or lack of thought on the part of geographers but from its very complexity. In the past it has been variously classified as an '*earth science*' (at Cambridge it is part of the Faculty of Geography and Geology, which include geophysics, mineralogy and petrology), a '*social science*' (as in most universities in the United States), and less commonly as a '*geometrical science*', a position it held in Greek times and which a few workers, notably members of the Michigan Inter-University Community of Mathematical Geographers, would like it to resume. These alternative placings arise largely from the different growth of the subject in Germany (the *landschäft* school), in France (the *human ecology* school), and in the United States (the *locational* school), although of course its evolution has been more complex than these facile associations of nation and school would suggest.

An Attempt at Fusion

An attempt to fuse these alternative views using the basic approach of set theory and Boolean algebra has been made by Haggett (In Press). Briefly, each of the three sciences into which geography has been placed can be viewed as a *set*, and each separate subject is an *element* within that set. Three sets can be defined: an earth sciences set (α), a social sciences set (β), and a geometrical sciences set (γ). Set α contains geography (1), geology (2) and other earth sciences and can be written as

$$\alpha = \left\{ 1, 2 \right\}$$

Similarly we can define the other two sets:

$$\beta = \left\{ 1, 3 \right\}$$
$$\gamma = \left\{ 1, 4 \right\}$$

where 3 is demography together with other social sciences, and where 4 is topology together with other geometric sciences. This situation can also be shown diagrammatically by the use of Venn diagrams as in Figure 18.3.

We can show the relations between any two sets by overlapping the diagrams. Thus geography is by definition part of both α and β sets, and its position is shown in the overlapping area of Figure 18.3B. Overlap of the three sets in pairs also suggests the position of the human ecology view of geography ('man in relation to his environment') in 5, of geomorphology (6) and of surveying (7) at the overlap of the α and γ sets, and of locational analysis (8) at the overlap of the β and γ sets. We can write these intersections as

$$\alpha \cap \beta = \{1, 5\}$$

$$\alpha \cap \gamma = \{1, 6, 7\}$$

$$\beta \cap \gamma = \{1, 8\}$$

More complicated relationships between the three sets are shown in Figure 18.3C where geography (1) is seen to occupy the central position at the intersection of all three sets, that is,

$$\alpha \cap \beta \cap \gamma = \{1\}$$

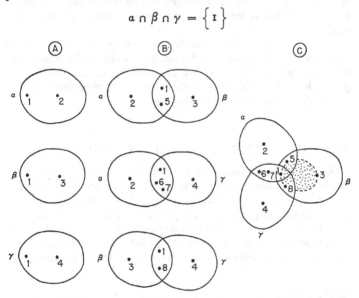

FIG. 18.3. *Set-theory approach to the location of geography within alternative definitions.*

with the cognate subjects, geomorphology, human ecology, surveying, and locational analysis occupying two-set intersections about it. The position of the newly emerging regional science (Haggett, Chapter 6) with its strong connections with geography locational studies, human ecology and the systematic social studies like economics is shown by the shaded area.

It is not suggested here that this type of analysis solves our problems of definition but it does suggest, if our analysis is correct, just why it is so difficult to 'locate' geography or to define it simply. To describe it as 'the study of the earth's surface', or 'man in relation to his environment', or 'the science of distribution', or 'areal differentiation' is to grasp only part of its real complexity. As in geomorphology (Chorley, Chapter 2) these are just some of the alternative views of the elephant. Geography can be defined not solely in terms of *what* it studies or of *how* it studies but by the intersection of the two. It is what Sauer (1952, p. 1) has called a 'focused curiosity' which has created techniques, traditions, and a literature of its own and which in our more hopeful moments we believe has contributed not a little to the understanding and the enjoyment of the planet we occupy.

In terms of our original classification problems we must learn to live with misclassification and timetable problems in school and university organization if we are to retain the essential identity of geography. Other subjects have equal burdens in other directions and we should cheerfully shoulder this one of our own.

EPILOGUE: THE PROBLEM OF INERTIA

Geography is a subject in which, despite great enthusiasm, large numbers of students, and growing post-graduate opportunities, progress continues to be slow. Most geographers welcome progressive changes and many geographers, in public at any rate, seem to want the same kind of changes, but somehow these aspirations are caught up in an academic log-jam. We suggest here some areas where this jam is being created and make some tentative suggestions for dynamiting it.

On the Problem

Perhaps the most immediate inertial problem faced by geographers is the constriction imposed by geography's past growth. We inhabit a Victorian academic structure every bit as solid and constraining as its architectural counterpart. Despite the vigorous growth of the subject in universities and schools over the last fifty years, the everyday popular image of geography is an antique one dogged by 'exploration, description and capes and bays' and this in turn influences the character of our intake of young minds, together with the amount, sources and destination of research funds. The influences of earlier models (of cyclic erosion, of environmental determinism, or of regional 'character') have been taken up in earlier chapters, but perhaps the saddest aspect of this apparent reverence for the past is that this has too infrequently been accompanied by genuine historical research. Old models, like old myths, have been handed down to us, while the original intentions and aims of their constructors lay, reverenced but unread, in some corner of the geographical library (see Wrigley on Vidal de la Blache, Chapter 1).

A second major problem is that of conflicting objectives. Paradoxically, this may stem in reality from the popularity of geography as a school and university subject in an era of educational expansion. Disregarding those students attracted to geography through a belief in its elementary character, much of its popularity derives not from its possessing any satisfying basic academic discipline as from the valuable 'side-products' which are believed to spring from its study. We would not deny that a deeper understanding of international affairs, of planning problems, or of the 'bridge' between the 'two cultures' may stem from a study of geography. We would argue, however, that if geography trims its sails to the vicissitudes of every profitable wind of social and educational demand that blows it is likely to lose any sense of distinctive intellectual purpose, will fail to attract its most necessary growth ingredient (the research student) and is likely to be eventually replaced by or amalgamated with other subjects which serve the purposes of society as well as possessing some intellectual identity. Geography's most important contribution to society will in the long-run result from its producing good geographical research, not by over-extending itself in fields of immediate educational profit.

A third inertial problem unquestionably results from the kind of academic isolationism which has been fostered by the idiographic and artistic preoccupation of so much past geographical work (see Ackerman, 1963, and Chorley, Chapter 8). There is no doubt that the most sterile aspects of present geography are the result of such academic in-breeding, whereas the most virile work is proceeding in fields which have been most willing to draw upon general intellectual advances (which, of course, are today mainly the scientific ones). It is not sufficient to dismiss this reality by the belief that 'the best research always goes on along inter-disciplinary boundaries'. Worthwhile disciplines do not develop like a coral atoll by the outward growth of active margins around a dead or atrophied centre. What should characterize the living heart of geography, and thereby justify its academic identity, is neither any single simple methodology nor any immutable body of subject matter, but the kind of physical/social/geometrical fusion which we have attempted to explain in terms of set theory.

On the Solution

Easy solutions are rarely the optimum ones. There has been no lack of thought or effort in British geography in the last half-century and if basic solutions had been at hand they would long since have been applied. Perhaps, with Chesterton's Father Brown, we find our greatest difficulty in recognizing the exact nature of the problem rather than in solving it. Here, however, we must be willing to commit ourselves on both the problem and the solution. On the problem, we suggest that the features of inertia recognized above are symptoms of a deeper malaise – the failure to recognize the multivariate nature of geography (as shown in Fig. 18.3C). In particular there has been the neglect of the strong geometrical tradition in geography.

The geometrical tradition was basic to the original Greek conception of the subject, and many of the more successful attempts at geographical models have stemmed from this type of analysis. The geometry of Christaller hexagons, of Lewis' shoreline curves, of Wooldridge's erosion surfaces, of Hägerstrand's diffusion waves, of Breisemeister's projections come vividly to mind. Indeed from one point of view, much of the new statistical work relating to regression analysis (Chapter 8) and generalized surfaces (Chapter 9) represent merely more abstract geometries. Much of the most exciting geographical work in the 1960's is emerging from applications of higher-order geometries; for example, the multidimensional geometry of Dacey's settlement models and the graph-theory and topology of Kansky's network analysis. It is an interesting reflection that the increasing separation of geomorphology and human geography may have come just at the time when each has most to offer to the other. Sauer (1925) in his *Morphology of Landscape* drew basic parallels between the two, but it was unfortunate, as Board (Chapter 10) so clearly shows, that 'landscape' was seized upon and 'morphology' neglected by those who drew inspiration from this important paper. The topographic surface is only one of the many three-dimensional surfaces that geographers analyse and there is no fundamental reason why, for example, the analysis of landform and population-density surfaces should not proceed along very similar lines. Geometry not only offers a chance of welding aspects of human and physical geography in a new working partnership, but revives the central role of cartography in relation to the two.

Our immediate solution then is to press for a re-establishment of the tripartite balance in geography by building up the neglected geometrical side of the discipline. Research is already swinging strongly into this field and the problem of implementation may be more acute in the schools than in the universities. Here we are continually impressed by the vigour and reforming zeal of 'ginger groups' like the School Mathematics Association which have shared in a fundamental review of mathematics teaching in schools. There the inertia problems – established textbooks, syllabuses, examinations – are being successfully overcome and a new wave of interest is sweeping through the schools. The need in geography is just as great and we see no good reason why changes here should not yield results equally rewarding. Better that geography should explode in an excess of reform than bask in the watery sunset of its former glories; for, in an age of rising standards in school and university, to maintain the present standards is not enough – to stand still is to retreat, to move forward hesitantly is to fall back from the frontier. If we move with that frontier new horizons emerge into our view, and we find new territories to be explored as exciting and demanding as the dark continents that beckoned an earlier generation of geographers. This is the teaching frontier of geography.

References

ACKERMAN, E. A., 1963, 'Where is a Research Frontier?', *Ann. Assn. Amer. Geog.*, **53**, 429–40.

BUNGE, W., 1962, *Theoretical Geography* (Lund).

CHORLEY, R. J., 1964, 'Geography and Analogue Theory', *Ann. Assn. Amer. Geog.*, **54**, 127–37.

CHRISTALLER, W., 1933, *Die zentralen Orte in Süddeutschland* (Jena).

HAGGETT, P., CHORLEY, R. J. and STODDART, D. R., 1965, 'Scale standards in geographical research: A new measure of a real magnitude', *Nature*, **205**, 844–47.

HAGGETT, P., In Press, *Locational Analysis in Human Geography* (London).

HARTSHORNE, R., 1939, *The Nature of Geography* (Lancaster, Pa.).

— 1959, *Perspective on the Nature of Geography* (London).

ISARD, W., 1960, *Methods of Regional Analysis* (New York).

POSTAN, M., 1962, 'Function and Dialectic in History', *Econ. Hist. Rev.*, 2nd Series, **14**, 397–407.

SAUER, C. O., 1925, 'The Morphology of Landscape', *Univ. of Calif. Pubs. in Geog.*, **2**, 19–53.

— 1952, *Agricultural Origins and Dispersals* (New York).

THOMPSON, D'ARCY W., 1917, *On Growth and Form* (Cambridge).

TOULMIN, S., 1953, *The Philosophy of Science* (London).

ZIPF, G., 1949, *Human Behaviour and the Principle of Least Effort* (New York).

Selective Index

It has seemed to the editors that the detailed table of contents (pp. *v–x*) and chapter references have obviated the necessity for extensive general and author indexes. There follows a short index in which some major methodological themes running through the volume are indicated. Where the whole of a chapter is concerned with such a theme it is shown by Roman numerals.

Printed in the United States
by Baker & Taylor Publisher Services